AF723088

OIL AND THE GREAT POWERS

Oil and the Great Powers

Britain and Germany, 1914–1945

ANAND TOPRANI

OXFORD
UNIVERSITY PRESS

OXFORD
UNIVERSITY PRESS

Great Clarendon Street, Oxford, OX2 6DP,
United Kingdom

Oxford University Press is a department of the University of Oxford.
It furthers the University's objective of excellence in research, scholarship,
and education by publishing worldwide. Oxford is a registered trade mark of
Oxford University Press in the UK and in certain other countries

© Anand Toprani 2019

The moral rights of the author have been asserted

First Edition published in 2019

All rights reserved. No part of this publication may be reproduced, stored in a retrieval system, or transmitted, in any form or by any means, without the prior permission in writing of Oxford University Press, or as expressly permitted by law, by licence or under terms agreed with the appropriate reprographics rights organization. Enquiries concerning reproduction outside the scope of the above should be sent to the Rights Department, Oxford University Press, at the address above

You must not circulate this work in any other form
and you must impose this same condition on any acquirer

Published in the United States of America by Oxford University Press
198 Madison Avenue, New York, NY 10016, United States of America

British Library Cataloguing in Publication Data
Data available

Library of Congress Control Number: 2018961442

ISBN 978–0–19–883460–1

Links to third party websites are provided by Oxford in good faith and for information only. Oxford disclaims any responsibility for the materials contained in any third party website referenced in this work.

The manufacturer's authorized representative in the EU for product safety is Oxford University Press España S.A. of Parque Empresarial San Fernando de Henares, Avenida de Castilla, 2 – 28830 Madrid (www.oup.es/en or product.safety@oup.com). OUP España S.A. also acts as importer into Spain of products made by the manufacturer.

Acknowledgments

My debts are many, but rather than test the patience of my editor, I shall live by the adage that less is more and keep my acknowledgments brief. The first group deserving of thanks are the archivists in the United States, Britain, and Germany who facilitated access to the primary sources upon which this book depends. The second are my many friends and colleagues in Washington, New Haven, Cambridge, and Newport who critiqued drafts, supplied novel insights, and challenged me to sharpen both my thinking and prose—you know who you are. The third are the anonymous reviewers at Oxford University Press, who supplied impressively detailed feedback that improved this book immeasurably.

Two particular people, however, are deserving of special mention. The first is my *Doktorvater*, David Painter. I still recall the look of shock on colleagues' faces when I told them that my graduate supervisor had read several drafts of my dissertation before my defense. What is even more remarkable is that he remained generous with his time even after I had graduated. Transforming the dissertation into a book took longer than I expected, requiring substantial excisions, additions, and revisions. Throughout this process, Dr. Painter provided constant encouragement, never losing faith even as my frustration mounted and my confidence ebbed. The result is a testament to his efforts as much as my own, but for the sake of my tenure application, I hope he will forgive me for retaining sole authorship.

The second is my wife, Maria. During our courtship, she often had to compete with the manuscript for my attention. Some partners might have resented sharing first billing with a book—but not Maria. She endured my divided attention with grace and supplied assurance during any setbacks. When I finally passed peer review, she was the only one of us who shed tears of joy. As a wholly inadequate token of my love, I dedicate this book to my wife and pay her the greatest compliment any historian can give—she is truly more precious to me than any document I ever found in the archives.

Some obligatory housekeeping. Portions of Chapters 7 and 8 were first published as "Germany's Answer to Standard Oil: The Continental Oil Company and Nazi Grand Strategy, 1940–1942," *Journal of Strategic Studies* 37: 6–7 (2014) 949–73. Parts of Chapters 1 and 2 also appeared initially in "An Anglo-American 'Petroleum Entente'? The First Attempt to Reach an Anglo-American Oil Agreement, 1921," *The Historian* 79:1 (2017) 56–79. They are reprinted with the permission of Taylor & Francis and John Wiley & Sons. Finally, this book relies on publicly available records in the United States, United Kingdom, and Federal Republic of Germany. The views expressed within it are my own and not necessarily those of the U.S. government.

Contents

List of Illustrations

List of Maps

Introduction
Oil and Strategy

The struggle for oil has been at the center of international politics since the beginning of the twentieth century. Securing oil—or, more precisely, access to it—has also been at the heart of many great powers' grand strategies during that time, particularly those in oil-poor Europe. The Continent's geographical and geological endowments, particularly its rich coal seams, had facilitated its rise to global predominance following the conquest of the New World and the start of the Industrial Revolution, but they conspired against it during the Age of Oil. Rather than accept their relegation to second-tier status, Britain and Germany developed elaborate strategies to restore their energy independence. These efforts wound up compromising their security by inducing strategic overextension—for Britain in the Middle East, and for Germany in the Soviet Union—thereby hastening their demise as great powers. For these reasons, the history of oil is also a chapter in the story of Europe's geopolitical decline.

While the control of oil is often an objective in great power politics, it is easy to overlook how the varying availability of oil among the great powers affected both the development of strategy and its execution. Of course, no serious scholar or analyst can "support the proposition that only natural resources structure the underlying competition among nations."[1] There is little evidence that nations fight wars over oil simply for its own sake.[2] Nevertheless, oil commands our attention in ways that no other commodity does, because ample supplies are indispensable for every nation's war machine and civilian economy. When nations go to war for oil, what they are actually fighting for is the capability to accomplish tasks that require oil. Great powers—no matter how brutal—cannot hope to translate their political ambitions into reality without oil. In 1942, to cite one example, "as Nazi planners worked furiously to realize the economic and racial goals associated with Lebensraum," the fate of the German war effort depended upon seizing the oil of the Caucasus, "without which the grand Nazi schemes would be mere chimeras."[3]

[1] Alfred Eckes, *United States and the Global Struggle for Minerals* (Austin: University of Texas Press, 1979), ix. Emphasis in the original.

[2] Emily Meierding, "Dismantling the Oil Wars Myth," *Security Studies* 25: 2 (2016): 258–88.

[3] Stephen Fritz, *Ostkrieg: Hitler's War of Extermination in the East* (Lexington: University Press of Kentucky, 2011), 230–9 (quotation from p. 236).

The study of the geopolitics of oil begins with World War I.[4] One of the factors that made this conflict so destructive was the blending of two earlier developments in warfare. The first was the "people's war" that first emerged during the Wars of the French Revolution, when states mobilized nationalist passions to wage unlimited war. The second was the "industrial war" of the second half of the nineteenth century, when soldiers grasped the fruits of the Industrial Revolution, including railways, telegraphs, and mass-produced weapons, to enhance the effectiveness of large conscript armies. During the latter period, coal was king. It was essential to the production of intermediate goods such as steel and dominated the market for propulsion fuels both on land (railways) and at sea (ships). By contrast, oil's chief purpose was as a source of illumination. That changed with the development of the internal combustion engine around the turn of the twentieth century and the decision of the Royal Navy, followed shortly thereafter by the United States, to convert its battle fleet to burning oil exclusively in 1912, both of which paved the way for oil to supplant coal as the world's premier propulsion fuel.

Between 1914 and 1918, oil revolutionized the conduct of war. Although the Allies won World War I using weapons dating from the nineteenth century, the conflict inaugurated a "military revolution" at the social, political, and technological levels that spawned "revolutions in military affairs" at the operational and tactical levels. Before the war had even ended, oil had transformed military operations and strategy, and compelled nations to adjust their force structures, military plans, and strategies.[5] Many of the weapons platforms we associate with modern conventional warfare emerged during this conflict, including tanks, trucks, aircraft, and submarines. All of these platforms required petroleum products either as a fuel or for lubrication. Oil was not yet vital to the civilian economy, and it did not replace coal as the world's largest source of energy until the 1960s.[6] Nevertheless, by 1918, modern war economies could not function without sufficient supplies of oil for sectors such as heavy industry, transportation, petrochemicals and even agriculture, where mechanization compensated for the manual labor drafted into the military or the factories. While oil played a contributing rather than a decisive role in the outcome of the conflict—horses were at least as significant, as the British shipped more fodder to their army in France than any other item—all of the participants understood that a profound change was occurring.[7]

There was considerable debate over the wisdom of embracing oil before 1914, but few expressed any hesitation after the Great War. On land, oil offered soldiers the tools to transcend the tyranny of firepower and fixed fortifications by restoring

[4] The following discussion draws from: Williamson Murray and MacGregor Knox, "Thinking about Revolutions in Warfare," in: *The Dynamics of Military Revolution, 1300–2050*, ed. Knox and Murray (New York: Cambridge University Press, 2001), 1–14.

[5] David Deese, "Oil, War, and Grand Strategy," *Orbis* 25: 3 (1981): 526ff.

[6] Bruce Podobnik, *Global Energy Shifts: Fostering Sustainability in a Turbulent Age* (Philadelphia: Temple University Press, 2006), 4–5; Raymond Stokes, "Oil as a Primary Source of Energy," in: *1956: European and Global Perspectives*, ed. Carole Fink, Frank Hadler, and Tomasz Schramm (Leipzig: Leipziger Universitätsverlag, 2006), 245–64.

[7] James Laux, "Trucks in the West during the First World War," *Journal of Transport History* 6 (1985): 69.

mobility to the battlefield using combined arms to shatter an enemy's lines and exploit a breakthrough. In the air, oil enabled the creation of an entirely new domain of warfare through the invention of heavier-than-air aircraft, which could support operations on land and sea or operate independently by striking an enemy's airfields, cities, and economic infrastructure. At sea, oil propulsion enhanced the range and lethality of surface combatants, enabled the development of submarines, and facilitated the development of naval aviation using aircraft carriers. The net effect across all of these domains of war was the enhancement of countries' ability to project power rapidly over vast distances.

The emergence of this internal combustion-propelled triad of vehicles, aircraft, and naval vessels ensured that the next war would be an "oil war." Even as the Great War reached its bloody crescendo in 1918, the leaders of the great powers adopted strategies to secure future oil supplies. It is not clear that they had a choice. Absent oil, as one German official remarked in 1918, "there is no militarily secure independence, no global relevance."[8] Success in war, other influential analysts surmised afterward, would hinge upon having ample supplies of oil from the start of hostilities.[9] The prize was nothing less than "world hegemony," which, the German Foreign Office concluded in 1921, depended on having "oil hegemony" as well.[10]

Throughout the interwar period, oil permeated geopolitical discourse.[11] Soviet dictator Josef Stalin declared in 1925 that "oil is the vital nerve of the struggle among the world states for supremacy both in peace and in war."[12] His fascist opponents concurred: "The struggle of world politics," one German military journal concluded in 1934, "is today more or less a struggle over oil."[13] Another declared the following year that "today, all of European politics, even world politics for that matter, is geared decisively toward oil interests."[14] These were obviously oversimplifications, since oil did not create interstate or ideological competition after the Great War. Nevertheless, oil was transforming the balance of power in subtle yet profound ways. After World War I, one scholar argues, "[the] global balance of power began to reflect the distribution of indigenous oil resources, access to foreign oil... and competition and potential conflicts over oil-bearing territory." Oil was now at the heart of the most important tasks of any strategist, including "[defining] vital interests... [developing] alliance and arms-transfer policies, the selection and design of national force structures, and the decision to initiate or enter international conflict of war."[15]

[8] Stuchlik, "Die Wirtschaftliche Bedeutung [...]," 1918, Bundesarchiv, Berlin-Lichterfelde (BA-B), R 3101/884.

[9] Fritz Fetzer, *Ölpolitik der Großmächte unter kriegswirtschaftliche Gesichtspunkten: Das japanische Beispiel* (Hamburg: Hanseatische Verlagsanstalt, 1935), 7–8.

[10] Note to Wirkl and Bücher, January 20, 1921, Politisches Archiv des Auswärtigen Amtes (PAAA), R 118173.

[11] "Kriegsfolgen [...]," *Vierjahresplan*, 1942/III.

[12] "Political Report [...]," December 18, 1925, accessed August 15, 2018, <https://www.marxists.org/reference/archive/stalin/works/1925/12/18.htm>.

[13] Roesner, "Die Treibstoff-Frage [...]," *Militär-Wochenblatt* (May 25, 1934).

[14] Bronk, "Deutschlands Erdöl-Selbstversorgung [...]," *Militär-Wochenblatt* (October 25, 1935).

[15] Deese, "Grand Strategy," 526–7.

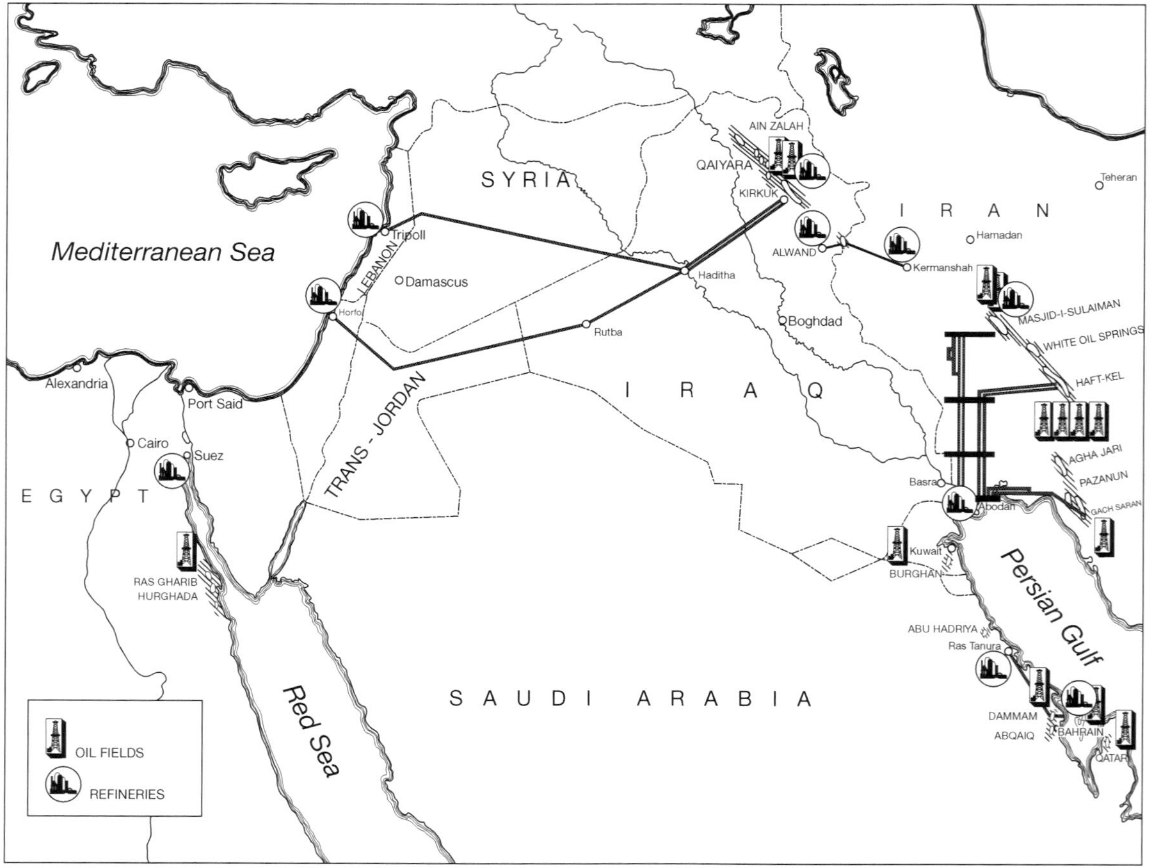

Map 1. Oilfields, Pipelines, and Refineries of the Middle East, no date (circa 1943)

Source: NARA, RG 107, Entry 141, Box 251.

The century before, the introduction of coal-powered steam engines aboard ships had radically affected geopolitics by freeing naval and merchant vessels from the vagaries of wind. This, in turn, restored the historical "transcendence" of some regions (the Mediterranean), elevated backwaters to prominence while relegating once significant arteries to secondary status (the Strait of Malacca at the expense of the Sunda Strait), opened new trade routes (Cape Horn), and sparked another global spasm of imperialist competition.[16] Oil prompted a similar rewriting of global political geography that has persisted to this very day. Besides benefiting oil-rich powers such as the United States and the Soviet Union, the rise of oil also gave a few developing nations in what came to be known as the "Global South" immense geopolitical significance as oil producers. No region was affected more than the Middle East. Already the "'Clapham Junction' [hub] of British imperial communications" since the nineteenth century, the Middle East became doubly important as an oil supplier and transit route for tankers or pipelines (see Map 1).[17]

WHY OIL?

Oil is not the only commodity crucial to a modern economy or war machine, but its peculiar characteristics—especially the circumstances under which the oil industry operates and the nature of production and consumption—sometimes make acquiring adequate and secure supplies more challenging than is the case with many other key commodities. Consider, by way of example, a bulk ore such as bauxite or metals such as chromium, copper, and nickel. As with oil, industrialized countries require vast quantities of bauxite for aluminum production that are too large to stockpile. Unlike oil, however, during the 1930s, supplies of bauxite were both plentiful and dispersed worldwide, including Europe, and traded for either hard or soft currency (currencies that are or are not freely convertible). Supplies of chromium were geographically concentrated like oil and purchasing them required hard currency, but unlike either oil or bauxite, chromium could be stockpiled for military purposes, and Germany entered World War II with a two-year supply.[18] As for nickel or copper, there was no need to stockpile—the Germans could instead recycle prefabricated goods made with copper to supplement their domestic production. Nickel was trickier—90 percent of prewar production was unavailable to Germany due to blockade. Berlin devised an ingenious response—it pulled its nickel coins from circulation after World War II began. Germany thereby doubled its prewar stockpile and, by combining this with conservation measures, survived the conflict without any major shortage.[19]

16 Bernard Brodie, *Sea Power in the Machine Age* (New York: Greenwood, 1969), 105–10.

17 John Darwin, "An Undeclared Empire: The British in the Middle East, 1918–39," in: *The Statecraft of British Imperialism: Essays in Honor of Wm. Roger Louis*, ed. Robert King and Robin Kilson (London: Frank Cass, 1999), 160.

18 Alan Milward, "The Reichsmark Bloc and the International Economy," in: *Aspects of the Third Reich*, ed. H.W. Koch (New York: St. Martin's, 1985), 346–9.

19 John Perkins, "Coins for Conflict: Nickel and the Axis, 1933–1945," *Historian* 55 (1992): 85–100.

Neither Britain nor Germany could resort to such gimmicks to satisfy their oil requirements in wartime. Before the discovery of North Sea oil in the 1969, Britain's only significant source of petroleum was Scottish oil shale, production of which was hideously inefficient. Shale production during World War I averaged 3,000,000 tons per annum, but since the output of crude oil per ton of shale had declined by a quarter between 1886 and 1918 (from 31.2 imperial gallons to 23.0 gallons), the total quantity of refined petroleum was under 300,000 tons.[20] As for Germany, it produced only 8 percent of its total oil consumption during World War I (418 tons versus 5,290 tons)—40 percent of which came from Alsace-Lorraine, which returned to France under the Treaty of Versailles, with the remainder from around Hannover.[21]

Oil shortages were not the only similarity. Britain and Germany also suffered from current account deficits—the sum of their trade balance plus net foreign investment income and transfers. Britain began running trade deficits in the 1880s, and World War I depleted its reserves of foreign exchange (convertible, "hard" currencies such as U.S. dollars or gold) and securities (stocks and bonds). In Germany's case, rearmament after 1933 stimulated demand for imports while reducing export surpluses, thereby starving Germany of the foreign exchange it needed to import oil. Even if both nations abjured any aspirations to great power status, in a world where the United States dominated the global oil trade, their capacity to import oil was still constrained by their foreign exchange position. Not surprisingly, independence for both Britain and Germany also meant purchasing oil whose price or currency they controlled.

Perhaps most striking, however, was both nations' persistence in pursuing independence irrespective of the costs. Changes in government or regime type sometimes affected the tactics but never the underlying aims or strategies. Although the Conservative Party dominated British politics during the twentieth century, a Liberal government first embraced oil as a matter of national security before 1914 and established the framework for Britain's postwar strategy in 1916. The interwar Labour governments avoided confrontation with the United States and France but they did not question the assumptions of British oil policy or challenge their predecessors' strategy. Meanwhile, the first tentative steps toward building a synthetic fuel industry to make Germany independent of oil imports took place during the Weimar Republic. The rise of the Third Reich led to a readjustment of German aims—from domestic stabilization to a genocidal war of conquest—but not to the mechanisms of Germany's energy strategy.

What explains this remarkable consistency? The answer, of course, is World War I. Britain and Germany drew the same lessons from that conflict with regard to oil. Not only did they need it, and in much greater quantities than heretofore, but they had to look for it beyond their borders. The war had revealed that a "general conflict could no longer be resolved with reference to European resources alone," and victory had gone to the side that had made greater use of assets beyond

[20] Department of the Interior, *Oil Shale: An Historical, Technical, and Economic Study* (Washington, DC: GPO, 1924), 55–7.

[21] Ferdinand Friedensburg, *Erdöl im Weltkrieg* (Stuttgart: Ferdinand Enke Verlag, 1939), 73.

Europe, whether from their overseas empires or the budding superpower on Europe's periphery—the United States.[22]

British and German aims—energy independence—were identical, but the divergent strategies they adopted were byproducts of their peculiar circumstances as either a maritime or a continental power. Although a disruptive enterprise in general, the Great War had validated a set of prewar assumptions in both countries. In Britain's case, the need to preserve secure sea lines of communication remained paramount. In Germany, it became clear that the only surefire way to augment the nation's economic strength was through "*continental* expansion and economic integration" of adjacent regions.[23]

Britain's primary instruments were its navy and major oil companies—specifically, the Anglo-Persian Oil Company (Anglo-Iranian Oil Company after 1935) and the Royal Dutch/Shell Group.[24] Whitehall had a 51 percent stake in the former since 1914, while the latter was the product of a 1907 merger between the Royal Dutch Petroleum Company and the Shell Transport and Trading Company. The resulting group was 60 percent Dutch-owned but still British in outlook.[25] Together, Anglo-Persian and Shell possessed valuable concessions throughout the Western Hemisphere, the Middle East, and East Asia, as well as sophisticated marketing and transportation networks. As of 1927, Shell produced 50 percent more oil than either of its two closest competitors, Standard Oil Company of New Jersey (Jersey) and Gulf Oil: 314,200 barrels per day (bpd) versus 214,700 and 212,500 bpd, respectively. It also had more tankers than Jersey and Gulf combined (156 versus 123).[26] Anglo-Persian was the smallest of the major oil companies, but it had a monopoly over some of the richest oilfields in the world. Its assets more than doubled between 1920 and 1927 from $110 million to $248 million, whereas those of Shell increased by only 50 percent (from $320 million to $480 million).[27]

Until World War II, Shell and Anglo-Persian/Anglo-Iranian produced more oil beyond the United States than every U.S. oil company combined did. In 1929, the former possessed 41 percent of world oil production outside of the United States (541,200 bpd), while the U.S. companies produced only 29.7 percent (392,000 bpd). Ten years later, Shell and Anglo-Persian still controlled 35.7 percent of global production beyond the United States (795,300 bpd), while the U.S. share had declined to 24.9 percent (554,800 bpd).[28]

[22] Ian Lesser, *Resources and Strategy* (Basingstoke: Macmillan, 1989), 41–6 (quotation from p. 46).

[23] Lesser, *Resources*, 46. Emphasis in original.

[24] Major companies are vertically integrated firms with their own oilfields, transportation infrastructure, refineries, and marketing networks. Independent companies focus on one or more of these business areas.

[25] Shell, "Memorandum," January 16, 1940, British National Archives (BNA), CAB 63/117.

[26] Jersey, which lost most of its upstream assets after the U.S. Supreme Court broke up the Standard Oil Company in 1911, caught up with Shell in 1935 at 560,000 bpd. Joost Jonker and Jan Luiten van Zanden, *From Challenger to Joint Industry Leader, 1890–1939*, vol. 1 of *History of Royal Dutch Shell* (Oxford: Oxford University Press, 2007), 224–5, 437.

[27] R.W. Ferrier, *Developing Years, 1901–1932*, vol. 1 of *History of the British Petroleum Company* (Cambridge: Cambridge University Press, 1982), 543.

[28] Mira Wilkins, *Maturing of Multinational Enterprise: American Business Abroad from 1914–1970* (Cambridge: Harvard University Press, 1974), 241.

Considering the tools at its disposal—aggressive companies, a sprawling empire, and a powerful navy—it should come as no surprise that Whitehall saw energy independence as a viable objective. Even if Britain had to import all of the oil it consumed, few countries were better equipped to import large quantities from overseas. Assuming secure sea lines of communication, Britain's supply position was enviable comparted to its most likely adversaries after 1918.[29]

That did not mean that Whitehall's task was a simple one. First on the list of concerns was the question of whence Britain should draw its oil. The menu of options was rather limited in the aftermath of World War I—the only significant sources were the United States, Russia, and Mexico (see Maps 2 and 3). War-torn, revolutionary Russia was out of the question. Another option was the Western Hemisphere, especially Latin America, where British firms had a long history, but it was not the ideal source of supply politically. Mexico was also in the throes of revolution, and Latin America now lay within the U.S. orbit. Starting in 1914, Whitehall expressed a preference for sources under British political control. Ideally, this would have meant from areas within the British Empire, but unfortunately imperial oil production was negligible, too far away from Britain to be transported at a reasonable cost, and often indefensible in the event of war.

There was, however, one further alternative that offered the prospect of ample oil supplies under British control—the Middle East. Thanks to the efforts of generations of civil servants, soldiers, academics, and merchants, the region was not exactly terra incognita—Britain had already established an extensive presence there before World War I to support the Ottoman Empire, contain rival powers such as Russia, and preserve lines of communication to India and the Far East, especially the Suez Canal. Drawing oil from the Middle East therefore supplemented rather than complicated Britain's strategic posture. If anything, oil kept Britain ensnared in the Middle East even as the imperial rationale for being there evaporated with the waning of British trade to India in the 1930s.[30]

In hindsight, Britain's decision to seek the bulk of its oil supplies from the Middle East appears a foregone conclusion, but at the time, it was a gamble. We should not confuse today's oil industry with the one that existed before 1945. At that time, the center of production was North America, which accounted for more than 80 percent of the world's oil in 1918—and by North America, we mean the United States. While this book focuses on the challenges that confronted Britain and Germany, the United States was never entirely out of the picture, since it provided the benchmark against which rival great powers had to measure themselves in the Age of Oil. No country or even group of countries, such as the Organization of Petroleum Exporting Countries (OPEC), has ever possessed the sort of commanding position the United States enjoyed during the first half of the twentieth century (see Illustration 1). Its land mass dwarfed the potential oil-producing regions of the Middle East (see Map 4). Before World War II, the U.S. share of annual global output and accumulative production throughout world history was

[29] Brodie, *Sea Power*, 117–18.

[30] B.R. Tomlinson, *Political Economy of the Raj, 1914–1947* (London: Macmillan, 1979), 44ff.

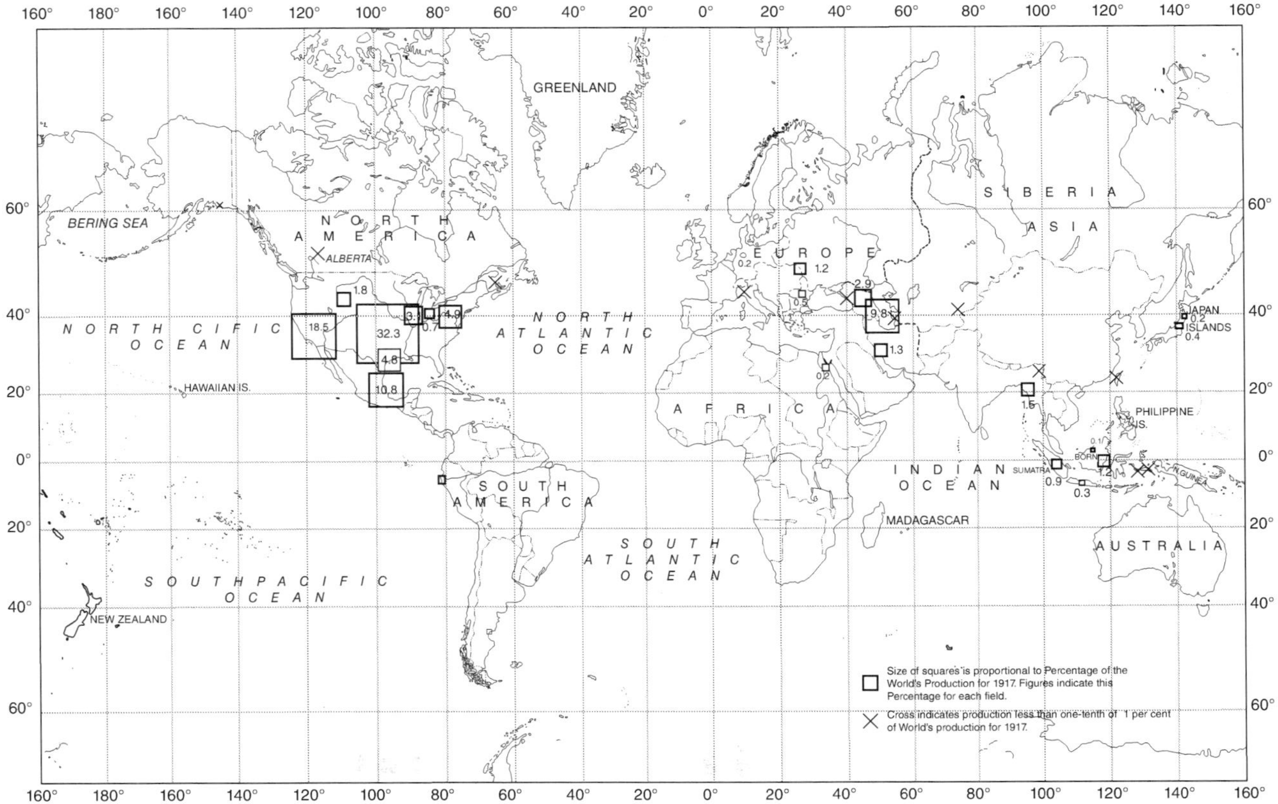

Map 2. Shares of World Oil Production, 1917

Source: NARA, RG 107, Entry 191, Office of the Assistant Secretary for War, Planning Branch (Entry 191), Box 100.

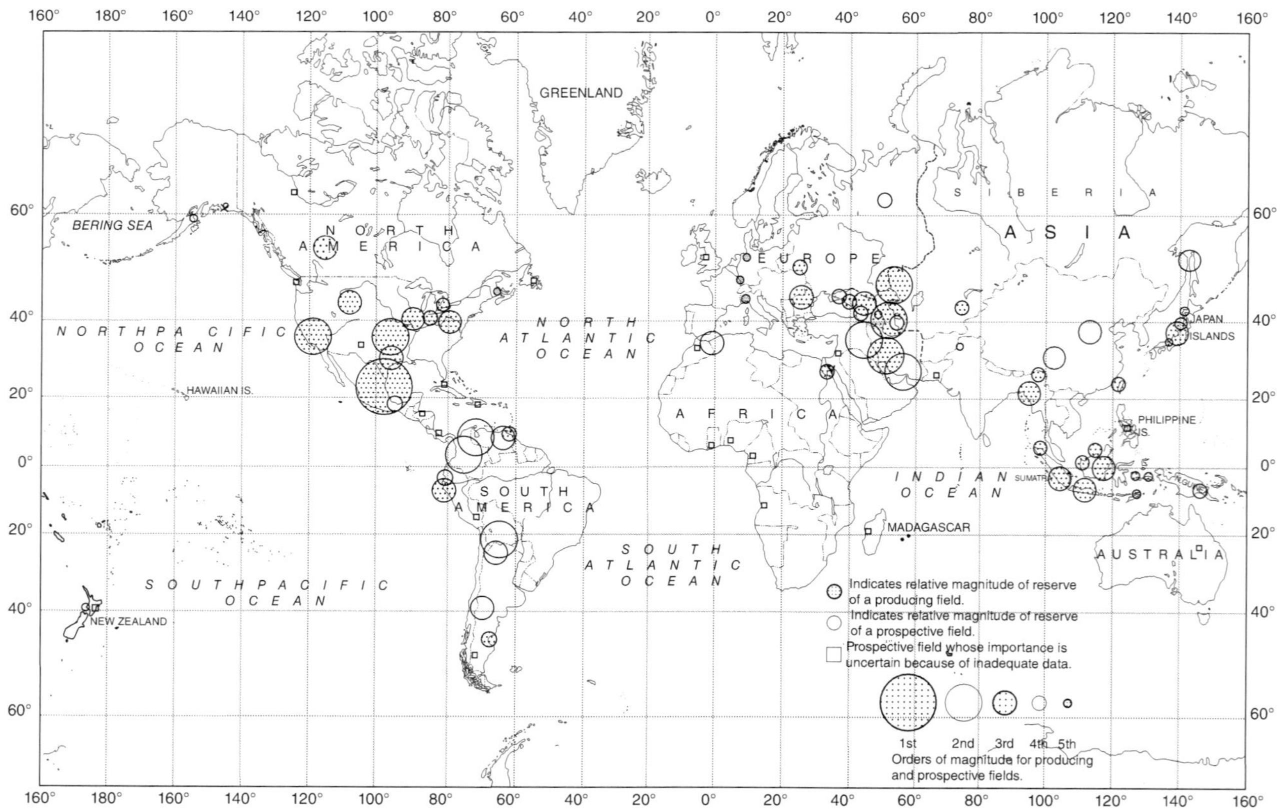

Map 3. Shares of World Oil Reserves, 1917

Source: NARA, RG 107, Entry 191, Box 100.

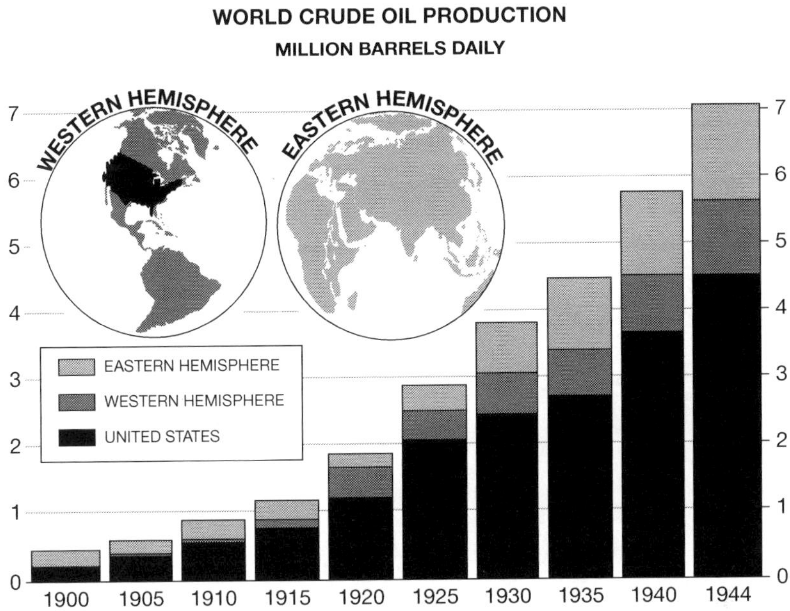

Illustration 1. "World Crude Oil Production," 22 February 1945

Source: Library of Congress (LOC), Harold L. Ickes Papers, Box 221.

more than 60 percent (see Illustration 2). The United States was equally strong in refining, with a capacity twice as great as the rest of the world combined (see Illustration 3). By comparison, the Middle East was a mere sideshow, with only 6 percent of world production and 7 percent of refining capacity before 1939 (see Illustration 4).[31] The British were nevertheless convinced that the future of the oil industry lay in the Middle East (see Illustration 5). By the late 1930s, British planners expected that Iran alone would supply roughly 40 percent of the empire's annual oil requirements in wartime. The security environment also looked promising. Although British commitments in the Middle East had expanded considerably, World War I had eliminated every great power rival to British predominance in the region except the United States, while Britain had little to fear from any local power.

Of course, controlling oilfields was useless if Britain could not get the oil to consumers. The British learned the hard way that, in wartime, "the only oil reserve worth defending is that which can be held with a minimum of defensive military commitments."[32] After 1935, after Italy became a likely adversary, Britain could neither guarantee the security of its sea lines of communication in the Mediterranean, nor procure enough tankers to redirect supplies around the Cape of Good Hope.

[31] Besides the various illustrations, the preceding statistics are drawn from: DeGolyer and MacNaughton, *Twentieth Century Petroleum Statistics: Historical Data* (Dallas: DeGolyer and MacNaughton, 2004), 3–4, 9, and 12.

[32] Bernard Brodie, "American Security and Foreign Oil," *Foreign Policy Reports* 23: 24 (March 1, 1948): 301.

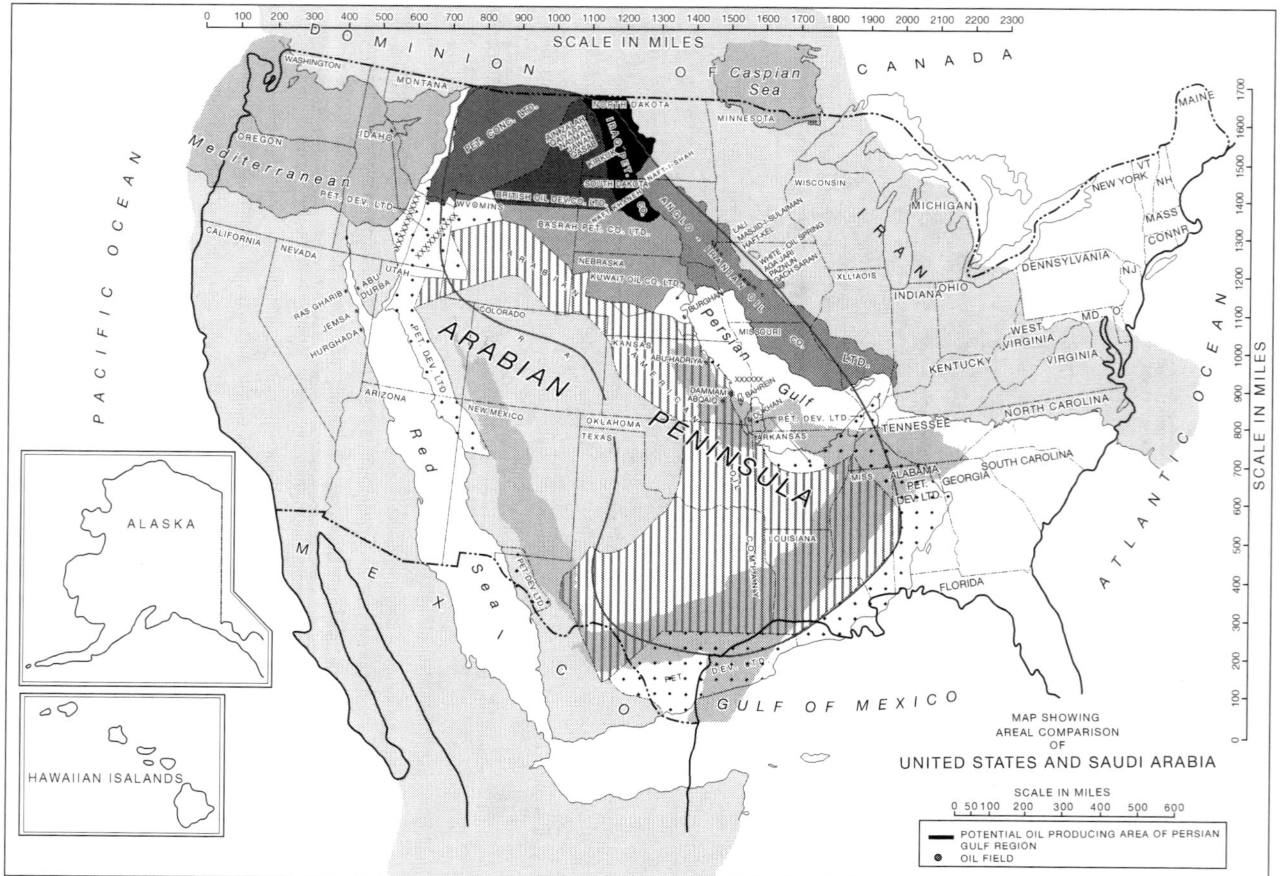

Map 4. Aerial Comparison of Major Oil Concessions in the Middle East against the United States, no date (circa 1939/41)

Source: NARA, RG 107, Entry 141, Box 251.

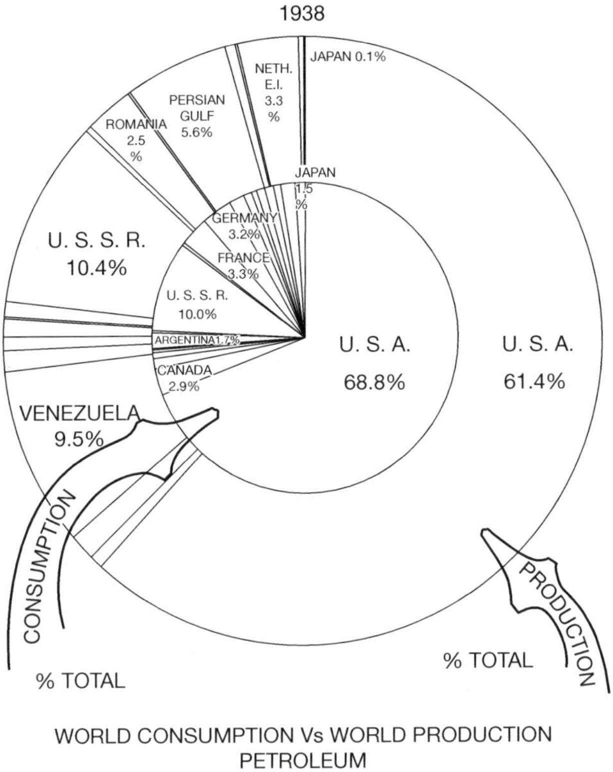

Illustration 2. "World Consumption vs. World Production [of] Petroleum," 1938

Source: National Archives and Records Administration (NARA), Record Group 107: Records of the Office of the Secretary of War (RG 107), Entry 141, Office of the Under Secretary for War, Administrative Office, Classified Decimal File (Entry 141), Box 251.

The only alternative was to import from the Western Hemisphere, primarily from Venezuela and the United States. This imposed ruinous economic costs since Britain was short of foreign exchange to pay for these imports.

Germany's rationale for embracing energy independence was a reflection of the dilemmas it confronted as a resource-poor, continental power. Although it had been at the forefront of the second Industrial Revolution, Germany was short of many important raw materials, including oil, copper, rubber, nickel, manganese, tungsten, chromium, and bauxite. Following the loss of Lorraine in 1918, it was also bereft of high-quality iron ore.[33] One of the few natural resources it did have in abundance, however, was coal. Before World War I, the discrepancy between its industrial might and weak raw materials position contributed to the development of both a dynamic export sector and an aggressive foreign policy.[34] Wartime defeat and the swift

[33] R.L. DiNardo, *Germany's Panzer Arm* (Westport: Greenwood Press, 1997), 6–7.

[34] Peter Hayes, "Carl Bosch and Carl Krauch: Chemistry and the Political Economy of Germany, 1925–1945," *Journal of Economic History* 47: 2 (1987): 353.

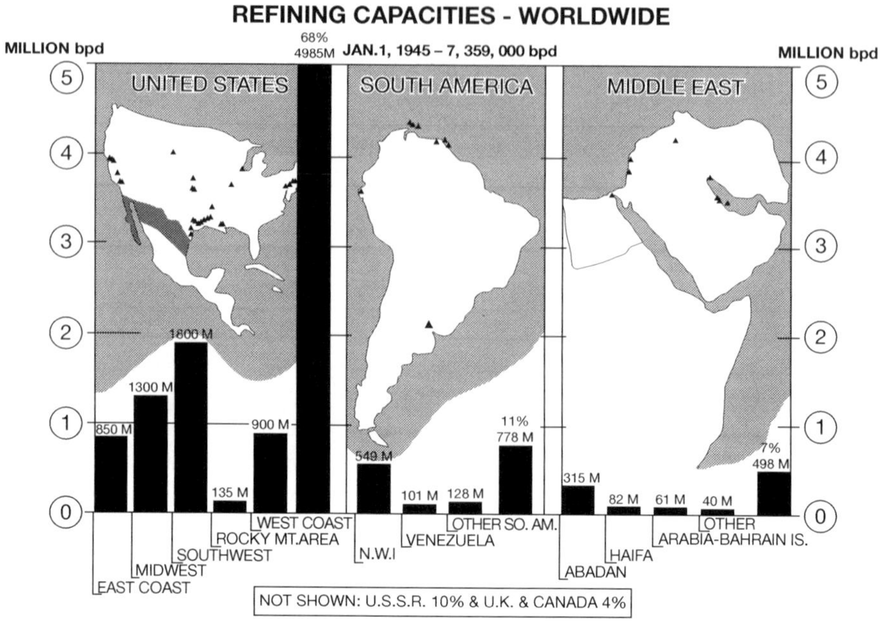

Illustration 3. "Refining Capacities—Worldwide," 22 February 1945
Source: LOC, Harold L. Ickes Papers, Box 221.

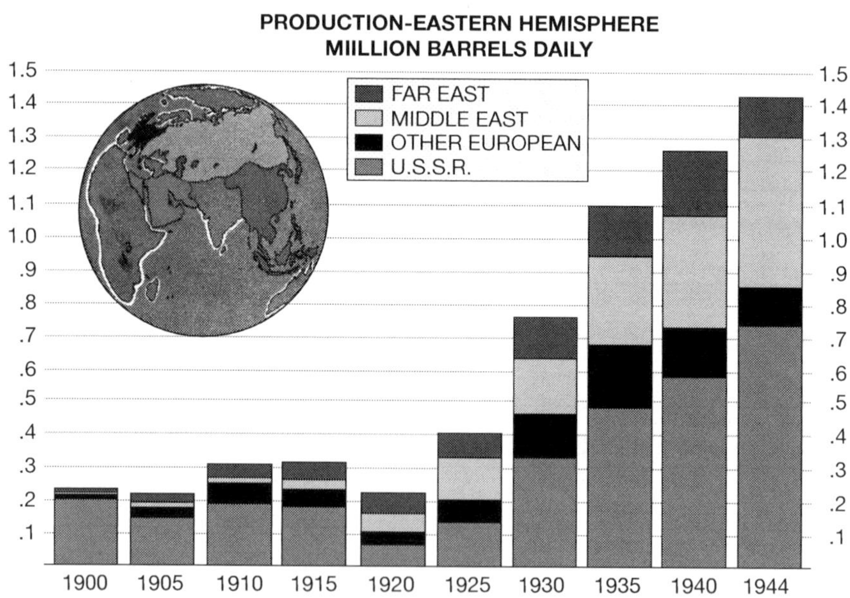

Illustration 4. "Production—Eastern Hemisphere," 22 February 1945
Source: LOC, Harold L. Ickes Papers, Box 221.

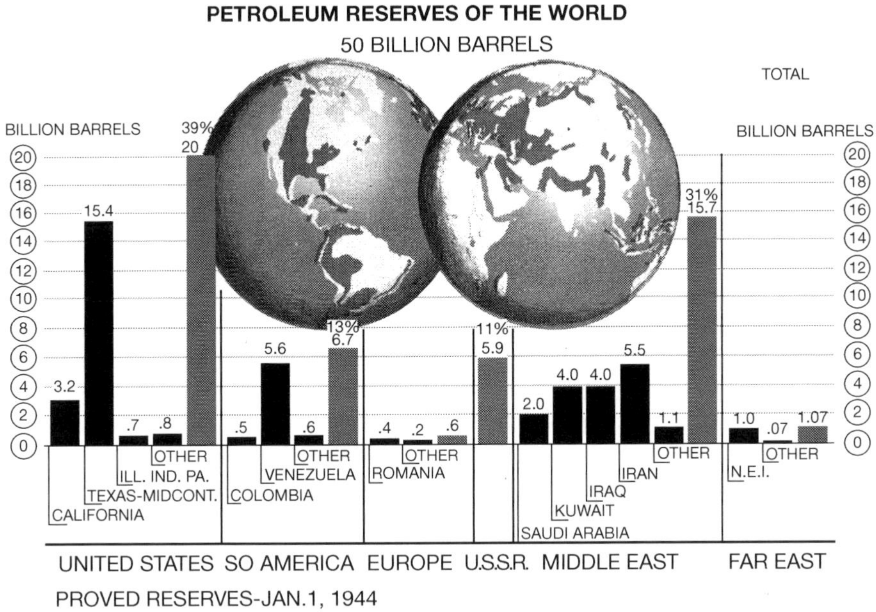

Illustration 5. "Petroleum Reserves of the World," 22 February 1945
Source: LOC, Harold L. Ickes Papers, Box 221.

collapse of global trade during the Great Depression meant that the latter instrument of national power came to assume disproportionate importance.

It would be a mistake to presume that Germany only decided to limit its dependence on imported oil following the rise of the Third Reich in 1933. The Great War had revealed the folly of relying on an Anglo-American-dominated global order, which provided Germany with economic security but only at the cost of forfeiting its strategic autonomy. The shortages imposed by the Allied blockade encouraged German officials to experiment with various methods of economic plunder in conquered territories such as Romania that they would perfect after 1939.[35] Defeat and the loss of their empire did not end the Germans' preoccupation with self-sufficiency, but after 1918, the basis would have to exist within Germany's borders with regard to oil. The scientific basis for converting coal into petroleum existed prior to World War I, and German firms were busy improving upon existing methods or developing new ones.

The company with the greatest stake in synthetic fuel, the chemical conglomerate IG Farben, had gotten into the business for commercial reasons, but it did not hesitate to extol synthetic fuel's security benefits to win financial and political backing. The firm's message found a receptive audience, for there was a widespread belief in Germany that synthetic fuels allowed the Second Reich to prolong the

[35] Daniel Hamlin, *Germany's Empire in the East: Germans and Romania in an Era of Globalization and Total War* (Cambridge: Cambridge University Press, 2017), 1–23 and 323–9.

Great War long after Germany lost access to overseas imports.[36] The German Army, in particular, became a vocal advocate for establishing a synthetic fuel industry from the early 1920s onward. Throughout the interwar era, the military was fascinated by the operational possibilities afforded by the internal combustion engine and convinced that synthetic fuel was Germany's only truly reliable source of petroleum. The military was not the only institution to grasp the potential of synthetic fuel to transform Germany's entire military and economic position. Civilian bureaucrats recognized that a synthetic fuel industry could boost employment by stimulating demand for domestic coal and steel. Reducing the amount of imported oil would also improve Germany's balance of payments.

Unlike in Britain, however, in Germany the strategic aims behind the quest for energy independence evolved over time. Before 1933, the various interest groups, whether military or civilian, public or private, were motivated by their own parochial concerns and did not integrate their desired policies into a wider strategic framework. For the National Socialists, however, synthetic fuel became a means to a far greater end than job creation and national solvency. The regime jettisoned the notion that synthetic fuel could make Germany energy independent due to the immense cost in raw materials, intermediate goods, and labor. Rather, it believed that synthetic fuel could lay the foundation for future expansion by giving the Third Reich the means to take by force what Germany needed over the long run.

Although woefully inadequate by U.S. or British standards, by the outbreak of World War II Germany's petroleum supply sufficed to meet Adolf Hitler's plan to wage short, sequential campaigns against individual adversaries. Although limited in scope, these conflicts would be unlimited in effort and aims. Germany was planning a war of conquest, exploitation, and genocide across Europe, the Soviet Union, and eventually the Middle East. Not surprisingly, the acquisition of oilfields in all of these regions was an explicit German war aim during World War II. Had the Third Reich succeeded, Caucasian and Middle Eastern oil, as well as synthetic fuel, would have provided the energy foundation to challenge Anglo-American predominance. In that case, as one historian concludes, "World War II can be seen above all as a war for oil, with those lacking it (the Axis) seeking to defeat those who controlled it (the Allies)."[37]

For all their differences, Britain and Germany had to learn the same lesson—energy independence is not synonymous with energy security. Politicians and pundits use terms such as "energy independence" and "energy security" interchangeably, but this is a grave conceptual error. Energy independence usually means self-sufficiency but can also refer to freedom from imported supplies, drawing supplies only from sources under the control of one's nationals, or privileging one form of imports over another (overland versus overseas). Energy security, by contrast, means secure access to supplies at stable prices.[38] Strategies to achieve energy security may include

[36] Hoffmann, "Das Oel [...]," *Wirtschaftlicher Teil der Deutschen Zeitung* 162 and 164 (April 8 and 9, 1924).

[37] Fritz, *Ostkrieg*, xxii.

[38] My definition differs from those of organizations such as the International Energy Agency (IEA), which defines it as "the uninterrupted availability of energy sources at an affordable price." IEA,

insuring the physical security of infrastructure, preserving access, and building robust systems to distribute supplies.[39] Seeking independence does not necessarily contribute to energy security, since a country may enjoy greater access to energy at a reasonable cost by importing it rather than producing it internally, and vice versa.

Another difference between energy independence and security is their relationship to prices.[40] For energy independence, the nominal price of oil is immaterial compared to the opportunity costs of the necessary industrial inputs such as coal, steel, and labor to produce petroleum internally. For energy security, price assumes primary significance. Extractive industries operate under a unique set of conditions. Since firms do not manufacture oil so much as discover it, the market supply at any moment may not correspond to demand. This fact, combined with various financial machinations and government regulations, contributes to sharp short-term fluctuations in price.

Over the long run, the price of oil rather than demand for it—which is inelastic in the short run—is the most important factor determining supply. As the price increases, so too does the volume of recoverable reserves that it is economical to extract. Higher prices will also stimulate the search for new reserves and technologies to recover previously inaccessible deposits—as evidenced by the development of Alaska and deep-water drilling in the 1970s and the recent application of horizontal drilling and hydraulic fracturing to extract oil trapped in shale formations. The former was a response to the First Energy Crisis of 1973–4 while the latter derived much of its impetus from the steady rise in oil prices between 2002 and 2008.

Provided they do not rise too much, thereby spurring inflation or incentivizing lower consumption and greater efficiency, high prices will increase supplies. Low prices, meanwhile, discourage production from high-cost fields and exploration even as they stimulate consumption, which may eventually create shortages. Since the 1920s, producers—initially firms, but later governments—have colluded to regulate supplies, usually by suppressing excess production. The purpose of manufacturing "scarcity" was to sustain a price that stimulates consumption without reducing incentives to hunt for new reserves.[41]

Pursuing and preserving energy independence requires a more active role for the state, but that is not the only reason why it is appealing to so many governments. For some, independence carries with it the promise of greater autonomy and geopolitical relevance. Opting for energy security as part of a strategy of interdependence could

"Energy Security," accessed August 15, 2018, <https://www.iea.org/topics/energysecurity/>. This definition is problematic because consumers and producers may have different ideas about what constitutes an "affordable" price. Whereas consumers want the lowest possible prices, companies prefer stable prices when making investment decisions.

[39] Daniel Yergin, "Energy Security and Markets," in: *Energy and Security: Strategies for a World in Transition*, ed. Jan Kalicki and David Goldwyn (Washington, DC: Woodrow Wilson Press, 2013), 69–87.

[40] The following discussion draws from: Jonathan Chanis, "U.S. Foreign Policy and Petroleum," in: Foreign Policy Association, *Great Decisions 2017* (New York, 2016), 69–71.

[41] Timothy Mitchell, *Carbon Democracy: Political Power in the Age of Oil* (New York: Verso, 2011), 39–65 and 205–9; Jonathan Nitzan and Shimson Bichler, *Global Political Economy of Israel* (Pluto: London, 2002), 219ff.

have provided Britain and Germany with ample supplies of energy and perhaps guaranteed their economic prosperity, but only at the cost of their political freedom. This was unacceptable to elites in either country. All of them embraced a geopolitical outlook that focused on relative gains and were keenly aware of how geography affected the balance of power. Whether policies that increased their state's relative power also raised the living standards of their subjects was something elites of both nations rarely considered beyond narrow issues such as employment—grappling with declining utility of force or empire was even less thought about. Elites were only willing to accept hard truths when confronted by the reality of national bankruptcy or catastrophic military defeat. True wisdom would have required acknowledging that European pretensions to an "autonomous, great power policy" were a "fiction."[42] Britain came to this realization too late to preserve its geopolitical standing. Germany refused to acknowledge the facts, and, after two failed bids for European hegemony, forfeited its existence as a nation state.

OUTLINE

Europe lost as much from geography during the Age of Oil as it had gained during the Age of Coal. None of Europe's great powers escaped the consequences, but only Britain and Germany opted for the risky and expensive strategy of independence. France, by contrast, sought to make the best of its circumstances after World War I. In many respects, French strategy mirrored that of Britain, but in some, the French were even more ambitious. French officials were determined to build a vibrant domestic refining industry; draw supplies from French suppliers using French ships; and use their limited share of Middle Eastern oil production to cement France's control of its longstanding sphere of influence in Lebanon and Syria (see Chapters 1 and 2). Independence, however, was never a serious prospect. French access to overseas oil depended on the assistance—technical, financial, and military—of Britain and the United States. In that sense, France's oil strategy was similar to its strategy for containing Germany—another vital French interest beyond France's means to accomplish independently.

Why, then, did Britain and Germany, despite their differences—one a status quo, maritime power, the other a revisionist, continental one—choose independence? There is a substantial body of literature examining why and how oil became an important factor in British strategy after World War I. Several scholars have identified energy independence from the United States as Whitehall's primary objective, and the vital role of the Middle East as an issue in Britain's strategy.[43] Unfortunately, their works are better at explaining what Britain desired than what it achieved. Most end their stories in the 1920s, when the British faced no

[42] Michael Geyer, *Aufrüstung oder Sicherheit: Die Reichswehr in der Krise der Machtpolitik* (Wiesbaden: Steiner, 1980), 177–88 and 495–7 (quotation from p. 497).

[43] Rosemary Kelanic, "The Petroleum Paradox: Oil, Coercive Vulnerability, and Great Power Behavior," *Security Studies* 25: 2 (2016): 197–202.

compelling threats to their sea lines of communication or control of foreign oilfields. Consequently, they cannot explain why Britain entered World War II just as dependent on the United States for oil as it was during World War I.[44] As for Germany, many historians offer an incomplete picture of the Third Reich's petroleum strategy that fails to reflect its true complexity. Specifically, they have overstated the significance of synthetic fuel, while devoting relatively little attention to how German aims evolved and expanded under the exigencies of war.

Besides rectifying such deficiencies, this study seeks to break down the compartmentalization between the history of oil and the study of international relations, by clarifying how the broader strategic context shaped British and German views about oil, and how the availability of oil (or lack thereof) in turn circumscribed their strategic options. To accomplish that, the study takes a focused, comparative approach that considers the British and German case studies in tandem. Such studies are rare, both within the field of oil history and in strategic studies, due to their archival and linguistic demands, not to mention the methodological challenges of amalgamating insights from different subfields or even whole disciplines. Case selection is a subjective endeavor, but as one historian of Fascist Italy and the Third Reich observes, when it is done well, comparative history can "clarify the unique causes and consequences of historical phenomena by comparing them to apparently similar phenomena, and . . . derive general patterns and potential explanations for those patterns from the analysis of groups of comparable cases."[45]

The comparative approach is also invaluable for identifying differences between evidently analogous case studies. Although there is some overlap in chronology and policy aims, the strategies adopted by Britain and Germany were as different as the reasons for their failure. Doing justice to the specific contexts in which British and German officials operated would be difficult if the two case studies were merged into a single narrative. Britain emerged from World War I as one of the world's two genuine global powers. Its strategy emerged out of its victory and blossomed in an era of peace before disintegrating under the threat of war. Germany started on the path toward independence at the same time, but with the handicaps brought by defeat. Its relative power was a fraction of Britain's and its strategy was, in part, a response to circumstances shaped by its Anglo-Saxon rivals.

Part I examines Britain's gambit to achieve energy independence from the United States after World War I and restore the political independence and strategic advantage it enjoyed during the Age of Coal. Chapter 1 focuses on the origins of Britain's oil strategy during the Great War, when officials began pursuing two of the key objectives of British strategy—imposing British control over Shell and the oilfields of the Middle East. The superficial majesty of the British Empire masked serious weaknesses, not the least of which was the fact that the empire had only a small fraction of global oil production. Britain would have to look elsewhere for

[44] William Gibson, *Britain's Quest for Oil: The First World War and the Peace Conferences* (Solihull: Helion, 2017), 195–6.

[45] MacGregor Knox, *To the Threshold of Power, 1922/33: Origins and Dynamics of the Fascist and National Socialist Dictatorships* (Cambridge: Cambridge University Press, 2007), 14–15.

oil over which it could exercise political control, and this meant the Middle East. The chapter also examines how oil influenced Britain's war aims in the Middle East and sparked Anglo-American competition over the region's oil.

Chapter 2 outlines how Britain's postwar financial constraints influenced internal government debates concerning naval oil stockpiling, before shifting to Iraq, which, along with Iran, was one of the centerpieces of Britain's oil strategy. It also discusses the formation of the Oil Board of the Committee of Imperial Defence, which was responsible for assessing Britain's wartime oil position. Britain's mixed record in the Middle East is contrasted with the development of the oil industry in Venezuela, which by the early 1930s was on its way to becoming Britain's most important oil supplier.

Chapters 3 and 4 demonstrate how and why Britain failed to achieve energy independence. Although the British did have some success in boosting output in Iran and Iraq, their political position in both nations was tenuous. The primary threat was from resource nationalism, which reared its head in Persia in 1932–3 and a few years later in Mexico. The second and most challenging threat was from Italy following the Abyssinian crisis of 1935–6. Although Italian hostility would doom Britain's plan to achieve oil independence through the Middle East, during the deliberations concerning their response to Italian aggression against Abyssinia British officials gave little attention to the risk to their oil lifeline until it was too late—yet another instance of policymakers, strategists, and planners ignoring the mundane topic of logistics, with grave consequences. While inferior to the Royal Navy, the Italian Navy, singly or with German assistance, could make the cost of importing Middle Eastern oil through the Mediterranean prohibitive for Britain. Across the Atlantic, Britain dodged a bullet when the United States did not ban oil exports under the Neutrality Acts, but U.S. restrictions imposed further strain on Britain's limited supplies of tanker tonnage and foreign exchange.

Chapter 4 provides a critical assessment of Britain's post-World War I oil strategy and details the strategic consequences of its failure during the early years of World War II. After British control of the Mediterranean came into question following Italy's entry into the war in June 1940, Whitehall was caught in a trap. Britain could only afford oil from the Middle East, which was priced in sterling, but it lacked enough tankers to reroute supplies around the Cape of Good Hope. Importing oil from the Western Hemisphere solved the tanker shortage, but it also strained British finances because it required payment in dollars. The British were unable to resolve the tradeoff between their logistical and financial difficulties, with the result that Britain's survival after 1940 would again depend upon the assistance of the United States.

Part II analyzes how oil shaped the Third Reich's plans for war, conquest, and extermination. Chapter 5 reveals that German politicians, soldiers, bureaucrats, and executives all placed enormous faith in Germany's "creative genius" to overcome its shortages of raw materials using technology.[46] Foremost in their minds was the experience of World War I—not simply the Allies' economic and financial

[46] "Ölstrategie," *Militär-Wochenblatt* (January 15, 1937).

advantage, but also Germany's vulnerability to economic strangulation through a British blockade. The National Socialists' "seizure of power" in 1933, therefore, did not represent a significant break in terms of energy policy. Despite its archaic values, National Socialism aimed to modernize Germany socially and economically. The regime reaffirmed Weimar-era measures supporting synthetic fuel but also went further by championing domestic oil exploration and the stockpiling of imports.

Chapter 6 sketches the origins and progress of the Four-Year Plan of 1936 to achieve economic self-sufficiency. Self-sufficiency was never an end in and of itself—rather, Germany's petroleum strategy aimed to provide its armed forces and economy with sufficient supplies to wage a war of conquest and resource acquisition. Germany did not have to be completely ready for war—which was impossible considering its economic deficiencies—only relatively better prepared than its adversaries.[47] By 1938–9, however, even this narrow objective was falling out of reach due to material and labor shortages. Time was also running out due to Germany's bellicose foreign policy. To bridge the remaining deficit between consumption and supplies, Germany sought to repair its relationship with Romania. By the autumn of 1939, there was a broad consensus within Germany that its petroleum supply could, assuming access to Romania and unimpaired domestic production, sustain campaigns of limited duration against a specific adversary.

Chapter 7 reassesses Germany's wartime petroleum strategy. Fuel consumption during Germany's early campaigns (1939–40) was lower than expected but acquiring imports from Romania and the Soviet Union proved logistically challenging. The conquest of Western Europe resolved the short-term difficulties but left Germany with a long-term crisis. Before the war, Europe had imported most of its petroleum from overseas. Germany's prewar efforts had only aimed to make the Third Reich self-sufficient—they could not hope to replace the supplies other European nations had imported from abroad. Since Axis Europe's immediate and long-term energy needs vastly exceeded the capacity of either the synthetic fuel or Romanian oil industries, unless Germany could exploit of the oil resources of either the Soviet Union or the Middle East, fuel shortages would soon derail the entire war effort. This looming energy crisis is not a sufficient explanation for Hitler's desire to invade the Soviet Union, whose conquest was central to his worldview, but it does clarify why he was willing to risk a two-front war in 1941 (Operation Barbarossa), and why he could not afford to wait until after he had subdued Britain.

Chapter 8 places oil at the center of the National Socialists' wider plans for the economic exploitation of the Soviet Union. Time was running out for the Germans—their supply position was already tenuous by the spring of 1941. By the autumn, Germany had burned through its entire operational reserve, even as Soviet resistance to Operation Barbarossa stiffened. Efforts to boost production in Romania foundered in the face of Romanian intransigence, while logistical

[47] Williamson Murray, *Change in the European Balance of Power, 1938–1939: The Path to Ruin* (Princeton, Princeton University Press, 1984), 3–49.

difficulties starved Germany's offensive spearheads in the Soviet Union of fuel. The failure of Operation Barbarossa not only ended Germany's best chance of winning World War II outright but also sealed its fate as a great power.

TERMINOLOGY AND MEASUREMENTS

The oil industry divides its work between upstream and downstream operations. The former deals with prospecting for and developing new sources of oil, whereas the latter refers to the process of getting it to consumers—refining, transportation, and distribution. The chemical properties of various kinds of crude oil determine the volume of space it occupies. One ton of oil can fill between six to eight (42 U.S. gallon) barrels (bbl) depending upon its specific gravity—how heavy or light it is compared to water—and viscosity—resistance to flow. For the sake of clarity, throughout this study, one ton of oil equals seven bbl of oil, and vice versa. Crude oil can also either be sweet or sour depending on its sulfur content. Sweet crude usually has a sulfur content of less than 0.5 percent and fetches a premium relative to sour crude because it is easier to transport and refine.

Americans enumerate oil in (42 U.S. gallon) bbl, whereas Europeans use metric tons. Rather than converting one to the other, I have used both measures depending on the source at hand. Unfortunately, many primary sources do not specify whether they are referring to "long," "short," or metric tons (2,240 lbs versus 2,000 lbs versus 2,204 lbs). The oil industry conducts business in "long" tons, and I have assumed that my sources do as well unless otherwise indicated.[48] Finally, I have simplified all figures greater than 100,000. Figures in the hundreds of thousands are rounded to the nearest thousand, those in the millions to the nearest ten thousand, and those in the billions to the nearest ten million. This means, for example, that 100,500 becomes 101,000; 1,015,000 becomes 1.02 million; and 1,015,000,000 becomes 1.02 billion.

[48] J.H. Bamberg, *Anglo-Iranian Years, 1928–1954*, vol. 2 of *History of the British Petroleum Company* (Cambridge: Cambridge University Press, 1994), 56–7.

PART I

BRITAIN

1

The Allure of Independence

1914–1921

During the first stage of the hydrocarbon era,[1] Britain was the "Saudi Arabia of coal."[2] Up to 2008, British coal mines produced as much energy as Saudi Arabia's oilfields.[3] Even the surge in global coal production during the second half of the nineteenth century—from 80 million to 1.3 billion tons between 1850 and 1914—still left Britain with a 25 percent share.[4] Oil was another story. While the sun may not have set on the British Empire, that empire produced only 1.52 million tons in 1913—2.5 percent of global production—against 4.7 million tons of consumption.[5] Whitehall had already recognized that oil had both strategic and commercial significance before the Great War due to its superiority over coal as a source of naval fuel. When war finally came, oil's impact extended far beyond the naval domain, transforming land warfare and creating an entirely new domain in the air.

Even as the war raged, British officials fretted over the precariousness of their oil supplies. The Allies could not have fought and won World War I without oil imported from the United States, which supplied 80 percent of their needs. Whitehall soon concluded that it must avoid depending on rival powers in future—but how exactly was Britain supposed to achieve independence? As an island nation with no appreciable domestic sources of oil, it was impossible to eliminate imports altogether. Rather, independence for Britain meant foregoing imports from areas not under British control—especially the United States, which also controlled access to the Western Hemisphere, the source of three-quarters of world oil production before World War II. British strategy required finding new sources of oil, developing ones already under British control, and patronizing British firms capable of supplying the empire. Contemporary journalists described this as "anglicising" Britain's oil supply. The stakes were potentially transformative—foreign observers hypothesized that a British success could tip the Anglo-American balance of power back in Whitehall's favor.[6]

[1] All primary sources cited in Chapters 1–4 are from the BNA unless otherwise indicated.

[2] Lesser, *Resources*, 25.

[3] Mitchell, *Carbon Democracy*, 14, n. 5.

[4] Jürgen Osterhammel, *Transformation of the World: A Global History of the Nineteenth Century* (Princeton: Princeton University Press, 2014), 655.

[5] These figures are drawn from: Petroleum Executive, "Negotiations…Volume I," March 1919, pp. 6 and 108–12, ADM 116/3452 (*PPHM*, i); "Petroleum Production," no date, attached to: Cadman to Amery, May 12, 1919, S.140/J., CO 323/813/50.

[6] E.H. Davenport and Sidney Russell Cooke, *Oil Trusts and Anglo-American Relations* (London: Macmillan and Co., 1923), 38–47 (quotation from p. 47); memorandum for Wirkl and Bücher, no date or author (circa 1921), PAAA, R 118173.

The focus of British strategy would be on the Middle East. Historians have been too credulous of British officials' "lusty denials" of oil's centrality to Britain's strategy in the Middle East.[7] By 1918, Whitehall was convinced that the road to energy independence ran through the Middle East. If all went according to plan, Britain would no longer depend upon the United States or U.S. companies for its oil needs during either peace or war. For a country that lived and died on imports, this was as close to energy independence as was possible. Moreover, by drawing supplies from regions already under British control and binding them closer to the metropole, British private interests would also be contributing to the wider public project of restructuring the empire as a single political, economic, and military commonwealth.

PREWAR DEVELOPMENTS, 1900–1914

At the turn of the century, most of Britain's demand for petroleum was limited to kerosene, which comprised 84 percent of British oil imports in 1900. By the eve of World War I, the situation had changed—gasoline and fuel oil accounted for 40 percent of British petroleum imports by 1913.[8] The government's attitude was also evolving, as evidenced by Whitehall's purchase of a 51 percent stake in Anglo-Persian in 1914. Ironically, the impetus had come from Anglo-Persian, which was following in the footsteps of Shell and Mexican Eagle (the latter was then the largest oil company in Mexico). Both firms had sought government support between 1902 and 1913.[9] British oil companies were confident that government support would provide commercial advantages without compromising their corporate autonomy, since British officials were loath to interfere with the free market.[10] Before 1914, however, Whitehall had preferred to lend its good offices rather than cash to troubled firms, such as William Knox D'Arcy's First Exploitation Company.[11]

D'Arcy won the oil concession to Persia in 1901, during the height of Anglo-Russian jockeying for influence in that country.[12] Russia was Persia's largest trading partner after 1910 and had a large market for its kerosene there. Russian oil

[7] Contra Clive Leatherdale, *Britain and Saudi Arabia, 1925–1939: The Imperial Oasis* (Abingdon: Frank Cass, 1983), 206–7 and 211–13, see: Peter Sluglett, *Britain in Iraq: Contriving King and Country* (London: I.B. Tauris, 2007), 65–6 and 75 (quotation from p. 75).

[8] T.A.B. Corley, "Oil Companies and the Role of Government: The Case of Britain, 1900–75," in: *Competitiveness and the State: Government and Business in Twentieth-Century Britain*, ed. Geoffrey Jones and Maurice Kirby (Manchester: Manchester University Press, 1991), 162–3.

[9] Geoffrey Jones, *State and the Emergence of the British Oil Industry* (London: Macmillan, 1981), 23–6; Cowdray to Churchill, "Mexican Oil Supply [...]," June 14, 1912; Fisher to Churchill, August 1913; CA, CHAR 13/9/71 and 13/21/31.

[10] Geoffrey Jones, "British Government and the Oil Companies, 1912–1924: The Search for an Oil Policy," Historical Journal (1977): 647–54.

[11] The following discussion draws from: Jones, *Industry*, 128–41; Ferrier, *History*, 27–113; Mostafa Elm, *Oil, Power, and Principle: Iran's Oil Nationalization and its Aftermath* (Syracuse: Syracuse University Press, 1992), 1–22.

[12] For the early history of the Persian oil industry, see: Ferrier, *History*, 15–27, and "The Iranian Oil Industry," in: From Nadir Shah to the Islamic Republic, ed. Peter Avery, et al., vol. 7 of *Cambridge History of Iran* (Cambridge: Cambridge University Press, 2008), 639ff.

companies also hoped to construct a pipeline through Persia to the Gulf that would allow them to market kerosene in India. The D'Arcy concession, while excluding the provinces bordering Russia, compelled Tehran to reject Russian efforts to build a pipeline across Persia.[13] Drilling for oil proved more expensive than planned, and to dissuade D'Arcy from selling his concession to either the Russians or French Rothschilds, Whitehall arranged for financing through the Burmah Oil Company in 1904/5.[14] Burmah dominated oil production and distribution within India, but its leadership was wary of the D'Arcy concession. It had no market for Persian oil but were worried some other company would use it to challenge Burmah's control of the Indian market.[15] Burmah took over the D'Arcy concession through a subsidiary, Concessions Syndicate Ltd, in 1905, and reorganized it as Anglo-Persian in 1909.[16]

Whitehall's brokering of a partnership between D'Arcy and Burmah reflected the government's awareness that it had to make a more energetic defense of British oil interests.[17] In Persia, that meant preserving British control of the oil concession. Unfortunately, Anglo-Persian was not making things easy. Although the company discovered oil in 1908 and constructed a pipeline to its refinery at Abadan three years later, its position was dire. For one thing, the refinery produced only substandard kerosene due to Persian oil's high sulfur content. The company was also in danger of being absorbed by Shell, with which it had signed a marketing agreement in 1912. Under the leadership of its managing director, Charles Greenway, Anglo-Persian began exploring ways to preserve its independence. The company lacked the downstream capacity to dispose of its Persian production and was having trouble arranging private financing.[18] In exchange for an annual subsidy, a naval-supply contract, and a deal to supply the Indian railways with fuel oil, Greenway offered Whitehall an annual supply of 500,000 tons at below market prices and government representation on Anglo-Persian's board.[19]

Anglo-Persian also wanted to leverage Whitehall's support to secure the oil concession to Mesopotamia, where it was competing with the Turkish Petroleum Company (TPC). The TPC started life in 1912 as an Anglo-Dutch-German

[13] Mitchell, *Carbon Democracy*, 49–50. The D'Arcy concession is reprinted in: Ferrier, *History*, 640–3.

[14] Daniel Yergin, The *Prize: The Epic Quest for Oil, Money, and Power* (New York: Simon & Schuster, 1991), 138ff. The most likely suitors were the Rothschilds, whom D'Arcy approached in 1904. Ferrier, *History*, 59–62; T.A.B. Corley, *History of the Burmah Oil Company, 1886–1924* (New York: Heinemann, 1983), 98–101.

[15] Warwick Michael Brown, "Royal Navy's Fuel Supplies, 1898–1939: The Transition from Coal to Oil" (Ph.D. Dissertation, KCL, 2003), 89–91; Mitchell, *Carbon Democracy*, 43–65; Yergin, *Prize*, 141–2.

[16] Corley, *Burmah*, 95–111 and 128–45.

[17] P.J. Cain and A.G. Hopkins, *British Imperialism: Innovation and Expansion, 1688–1914* (London: Longman, 1993), 408ff.

[18] See Greenway's speech of January 25, 1927 in: CA, CHAR 2/151/16–24.

[19] Marian Jack, "The Purchase of the British Government's Shares in the British Petroleum Company, 1912–1914," *Past & Present* 39 (1968), 142–3; Jones, *Industry*, 144ff; Yergin, *Prize*, 149 and 158–9; Brown, "Fuel Supplies," 91ff.

consortium to prospect for oil in the Ottoman Empire.[20] Before World War I, Britain was not the only great power that coveted Mesopotamian oil—Deutsche Bank had received a concession in 1904, and although it had lapsed in 1907, the bank had not given up its ambitions.[21] Whitehall determined to prevent Germany from acquiring control of Mesopotamia, since it assumed Deutsche Bank was acting as a proxy for the German Navy.[22] In fact, the Germans preferred to collaborate rather than compete—in 1912, Deutsche Bank and Shell agreed to partner with the National Bank of Turkey, a British-owned bank that financed development projects within the empire (see Chapter 5). Whitehall grasped the opportunity and applied pressure on Anglo-Persian, which decided to join the TPC to prevent its rivals from flooding the market with Mesopotamian oil.[23] In March 1914, the TPC, Anglo-Persian, and representatives of both the British and German governments signed the "Foreign Office" Agreement, which transferred the National Bank's shares to Anglo-Persian. The agreement also included a "self-denying" ordinance that enjoined signatories from operations anywhere within the Ottoman Empire except through the TPC or with the consent of the three major partners: Anglo-Persian (47.5 percent of the TPC's shares), Deutsche Bank (25 percent), and Shell (22.5 percent).[24]

Why at this point did Whitehall take the unprecedented step of purchasing majority control of a private firm? The answer was oil's relationship to British naval supremacy, which was under threat from Germany. Overcoming that threat required embracing technological innovations that preserved Britain's operational and tactical superiority at sea. First Sea Lord John ("Jackie") Fisher's re-conception of British naval strategy depended upon increased firepower and speed—four to five knots for battleships.[25] Coal consumption was increasing so quickly—50 percent per knot for vessels constructed in 1897 and 1909—that coal-fired ships would soon reach their ceiling in speed and range.[26]

Only replacing coal with oil could resolve the dilemma between speed and endurance, which had bedeviled navies since the introduction of steam. Securing oil was, in turn, "the main reason for the growth of the State's interests in the affairs of the oil companies," who now became "suppliers of a strategic commodity on a

[20] The following discussion draws from: Edward Mead Earle, "The Turkish Petroleum Company—A Study in Oleaginous Diplomacy," *Political Science Quarterly* 39: 2 (1924): 265–77; Gregory Nowell, *Mercantile States and the World Oil Cartel, 1900–1939* (Ithaca: Cornell University Press, 1994), 53–76, 116–32, and 183–91; Yergin, *Prize*, 184–206.

[21] Inge Baumgart and Horst Benneckenstein, "Die Interessen der Deutschen Bank am mesopotamischen Erdöl," *Jahrbuch für Wirtschaftsgeschichte* 1 (1988): 49–60.

[22] Fisher to Churchill, October 10, 1912, CA, CHAR 13/16/103.

[23] Jack, "Shares," 141–7; Fiona Venn, "In Pursuit of National Security: The Foreign Office and Middle Eastern Oil," in: *Foreign Office, Commerce and British Foreign Policy in the Twentieth Century*, ed. John Fisher et al. (London: Palgrave, 2016), 71–2.

[24] Reprinted in: D.J. Payton-Smith, *Oil: A Study of War-time Policy and Administration* (London: HMSO, 1971), 26–7.

[25] Winston Churchill, *World Crisis, 1911–1914* (Toronto: Macmillan, 1923), 125–33; Martin Gibson, "'Oil Fuel Will Absolutely Revolutionize Naval Strategy': The Royal Navy's Adoption of Oil before the First World War," in: *A Military Transformed? Adaptation and Innovation in the British Military, 1792–1945*, ed. Michael LoCicero et al. (Helion: Solihull, 2014), 116.

[26] Brown, "Oil Supplies," 26–7.

par with armaments."[27] The government took this step without any evidence that oil supplies would be sufficient for the next twenty-five years (the lifespan of the ships), which is why Fisher also pressed for the largest possible reserves.[28] The British hoped that output would keep pace with consumption, but this was a gamble—contrary to the transition from wood to coal in the United States, the Royal Navy expected that plentiful future demand for oil would create additional supplies, rather than vice versa.[29]

The Royal Navy had been experimenting with petroleum since 1865 but progress was slow until 1904.[30] Thereafter, the Admiralty's annual demand for fuel oil spiked from 1,200 tons at a price of 40/- per ton in 1902 to 279,000 tons at a cost (including freight) of more than 51/- by 1912.[31] Much of the impetus came from Fisher. A self-confessed "oil maniac" since 1886, Fisher was obsessed with oil's transformative possibilities, either in naval propulsion or through the development of submarines and aviation.[32] Although retired since January 1910, in the winter of 1911/12 Fisher bombarded the First Lord of the Admiralty, Winston Churchill, with missives on the value of oil and urged him to meet with the cofounder of Shell, Marcus Samuel. To Fisher, the superiority of oil was irrefutable in terms of efficiency and cost. Based on assurances from Samuel, he was also sanguine about long-term supplies.[33]

The Royal Navy already had fifty-six destroyers and seventy-four submarines burning oil exclusively when Churchill became First Lord in 1911, but it had shrunk from making oil its only source of naval fuel. Churchill took the "fateful plunge" the following April when he included construction of five oil-fired dreadnoughts (the *Queen Elizabeth* class) in that year's Naval Estimates.[34] Today, the shift to oil appears overdetermined, but if any nation could afford to buck the trend, it was Britain. The supply and cost of British coal were stable, its quality was high, and even allowing for oil's higher thermal content, coal was still cheaper (almost 50 percent in home waters). The coal industry also had a sophisticated infrastructure for distributing coal at home and abroad using overseas bases and colliers.[35] By contrast, thanks to oil, the Royal Navy's fuel costs as a share of expenditures

[27] Jones, *Industry*, 9. The best study of the Royal Navy's conversion to oil and Whitehall's purchase of shares in Anglo-Persian remains Jack, "Shares," 139–68; although Brown, "Fuel Supplies," 41–115, Ferrier, *History*, 158–201, and Jones, *Industry*, 9–31 and 144–76 are still useful.

[28] Fisher to Hopwood, August 31, 1912, CA, CHAR 13/16/88–92.

[29] Compare Fisher to Greene, August 1913, CA, CHAR 13/21/106 with Christopher Jones, "A Landscape of Energy Abundance: Anthracite Coal Canals and the Roots of American Fossil Fuel Dependence, 1820–1860," *Environmental History* 15 (July 2010): 453–4 and 461ff.

[30] Brown, "Oil Supplies," 41–3 and 47–9; Gibson, "Strategy," 111–14; David Allen Snyder, "Petroleum and Power: Naval Fuel Technology and the Anglo-American Struggle for Core Hegemony, 1889–1922" (Ph.D. Dissertation, Texas A&M University, 2001), 121–44.

[31] Brown, "Oil Supplies," 283–4.

[32] Fisher's role has been exaggerated. See Brown, "Oil Supplies," 59–60; Gibson, "Strategy," 115.

[33] Fisher to Churchill, December 3, 1911; December 7, 1911; January 1912; February 25, 1912; March 17, 1912; CA, CHAR 13/12/122–3 and 128; CHAR 13/16/1–3, 28–31, and 56.

[34] Brown, "Fuel Supplies," 65–70; Churchill, *World Crisis*, 133ff.

[35] "Tables [...]," no date or author (circa 1913), CA, CHAR 13/23/21. See also: Brown, "Oil Supplies," 12–40 and 259–60.

doubled between 1900/1 and 1912 following the 1904 decision to construct battleships and cruisers capable of burning either coal or oil.[36]

Supporters of oil retorted that scarcity and high prices were self-inflicted because of the Admiralty's stringent specifications.[37] For oil's boosters, its advantages were legion.[38] The most important was its caloric value: 33 percent greater than an equivalent amount of coal. One ton of oil required thirty-eight cubic feet of storage compared to forty-three for coal, and the former did not deteriorate in quality once extracted, unlike the latter, which lost 30 percent of its thermal content per year depending on the climate.[39] Oil's higher thermal content created a saving of space and weight, and enabled a 50 percent reduction in stokers. By 1934, the Royal Navy had slashed 15,000 stokers from its rolls—an annual saving of £30 million.[40]

Oil's chemical properties were less significant than the gains in "performance of the end-user technologies"—the engines.[41] More powerful engines provided superior performance and lethality by giving warships higher speeds, quicker acceleration, longer range, thicker armor protection, and heavier armament. The problem of refueling was also simplified. The record in the Royal Navy for refueling with coal (in port) was 1,450 tons in 3.5 hours, but an oil-fired ship could load the same amount of energy in fifteen minutes.[42] Unlike with coal, ships powered by oil could be refueled at sea. This further increased the range of oil-fueled ships and ensured that British fleets could go into battle with their full complement of vessels.[43] Oil also facilitated the blockade of enemy ports by allowing ships to remain on station for longer periods.[44]

The benefits were not limited to warships; merchant ships could also increase their efficiency. Whereas one ton of coal burned in a steam engine could move a ton of freight 60,000 miles, a ton of oil could move the same weight up to 90,000 miles or as much as 200,000 miles with an internal combustion engine. This meant that a 10,000-ton steamer sailing between Britain and Singapore (8,190 miles) at 11 knots required 1,365 tons of coal, or 1,020 tons of oil with a steam engine, and 545 tons with an internal combustion engine.[45] Finally,

[36] Jones, *Industry*, 13. Even before the shift, the Royal Navy expected its fuel oil consumption to double from 83,000 tons to 156,000 tons between 1910/11 and 1912/13. Admiralty, "Notes [...]," no date (circa November 1911), CA, CHAR 13/23/10.

[37] Churchill, Minute, CA, CHAR 13/6A/14–19.

[38] "Extracts [...]," attached to: Churchill, "Proposed Agreement [...]," May 11, 1914, CA, CHAR 13/31/24. See also: Brown, "Oil Supplies," 44–7.

[39] Brown, "Oil Supplies," 27.

[40] "Öl gegen Kohle, endgültig?" *Deutsche Wehr* (July 19, 1934); Waldener-Harz, "Kohle oder Öl," *Militär-Wochenblatt* (October 18, 1934).

[41] Volkan Ediger and John Bowlus, "A Farewell to King Coal: Geopolitics, Energy Security, and the Transition to Oil," Historical Journal (2018): 2 and 6.

[42] Clipping enclosed with: Fisher to Churchill, January 1913, CA, CHAR 13/21/131–133. For more on the difficulties of refueling with coal, see: Brown, "Oil Supplies," 16–21.

[43] Fisher to Churchill, July 24, 1912, CA, CHAR 13/16/73–76; Warwick Michael Brown, "When Dreams Confront Reality: Replenishment at Sea in the Era of Coal," International Journal of Naval History 9 (2010): 1–3.

[44] Churchill, "Oil," March 1914, CA, CHAR 13/6A/4–10; Ediger, "Geopolitics," 11–22.

[45] Petroleum Imperial Policy Committee (PIPC), "Oil Policy [...]," May 29, 1918, POWE 33/44. The British never converted the Royal Navy from steam turbines to diesel-powered internal combustion

supporters of oil observed that only Britain was energy independent with regard to coal—Royal Navy ships on deployment, however, depended on coaling stations supplied from overseas and vulnerable to interdiction.[46]

The fact other navies were considering a shift to oil convinced the British to break with their practice of not adopting technological innovations that eroded their numerical supremacy.[47] Fisher expected the Germans would try to keep pace but be unsuccessful.[48] He was right—the Germans went no farther than installing supplementary oil burners starting in 1909.[49] The country best suited to following Britain was the United States. The U.S. Congress had first appropriated money in 1866 to investigate oil-fueled engines. By the turn of the century, concerns over coal supplies had led to the creation of a Liquid Fuel Board that reported in favor of converting to oil in 1904. In 1910, the U.S. Navy switched to burning only oil in its destroyers and submarines, and the following year the Congress authorized construction of the first oil-fired U.S. battleships (the *Nevada* class).[50] Fisher used these developments as further justification for converting the Royal Navy to burning oil exclusively.[51]

In Britain, two committees considered the question of securing future requirements: the Pakenham Committee, which met between 1911 and 1912, and the Royal Commission on Fuel and Engines, established by Churchill in July 1912 and led by Fisher, which issued reports in November 1912, February 1913, and February 1914.[52] Churchill urged Fisher to solve the question of how the Britain should buy and store the necessary oil.[53]

In June 1913, based on the Royal Commission's findings, Churchill laid down three principles that served as the basis of Britain's strategy. The first was the geographic dispersal of supplies to avoid dependence on any single source. The second was promoting competition in the oil industry to avoid relying on any one company. The third was drawing supplies from within either the empire or areas of British

engines because of technical difficulties and weight issues—although the former burned three times as much fuel, they weighed less, which was an important factor due to the tonnage restrictions of the 1922 Washington Naval Treaty. Brown, "Fuel Supplies," 244–5; Gibson, "Strategy," 118.

46 " 'Milestones' and Motors," April 1913, CA, CHAR 13/21/107–117.

47 Holger Herwig, "The German Reaction to the *Dreadnought* Revolution," *International History Review* 13: 2 (1991): 276–7 and 282.

48 Fisher to Churchill, January 1912; Fisher to Hopwood, August 31, 1912; Fisher to Churchill, September 13, 1912 and January 18, 1913: Churchill Archive (CA), CHAR 13/16/1–3, 87, 99, and 13/21/1–3.

49 German ships had a coal-to-oil ratio of 2:1. Holger Herwig, *"Luxury" Fleet: The Imperial German Navy, 1888–1918* (London: Ashfield, 1980), 64–6, 80, 82–3, and 201.

50 The Taft Administration had decided in 1911 to construct the U.S.S. *Nevada* as an oil-burning battleship, but the decision was not finalized until April 1913. John DeNovo, "Petroleum and the United States Navy before World War I," *Mississippi Historical Review* 41: 4 (1955): 641–56; John Maurer, "Fuel and the Battle Fleet: Coal, Oil, and American Naval Strategy," *Naval War College Review* 34: 6 (1982): 69–74; Peter Shulman, " 'Science Can Never Demobilize': The United States Navy and Petroleum Geology, 1898–1924," *History and Technology* 19: 4 (2003): 365–72.

51 Fisher to Churchill, December 3, 1911, CA, CHAR 13/2/122–123.

52 *PPHM*, i: 6. The findings of both committees are summarized in: Brown, "Fuel Supplies," 67–76.

53 Churchill to Fisher, June 11, 1912; Fisher to Churchill, June 17, 1912: CA, CHAR 13/5/25–26 and CHAR 13/6/61–63. See also: Yergin, *Prize*, 156–7.

influence. The best way to safeguard the navy's requirements, Churchill argued, would be through long-term (forward) supply contracts that locked in prices. Before 1912, the price of fuel oil exclusive of shipping costs (22/-) had been equivalent to coal. In 1913, however, it rose to 39/- and was set to rise to 50/- by 1914, while freight rates had increased from 17/- to 27/- per ton. Although Anglo-Persian would be the largest recipient of government orders—starting at 200,000 tons in 1915 and rising to 500,000 tons if necessary—at Fisher's urging Churchill also supported contracts with Shell (100,000 tons annually), Union Oil of California (80,000 tons), and Mexican Eagle (200,000).[54] Churchill estimated that government purchases would rise from 664,000 tons in 1913/14 to 729,000 tons by 1917/18. He also advised accumulating a reserve equivalent to six months of wartime requirements—from 441,000 tons in 1913 to 1.4 million tons by 1917—instead of the four-year peacetime reserve suggested by the Royal Commission.[55] The reason for this is unclear. According to the Admiralty's "conservative" estimate, switching to oil and accumulating a four-year peacetime reserve would require an additional expenditure of £884,800 between 1913 and 1920, assuming an average fuel price of 50/- per ton.[56] Churchill probably considered a four-year reserve unnecessary, since the navy merely wished to replicate the security it enjoyed with coal.[57]

During a July 1913 speech, Churchill ventured that the government's "ultimate policy" ought to be to own a portion of its requirements, as well as refining capacity. Giving the Admiralty the flexibility to buy, transport, and refine its own oil would enable it to take advantage of favorable market conditions.[58] The question of whether or not it was wise to base Britain's security on an imported commodity missed the point: "If we cannot get oil, we cannot get corn, we cannot get cotton, and we cannot get a thousand and one commodities necessary for the preservation of the economic energies of Great Britain."[59] A wave of labor unrest across the coal

[54] Fisher was keen on an arrangement with Mexican Eagle. Fisher to Churchill, July 16, 1912 and August 6, 1912: CA, CHAR 13/16/60 and 77; Fisher to Hopwood, August 31,1912: CA, CHAR 13/16/88–92. He also referenced a deal with a Californian company for one million tons annually for ten years at 53/- including freight, but there is no record of such a contract. Fisher to Churchill, April 24, 1913, CA, CHAR 13/21/12. The details of the arrangement with Shell are in: Churchill to Deterding, September 3, 1913, CA, CHAR, 13/22B/195–198. The terms of the Mexican Eagle deal are in: Brown, "Fuel Supplies," 101–2.

[55] Churchill, "Oil Fuel [...]," June 16, 1913, CAB 37/115. See also: Yergin, *Prize*, 159–60. Churchill misremembered the commission's recommendations, claiming it had recommended a four-year war reserve. This would have amounted to 10.5 million tons according to estimated wartime consumption by 1917–18, as opposed to 2.08 million tons for four years of peacetime consumption by 1917. Contra Churchill, *World Crisis*, 179–81, see: Fisher to Marsh, November 27, 1912, CA, CHAR 13/16/109–110. Jack, "Shares Purchase," 152–3.

[56] Forsey, "Financial Effect [...]," February 4, 1913, CA, CHAR 13/23/30.

[57] Brown, "Fuel Supplies," 261–2.

[58] Churchill, Minute, 1913; Churchill to Hopwood, June 1, 1913, CA, CHAR, 13/22A/130 and 13/6A/25.

[59] "Navy Estimates...July 17, 1913," reprinted in: *Winston S. Churchill: His Complete Speeches, 1897–1963*, ed. Robert Rhodes James (New York: Chelsea House, 1974), ii: 2125–32 (quotation from p. 2128).

and transportation sectors between 1910 and 1914 called into question the reliability of domestic coal supplies.[60]

Churchill and his supporters were determined to break the dominance of foreign oil companies. The First Lord excoriated Shell in the House of Commons as the source of newspaper articles critical of the Admiralty for considering a supply contract with Mexican Eagle. The Royal Navy, he argued, could not afford to rely on any company "under foreign or potentially hostile control."[61] Instead, Whitehall should cultivate British alternatives such as Anglo-Persian or Burmah. Churchill's policy concerning Anglo-Persian went far beyond the recommendations of his own Royal Commission, but a spike in prices and lack of interest from any firms besides Mexican Eagle and Anglo-Persian suggested that the oil market was not a reliable supplier.[62]

The Admiralty already had a small supply contract with Burmah. Thanks to protection from the government of India, which imposed tariffs on imported petroleum between 1888 and 1911 and blocked attempts by Standard Oil and Shell to drill in Burma in 1902 and 1904, respectively, Burmah controlled oil production and distribution within India. By 1924, it was the world's tenth-largest oil company by capitalization, even though it had virtually no commercial presence beyond the subcontinent and accounted for only 1 percent of global production. Under their 1905 agreement, the Royal Navy could acquire up to 10,000 tons per year in peacetime and 100,000 tons in wartime. In fact, the Admiralty only purchased 11,000 tons a year during World War I, since the Royal Navy did not have any significant threats in the Indian or Pacific Oceans, and because Burmese oil was light and better suited for refining into kerosene than fuel oil.[63]

Churchill knew that he had a willing partner in Anglo-Persian. The Admiralty and the Foreign Office were supportive, but the Treasury was skeptical. To assuage any doubts about the productivity of the Persian oilfields, the Admiralty dispatched a team led by Vice Admiral Edmond Slade and John Cadman. Both men had impressive resumes—Slade was the former Commandant of the Royal Naval War College and Director of Naval Intelligence, while Cadman was a professor of petroleum geology at the University of Birmingham and previously served as a government mining inspector in England and Trinidad. The assignment was a turning point in both men's careers: Slade soon joined Anglo-Persian's board as one of the two government directors, while Cadman became Whitehall's chief adviser on oil policy in 1917, then an Anglo-Persian board member, and finally its chair in 1929.

None of that seemed likely when they visited Persia in December 1913. Construction of the Abadan refinery was behind schedule—output that year was only 33,000 tons.[64] The most pressing problem was the quality of Anglo-Persian's

[60] Mitchell, *Carbon Democracy*, 61–5. There is, however, no proof that such fears affected policy. Brown, "Fuel Supplies," 268, n. 26.

[61] Churchill speech, 1912, CA, CHAR 13/13/179–182.

[62] Brown, "Fuel Supplies," 74–7 and 94.

[63] Brown, "Oil Supplies," 58; Corley, *Burmah*, 1–3, 64–5, 78–94, and 229–30; Jones, *Industry*, 88–105.

[64] Ferrier, *History*, 278.

fuel oil, as its viscosity was too high. This was not a problem in warmer temperatures, but the Royal Navy's primary theater of operations was the North Sea, where the colder temperatures rendered Persian fuel oil unusable. The company failed to resolve this problem until after World War I began, to the point that the navy faced "considerable difficulties."[65] Slade's team nevertheless issued a favorable report in April 1914, which—having made no mention of the aforementioned difficulties—sufficed to pry the necessary funds from the Treasury.[66]

In retrospect, Whitehall took a gamble when it purchased a majority stake in Anglo-Persian. Although Fisher supported relaxing Admiralty specifications to take advantage of price differentials, he had opposed purchasing Mexican oil in 1912 due to its sulfur content, suggesting holding off until the Admiralty was certain it could resolve the technical difficulties.[67] Churchill, however, saw no reason to delay signing contracts with Mexican Eagle and instructed that the *Queen Elizabeth*-class dreadnoughts be modified to burn oil with a greater viscosity.[68] Fisher's Royal Commission also suggested that blending heavier and lighter oils might solve the problem.[69] Optimism and frugality seem likelier explanations for British decision-making than corruption. Fisher and Churchill were careful to avoid even the hint of impropriety.[70] Fisher sold any oil company shares he owned before joining the Royal Commission in 1913.[71] Churchill was also determined to control the spiraling costs of the shift to oil, which exceeded £10 million just for accumulating reserves, but this was only possible if the Admiralty could purchase cheaper Mexican or Persian oil.[72] Clearly, Whitehall's decision to bail out Anglo-Persian went beyond commercial considerations—thwarting Shell, assisting Burmah, finding a cheap supplier for the Indian railways, and maintaining British predominance in Persia and Mesopotamia were all factors.[73] For all Anglo-Persian's shortcomings, its oilfields seemed to satisfy Churchill's criteria of being "strategically and politically convenient and easy to protect"—but only in the absence of any credible military threat.[74]

In June 1914, following a contentious debate in the House of Commons, the government purchased a 51 percent share in Anglo-Persian for £2.2 million and

[65] Ferrier, *History*, 148–53 and 199–200; Corley, *Burmah*, 193–4; Brown, "Fuel Supplies," 141–2.

[66] Slade et al., "Admiralty Commission [...]," April 6, 1914, attached to: Churchill, "Proposed Agreement [...]," May 11, 1914, CA, CHAR 13/31/24. See also: Brown, "Fuel Supplies," 94–7; Jack, "Shares," 158–62; Yergin, *Prize*, 160.

[67] Fisher to Hopwood, August 31, 1912, CA, CHAR 13/16/88–92.

[68] Churchill to Moore, March 1913; Churchill, Minute, March 29, 1913; Churchill, note, April 6, 1913; CA, CHAR 13/6A/11–12, 14–19, and 22–3. See also: Brown, "Fuel Supplies," 84–6.

[69] Fisher to Greene, August 1913, CA, CHAR 13/21/106.

[70] Churchill to Crewe, May 1, 1913, CA, CHAR 13/19/93–94.

[71] Fisher to Churchill, November 21, 1912 and April 20, 1913: CA, CHAR 13/16/107 and 13/21/11.

[72] Churchill, Minute, March 29, 1913, CA, CHAR 13/6A/14–19; Hopwood to Churchill, CA, CHAR 13/19/125–128; Churchill, *World Crisis*, 137.

[73] Churchill, "Proposed Arrangement [...]," July 4, 1913, CA, CHAR 13/23/31. Churchill expected the India Office to guarantee the interest on any loan to Anglo-Persian. Churchill to Grey, July 9, 1913, CA, CHAR 13/22A/133–134. See also: Mitchell, *Carbon Democracy*, 59–61.

[74] Churchill to Slade, September 30, 1913, CA, CHAR, 13/22B/226–227.

concluded a secret agreement with the Admiralty.[75] Whitehall would also nominate two members of the company's Board of Directors, who would enjoy veto rights on matters affecting the company's relationship with the government.[76] These appointees (usually from the Treasury and Admiralty) often went native and—like Slade, for instance—could not distinguish between Britain's interests and the company's.[77] Under the terms of the supply contract, the Admiralty was guaranteed six million tons of fuel oil over the next twenty years at a cost of 30/- per ton excluding freight (roughly 25–30/- per ton), minus a maximum rebate of 10/- per ton based on the company's excess profits. The final cost was comparable to the market price in May 1914 (65/- per ton).[78] Deliveries would start at 50,000–70,000 tons in 1914–15, rising to 300,000–350,000 tons by 1917–18 and 500,000 tons thereafter.[79] After the war, however, the terms became more favorable—as Anglo-Persian's profits rose, so did the rebate, which reached its maximum level in March 1921. Including freight charges and its rebate, the Admiralty paid Anglo-Persian as little as 40/- per ton of fuel oil in 1922, whereas the market price was 65/- and rising. The two parties revised their contract in 1928, and Anglo-Persian agreed to fulfill the remainder of its obligation at a fixed price of 22/- per ton.[80] This was from the Admiralty's point of view a "bargain," but Anglo-Persian's leadership was also pleased to preserve its independence from Shell and secure a commanding position within the TPC, which was poised to begin prospecting within Mesopotamia.[81]

Two parties, however, did lose out. The first was Tehran, whose oil revenues were linked to Anglo-Persian's profits and suffered because of the Admiralty discount. The second was Shell, as Whitehall's decision thwarted its acquisition of Anglo-Persian. Samuel questioned why the Admiralty was privileging Persian oil to the detriment of imperial sources or protectorates such as Egypt. He warned Whitehall that it was setting a precedent for other nations while discouraging development within the empire.[82] Churchill scoffed at Samuel's "blackmailing," but this charge was unfair since Shell's sales price to the Admiralty at the time was only slightly higher than Anglo-Persian's (70/- including freight versus 62/-).[83]

[75] *Agreement* [...], Cd. 7419 (London: HMSO, 1914), CAB 27/180. See also: Yergin, *Prize*, 160ff.

[76] The terms of the veto are spelled out in: Bradbury to Anglo-Persian, May 20, 1914, reprinted in: Ferrier, *History*, 645–6. The government first exercised its veto in 1920, when it opposed Anglo-Persian's offer of 100,000 shares to Tehran to settle outstanding disputes over royalties and wartime damages. Ferrier, *History*, 198–9 and 352–5.

[77] Slade's appointment is a mystery—he had no expertise in the oil industry, and Fisher derided him as a "fool." Corley, *Burmah*, 192.

[78] Churchill, "Proposed Agreement [...]," May 11, 1914, CA, CHAR 13/31/24; Brown, "Fuel Supplies," 97–8; Jack, "Shares Purchase," 166, n. 82.

[79] Wartime deliveries peaked at 424,000 tons for the Admiralty in 1918/19—well above the 350,000 tons contracted—plus another 148,000 tons to British forces in Mesopotamia. Ferrier, *History*, 289. Except for a brief dip from 1930 to 1931, they increased every year between 1922 and 1937, rising from 434,494 tons to 1,300,829. Brown, "Fuel Supplies," 304.

[80] "Admiralty Agreement [...]," no date (circa April/May 1928), T 161/1206.

[81] Brown, "Fuel Supplies," 113–15; Jack, "Shares Purchase," 162–7; Jones, *Industry*, 154–5. The price paid by the Admiralty averaged 22/-. British Petroleum Archive (BP), Arc. Ref. 78504. The 1914 supply contract expired in 1951. The Admiralty signed its last contract in 1959. See BP 63478.

[82] Samuel to Hopwood, May 29, 1914, attached to: Churchill to Cabinet, June 9, 1914, CAB 37/120.

[83] Brown, "Fuel Supplies," 99–100.

The shares purchase and the supply contract were complementary since the latter depended upon the commercial viability of Anglo-Persian, which was in danger until it received the infusion of public capital.[84] Churchill had no wish to interfere with the operations of a private enterprise. British officials had an abiding faith in the free market, and if they occasionally departed from it, it was to preserve competition. The fact that oil was also becoming a "strategic" commodity that was in short supply within the empire forced the government to swallow its liberal sentiments.[85] The decision was not taken lightly, but it was "the lesser of two evils."[86] On the one hand, Whitehall had a guaranteed supplier for a significant portion of the Admiralty's fuel requirements. On the other, the technological shift from coal to oil had shifted the political calculus in Anglo-Persian's favor. Whitehall had every reason to fear that the company would exploit its "special relationship" and become a veritable "Frankenstein" monster.[87]

THE GREAT WAR, 1914–1918

These fears would come to life during the global conflict that commenced only weeks after Whitehall completed its purchase of Anglo-Persian shares. The British Expeditionary Force (BEF) that arrived in France in August 1914 possessed only 1,200 trucks, compared to 60,000 horses.[88] Four years later, British forces would field 56,000 trucks, 23,000 cars, and 34,000 motorcycles, not including tanks and armored cars, and the Royal Air Force 22,000 aircraft.[89] Oil had finally become a "vital necessity" when waging war.[90] The Allies' oil supplies also came increasingly from just one place. In 1913, the United States supplied 62 percent of the 1.81 million tons that Britain imported, followed by Romania (11.8 percent), Russia (7.8 percent), the Dutch East Indies (7.7 percent), and Mexico (4.2 percent).[91] Britain and France lost access to Russian and Romanian oil following the Ottoman Empire's entry into the war in October 1914. Russian exports to France as a proportion of U.S. exports collapsed from 51 percent in 1911 to 0.9 percent in 1915.[92] Of the alternatives, Persian production was negligible and East Indies oil too far away. That left the United States, which quickly became the Allies' largest supplier. Production in the United States increased from thirty-four million tons in 1913 to forty-eight million tons in 1918, while exports during the war exceeded thirty-four million tons. During the 1917/18 fiscal year, of the 8.7 million tons the United States exported, 4.01 million tons went to Britain, plus another 2.19 million

[84] Slade et al., "Anglo-Persian Oil," February 10, 1914, BP 71402.
[85] Bamberg, *History*, 174–5.
[86] Walker to Long, February 28, 1918, P. 84548/48009, POWE 33/42.
[87] Anthony Sampson, *Seven Sisters: The Great Oil Companies and the World They Shaped* (New York: Viking Press, 1975), 56.
[88] Laux, "Trucks," 64.
[89] The following discussion draws from: Yergin, *Prize*, 167–83.
[90] *Official History of the Ministry of Munitions* (East Sussex: Naval & Military Press, 2009), 134–56.
[91] PIPC, "Oil Policy [...]," May 29, 1918, POWE 33/44.
[92] Nowell, *Cartel*, 91.

tons to the empire as a whole.[93] Of the 21.2 million tons the Allies imported during the war, 17.9 million (84 percent) came from the United States.[94] British imports of U.S. oil increased from 1.18 million tons in 1914 to four million tons in 1918 (or 70 percent and 77 percent, respectively, of total imports).[95]

Mexico was also an important supplier for Britain, even though the sulfur content of its oil was too high for Admiralty specifications unless blended with sweeter crude oils.[96] At its peak, Mexico accounted for 10 percent of British wartime oil imports (399,000 tons in 1918). It also tripled its share of U.S. imports during the war to 95 percent.[97] These exports indirectly benefited Britain by allowing the United States to build up a strategic reserve and export a larger percentage of its own production to the Allies.[98] Persia became an increasingly important supplier as the war progressed, supplying one-eighth of the Royal Navy's wartime consumption. Its value remained limited, however, due to internal security problems such as the sabotaging of Anglo-Persian pipeline in 1915, Persian oil's high viscosity, and Britain's limited supply of tanker tonnage—a tanker that made six round-trips to the Gulf of Mexico annually could only make four to Abadan. As early as August 1916, the First Lord of the Admiralty, Arthur Balfour, identified the key dilemma confronting British oil strategy—Britain need not fear any shortage of oil so long as it had enough tankers to carry it and money to pay for it.[99] In both world wars, Britain turned out not to have enough of either.

The logistical challenge was daunting because of the growth in oil imports—British imports by 1917 were already double those of 1913 (four million tons versus two million).[100] The pace quickened further as the war reached its climax. During the last six months of 1918, Britain consumed 3.2 million tons of petroleum, 97 percent of which was imported (77 percent from the United States). Of this, 2.2 million tons was fuel oil primarily for the Grand Fleet, whose oil-burning units in 1917 required 450,000 tons of imports per month. Against the Royal Navy's estimated annual demand of four million tons of fuel oil by 1918, the

[93] Friedensburg, *Erdöl*, 105–13.

[94] "Der amerikanische Kontinent" *Vierjahresplan*, 1939/XVI.

[95] Although the high point in percentage terms (82 percent) came in 1917. Friedensburg, *Erdöl*, 90.

[96] Jonathan Brown, *Oil and Revolution in Mexico* (Berkeley: University of California Press, 1993), 104; Brown, "Fuel Supplies," 75 and 102.

[97] Imports continued to grow after the war, quadrupling between 1918 and 1922 (from 37.72 million bbl to 126.2 million bbl). DeGolyer, *Historical Data*, 60–1. By 1920, Mexican oil exports (78.08 percent of which went to the United States) equaled 20 percent of U.S. consumption and were valued at $61 million (compared to $11 million in 1914). Linda Hall, *Oil, Banks, and Politics: The United States and Postrevolutionary Mexico, 1917–1924* (Austin: University of Texas Press, 1995), 13 and 61.

[98] Friedensburg, *Erdöl*, 90 and 106–7; Paul Garner, *British Lions and Mexican Eagles: Business, Politics, and Empire in the Career of Weetman Pearson in Mexico, 1889–1919* (Stanford: Stanford University Press, 2011), 204–5. In 1920, the United States still accounted for 59.8 percent of Britain's imports of 3,310,587 tons, and Mexico for 28.5 percent. The only other suppliers of consequence were Persia (4.5 percent) and the East Indies (3.7 percent). Appendix I to CID, "Production [...]," November 29, 1937, 272-A, attached to: Inskip, "Report [...]," February 3, 1938, C.P. 19 (38), CAB 24/274.

[99] Brown, "Fuel Supplies," 140–2 and 156–7.

[100] Treasury, "H.M. Petroleum Executive," October 3, 1918, POWE 33/150.

French and Italian navies only consumed 90,000 and 120,000 tons, respectively. Britain was, however, responsible for supplying 100 percent of the French navy's requirements (20,000 tons originating from the United States) and two-thirds of Italy's (one-third from the United States). Finally, the Allies and the United States (officially an Associated Power since it never joined the Allied Powers by treaty) required 167,000 tons of tanker tonnage to carry the 8.3 million tons they expected to consume between April 1, 1918 and March 31, 1919. On land, the BEF consumed only 6,900 bpd, which could be transported on a single oil tanker.[101]

Most of the BEF's requirements were for its truck fleet, which increased from 100 in August 1914 to 60,000 by war's end.[102] Trucks were inefficient for moving troops for long distances but ideal for getting them from railheads to the front.[103] The importance of trucks also rose along with the army's supply requirements. In 1916, British divisions required fifty wagonloads of supplies each day, and trucks were preferable because each five-ton truck could haul the same amount of goods as three teams of horses.[104] Four-wheel drive trucks also replaced horses as the Allies' primary means of moving artillery, with the French using 8,600 for that purpose by war's end.[105] Consequently, Allied consumption of oil during the last six months of the war reached 4.59 million tons, 80 percent of which came from the United States.[106] Due to lagging motorization, gasoline comprised only 20 percent of Allied consumption, compared to 60 percent for fuel oil.[107] As a percentage of U.S. production in 1918, U.S. exports were trivial—less than 10 percent.[108] The cost was not—between 1914 and 1918, Britain alone spent £170,393,000 on petroleum imports.[109]

At the start of the war, Whitehall allowed the oil industry to operate on a voluntary basis and contented itself with a variety of measures to limit consumption until the supply situation began to deteriorate in 1916.[110] At the start of the year, stocks of refined petroleum stood at thirty-six million gallons. By the end of July, in the midst of the Battle of the Somme, they had declined by two-thirds. The government now introduced rationing, and stocks swelled to 37.5 million gallons by the end of December.[111] Whitehall also centralized government purchases of petroleum under the supervision of the Board of Trade and created an Interdepartmental Petroleum Committee in February 1917 to offer guidance on

[101] Paul Foley, "Petroleum Problems of the World War: A Study in Practical Problems," *United States Naval Institute Proceedings* 50: 261 (1924): 1802–32.

[102] Jones, "Oil Companies," 655.

[103] Laux, "Trucks," 67.

[104] Jonker and van Zanden, *History of Shell*, 162–3.

[105] Laux, "Trucks," 66.

[106] Or nine million tons for all of 1918, including 1.93 million tons of gasoline and 5.35 million tons of fuel oil. Average monthly gasoline consumption was divided between 88,000 tons for vehicles, 42,000 tons for tanks, and 31,000 tons for aircraft. Friedensburg, *Erdöl* 15. The Royal Navy's cumulative wartime consumption exceeded nine million tons. Miles, "Supply [...]," 14 April 1920, ADM 203/48.

[107] Jonker and Zanden, *History of Shell*, 163.

[108] DeGolyer, *Historical Data*, 5.

[109] Friedensburg, *Erdöl*, 120. The United States also supplied Britain and France with 35,000 trucks. Laux, "Trucks," 67.

[110] Brown, "Fuel Supplies," 139–40; Corley, "Oil Companies," 164–5.

[111] Jones, "Oil Companies," 655–6.

wartime and postwar policy.[112] Meanwhile, the Pool Board handled coordination between the companies. In 1917, a new Petroleum Committee superseded the Interdepartmental Petroleum Committee. This new entity was under the leadership of the Colonial Secretary, Walter Long, and reported to the War Cabinet.[113] Renamed the Petroleum Executive in December 1917, it became the primary agency coordinating oil policy.[114]

Long received this authority because the supply situation had again become grave. Due to losses to U-boats, increased consumption, and a shortage of tankers, the Royal Navy estimated in June 1917 that Britain would use up its fuel stocks by January 1918 unless the government reduced consumption and tanker losses.[115] Long insisted that the country build up reserves equal to three months' consumption, but he fell short of true centralization—the Admiralty still refused to cede its power to arrange U.S. purchases or deliveries to Allied navies.[116] Nevertheless, thanks to various reforms, the contribution of U.S. tankers, and the introduction of convoying, stocks also doubled from 874,000 tons to 1.81 million tons between May 1917 and October 1918 even as monthly consumption increased from 397,000 tons to 487,000 tons.[117]

Since Long had no prior experience with oil matters, it is fair to surmise that his views reflected those of his chief adviser, Cadman, "in whom he had so much trust and confidence."[118] In February 1919, Long (now First Lord of the Admiralty) appealed to his colleagues to preserve the Petroleum Executive as an independent and permanent department. Britain could not afford to regress to the status quo ante, when government departments handled oil-related matters separately. The defects of that system became evident following the resumption of German unrestricted submarine warfare in 1917, when departments squabbled over tanker tonnage. Long also argued that the importance of the agency's work would only increase after the war.[119] In June 1919, he appealed to Prime Minister David Lloyd George, warning him that the Petroleum Executive was losing its

[112] See memoranda enclosed with: Greene to Hankey, "Proposed [...]," February 23, 1917, G.T. 51, CAB 24/6; "Terms [...]," G.T. 107, no date or author (circa March 1917), CAB 24/7.

[113] "Petrol [...]," no date or author, enclosed with: Long to Lloyd George, "Control [...]," G.T. 1006, CAB 24/16.

[114] Board of Trade "Oil Control [...]," August 1927, O.B. 23, CAB 50/3; "Petroleum Policy," no date or author, BP 70025; Jones, "Oil Companies," 654–7; Marian Kent, *Oil and Empire: British Policy and Mesopotamian Oil, 1900–1920* (London: Macmillan, 1976), 185–8.

[115] Tothill, "Oil Fuel [...]," June 30, 1917, G.T. 1233, CAB 24/18.

[116] Geddes and Tothill, "Requirements [...]," July 11, 1917, G.T. 1433, CAB 24/20; Long, "Memorandum [...]," August 16, 1917, enclosed with: Long, "The Petroleum Position," August 16, 1917, G.T. 1756, CAB 24/23.

[117] Long, "Memorandum [...]," October 15, 1917, G.T. 2320, CAB 24/29; Long, "Memorandum [...]," December 13, 1917, GT. 3201, CAB 24/36; Treasury, "H.M. Petroleum Executive," October 3, 1918, POWE 33/150; Long to War Cabinet, October 5, 1918, G.T. 5916, CAB 24/66; Brown, "Fuel Supplies," 150ff.

[118] Lilian Cadman, Memoir, July 22, 1918. I am grateful to John Calvin for sharing with me biographical details concerning his great-grandfather, Cadman, as well as personal papers in the family's possession.

[119] Long, "Need [...]," February 27, 1919, G.T. 6960, CAB 24/76.

best staff and suggesting its reconstitution as an independent agency within the Ministry of Supply.[120]

After the Cabinet had failed to take action by the first anniversary of the Armistice, Long again chastened his colleagues.[121] The results were disappointing. Renamed the Petroleum Department, Long's old agency retained its advisory role but would not have its own minister.[122] The department was also shuffled between agencies—first to the Board of Trade in 1922 and then the Mines Department in 1928. By 1934, the Petroleum Department had a full-time staff of only three, supplemented by another three part-time employees. None of the departments with any equity—the Admiralty, the Treasury, the Foreign Office, and the Colonial Office—was willing to cede its authority.[123] Their bickering made it impossible to achieve unanimity on a strategy to accomplish what they all agreed was in Britain's long-term interest after the war—independence from the United States or any other foreign power.

Whitehall began postwar oil planning years before hostilities ended. Starting in 1916, various government agencies argued for the establishment of a major British company that would reduce the influence of foreign firms and encourage greater production within the empire.[124] In February, Foreign Secretary Edward Grey ruled out both Anglo-Persian and Burmah as candidates, as neither company possessed the skills or oilfields to handle the task. Luckily, Shell appeared receptive to an understanding with Whitehall. Britain would also need an aggressive firm to defend British interests in Mesopotamia, and Shell appeared the only viable candidate.[125] The Board of Trade produced its recommendations for Britain's postwar oil strategy in August 1916, although Slade—one of the two government directors on Anglo-Persian's board—caught wind of its contents in June.[126]

The author was Lewis Harcourt, a former Colonial Secretary. Harcourt was well versed in oil policy and had organized a supply contract between the Admiralty and Shell's Trinidadian subsidiary in 1913.[127] His paper represented "a landmark in official thinking," for it signaled the start of an eight-year effort, focused on Shell, to create an "all British" company that could challenge the largest U.S. firms.[128] Harcourt concluded that "petroleum is rapidly becoming as important as coal; in some directions it has already surpassed it." Foreigners currently supplied the bulk of the British Empire's requirements, and it was "essential" that Britain control imperial sources and expand British interests elsewhere. This was beyond the

[120] Long to Lloyd George, June 5, 1919, and Long to Inverforth, June 5, 1919, both in: MUN 4/6718.

[121] Geddes, "Petroleum Executive," November 10, 1919, C.P. 91, CAB 24/92; Long, "Memorandum [...]," November 11, 1919, C.P. 115, CAB 24/93.

[122] Hankey, "Petroleum Committee," November 24, 1919, C.P. 180, CAB 24/93.

[123] "Petroleum Department," no date (circa 1930), POWE 33/16; Bamberg, *History*, 175–6; Payton-Smith, *Oil*, 40.

[124] The following discussion draws from: Brown, "Fuel Supplies," 158–75 and 291–309; Ferrier, *History*, 235–61; Jones, *Industry*, 177–207.

[125] De Bunsen to the Admiralty, March 2, 1916, F. 36846, POWE 33/41.

[126] Slade to Samuel, June 23, 1916, BP 69537.

[127] B.S. McBeth, *British Oil Policy, 1919–1939* (London: Frank Cass, 1985), 10.

[128] Corley, *Burmah*, 250; Ferrier, *History*, 242–3.

capacities of the existing British firms. The most promising instrument would be a "combination" including Shell. Harcourt recommended using Burmah as bait to secure control of Shell Transport and Trading and the Anglo-Saxon Petroleum Company (Shell's British marketing subsidiary).[129] The companies would be reconstituted as a majority-British "Imperial Oil Company."[130]

The Admiralty had supported the Anglo-Persian shares purchase in 1913–14 due to suspicion of Shell's "foreign" ownership.[131] These doubts were without merit. Although Shell had collaborated with Deutsche Bank in Europe and Mesopotamia, this was for commercial reasons.[132] Fisher had warned Churchill not to antagonize Shell, as Britain needed its tankers, 75 percent of which flew British flags.[133] Any doubts about Shell's loyalties should have evaporated during the war. Although Royal Dutch was officially neutral, in practice it was an enthusiastic Allied supporter,[134] and leased tankers to the Admiralty at below market rates.[135] Shell also constructed two facilities for the production of toluol, a petroleum extract essential for TNT production, in Britain and even supplied British forces with gasoline from stocks in Holland in 1916 during a shortage. In addition, the Dutch partners no longer insisted on remitting their share of Shell's profits to The Hague. These reserves went to London, bolstering sterling. By 1918, Royal Dutch's reserves in Britain had reached £13 million, and the company bought Treasury securities and sold them to the Dutch government. The displacement of Shell's finances to London also enabled the Anglo-Saxon Petroleum Company to become the group's treasurer.[136] If Shell was expecting gratitude, it was mistaken. Rather than admit its supply contract with Anglo-Persian was useless, the Admiralty forced Shell to sell it lighter East Indies crude oil that it could mix with the heavier Persian oil.[137] The war also hurt Shell more than any other major oil company—it lost access to its Caucasian oilfields in 1914 and those in Romania two years later.[138]

Slade took up the task of killing a partnership with Shell. After leading the commission to Persia in 1913, he became one of the government representatives on Anglo-Persian's board in 1916 and vice-chairman a year later, while still on the Admiralty payroll.[139] Slade had a number of allies, including his superior, Balfour, whose support gave Slade cover to draft rebuttals of the Board of Trade's proposal, which Balfour circulated to the Cabinet in September.[140] Slade argued that a Shell–Burmah merger would hurt Britain, leaving it vulnerable to price hikes and harming its financial position after the companies repatriated their profits

[129] Shell broached a merger in 1915, which Burmah (with Whitehall's backing) rejected. Greene to Islington, July 28, 1915, T 161/142; Corley, *Burmah*, 238–40, 243–9, and 251.

[130] Harcourt, "Future Control [. . .]," August 1916, ADM 1/8537/240.

[131] Jack, "Shares Purchase," 143, 149–51, and 165.

[132] Nowell, *Cartel*, 92.

[133] Fisher to Churchill, July 13, 1913 and August 6, 1913; CA, CHAR 13/21/13–14 and 17.

[134] Yergin, *Prize*, 174–5.

[135] Nowell, *Cartel*, 92ff.

[136] Jonker and Zanden, *History of Shell*, 157–62 and 169–73.

[137] Brown, "Fuel Supplies," 145–7.

[138] Nowell, *Cartel*, 90–2; Jonker and Zanden, *History of Shell*, 185–90.

[139] His ascent within Anglo-Persian stalled in January 1918 following a failed coup against Greenway. Corley, *Burmah*, 239.

[140] Balfour, memorandum, August 18 or 19, 1916, CAB 37/154/3; Brown, "Fuel Supplies," 159.

(£8 million annually).[141] A merger of Anglo-Persian, Burmah, and Mexican Eagle would yield more oil than a Shell–Burmah combination—4.1 million tons compared to 3.37 million tons.[142] The government's aim should be to acquire not only "a predominating factor in Oil *Production*," but also "a controlling influence in the *Distribution* business of the Empire." The best way to accomplish the latter, Slade suggested, was to nationalize British Petroleum, a subsidiary of Deutsche Bank, which had been a major distributor of Romanian and Russian oil within Britain before the war.[143]

Slade's efforts came to naught—Lord Cowdray (Mexican Eagle's founder and chairman) was reluctant to merge with a state-owned firm, while the Admiralty feared offending Washington.[144] Anglo-Persian did not lose out completely—the following year it received permission to purchase three foreign-owned marketing firms: British Petroleum; its shipping subsidiary, the Petroleum Steamship Company; and the Homelight Oil Company, the marketing arm of independent Russian refiners. Control of British Petroleum's assets—850 depots and eleven tankers, at a cost of £2 million—was "the missing link" that completed the company's drive for vertical integration and allowed it to market its Persian crude oil without relying on its 1912 marketing agreement with Shell.[145]

Despite Slade's opposition, the Admiralty agreed to support a Shell–Burmah merger if it did not compromise Anglo-Persian's independence.[146] The Cabinet Committee on Oil approved the merger in November 1916, but it fell through the following month.[147] Although the Board of Trade had failed to create an "all-British" Shell, it had brought oil to the forefront of discussion, where it stayed thanks to the intrigues of Anglo-Persian and Shell, both of which sought to influence postwar policy. In the winter of 1917/18, Anglo-Persian began applying pressure on the government to establish an "all-British" major oil company with Anglo-Persian as its nucleus.[148] Greenway believed that Anglo-Persian's interests overlapped

[141] Slade, "Observations [...]," August 24, 1916, attached to: Balfour to Cabinet, September 6, 1916, CAB 37/154/16.

[142] Slade, "Strategic Importance [...]," August 24, 1916, attached to: Balfour to Cabinet, September 6, 1916, CAB 37/154/16. These figures are from an assessment Greenway prepared for the Admiralty. He argued that much of Shell's production was consumed beyond the British Empire, unsuitable, or uneconomical to transport. An "all-British" firm comprising Anglo-Persian, Burmah, and Mexican Eagle possessed 3,682,300 tons suitable for Britain, compared to Shell's 1.22 million tons. "Royal Dutch/Shell [...]," July 18, 1916, attached to: Greenway to Serocold, July 19, 1916, ADM 1/8537/240.

[143] Slade, "Petroleum Supplies [...]," August 24, 1916, attached to: Balfour to Cabinet, September 6, 1916, CAB 37/154/16. Emphasis in the original. For Slade's lobbying efforts, see: Jones, *Industry*, 182–6 and 197–200. Greenway was simultaneously pushing for Anglo-Persian to acquire British Petroleum. Greenway, "Proposal [...]," July 8, 1916, ADM 1/8537/240.

[144] Jones, *Industry*, 179–80 and 189–92.

[145] Ferrier, *History*, 217–19. For postwar developments, see: Alfred Chandler, *Scale and Scope: The Dynamics of Industrial Capitalism* (Cambridge: Harvard University Press, 1990), 298–304.

[146] Balfour, "Progress [...]," October 19, 1916, ADM 1/8537/240.

[147] *PPHM*, i: 6–7; "Views [...]," September 19, 1918, no author, BP 69537. Negotiations collapsed over the Admiralty's insistence that Anglo-Persian be released from its marketing agreement with Shell. Kent, *Oil*, 132.

[148] Long ordered a warning sent to Greenway. "Conference...December 21st, 1917," POWE 33/42.

with those of Whitehall—a claim Shell disputed.[149] Although Cadman reassured Shell that the government did not condone Anglo-Persian's behavior, the Admiralty found Shell's complaints disingenuous, since Samuel had long urged Whitehall to "to identify with the Shell interests."[150]

These differences over strategy were indicative of a wider uncertainty over postwar policy. To achieve some clarity, in 1918, Long (at Cadman's urging) formed the Petroleum Imperial Policy Committee (PIPC) under the leadership of Harcourt.[151] Long intended that the committee should develop a comprehensive policy for the government.[152] He also hoped it would tackle the future of Mesopotamia's oil.[153] He left the committee with no doubt as to the significance of its task: "The importance of the problems cannot be overstated and it is no exaggeration to say that the future of the British Empire depends on their satisfactory solution."[154] One area of concern that arose during the PIPC's first meeting was the United States, whose oil industry was "showing distinct signs of distress and cannot be depended upon to continue her supplies at the same rate as formerly." Prospects in Latin America were promising, but the United States would claim any increases in production there or in North America. The only alternatives were Persia, Mesopotamia, and India (Burma). Mesopotamia's reserves were still unproven and could not contribute to Britain's needs for several years. The onus therefore fell on Persia, which meant the government had to ensure the survival of Anglo-Persian and Burmah.[155]

Whitehall recognized that the Middle East was central to Britain's energy future during World War I, which belies the claim that Middle Eastern oil only became a vital British interest after 1945.[156] The defense of oil interests in Persia was at the forefront of British strategy in the Middle East from the start of the Great War, starting with the decision to occupy Basra in 1914.[157] There was, however, initially no agreement on whether to acquire the potentially oil-rich Ottoman province of Mosul. The War Office and the Admiralty pushed for the acquisition of Mesopotamia, plus Alexandretta, to deny it to either Russia or France, preserve British predominance in the Persian Gulf, and secure the Persian oilfields.[158] In 1915 the de Bunsen Committee, which studied Britain's war aims in the Middle

[149] Samuel to Cadman, December 6, 1919 (*sic*; 1917), POWE 33/42.

[150] Walker to Long, February 28, 1918, P. 84548/48009, POWE 33/42; Brown, "Fuel Supplies," 162–3.

[151] Long, "Warrant [...]," May 29, 1918, POWE 33/12. See also: Brown, "Fuel Supplies," 163–4.

[152] Long to War Cabinet, May 15, 1918, G.T. 4564, CAB 24/51; Long to Balfour et al., May 16, 1918, POWE 33/12.

[153] Cadman to Long, May 7, 1918, POWE 33/41.

[154] Long, memorandum, May 29, 1918, POWE 33/12. See also: Long to War Cabinet, "Petroleum," July 31, 1918, G.T. 5275, CAB 24/59.

[155] PIPC, "Oil Policy [...]," May 29, 1918, POWE 33/44.

[156] Elizabeth Monroe, *Britain's Moment in the Middle East, 1914–1971* (Baltimore: Johns Hopkins Press, 1981), 11.

[157] V.H. Rothwell, "Mesopotamia in British War Aims, 1914–1918," *Historical Journal* 13: 2 (1970): 288–9; Sluglett, *Iraq*, 3–4.

[158] Kitchener, "Alexandretta [...]," March 16, 1915, G-12; Jackson, "Remarks [...]," March 18, 1915, G-15; both in: CAB 24/1.

East, concluded that oil "makes it commercially desirable for us to carry on to Mosul" but rejected annexation because of the cost.[159] Mosul waned in importance following the disaster at Kut in April 1916 and pressure to cease diverting resources to peripheral theaters at the expense of the Western Front.[160] In May, Britain ceded control of Mosul to France under the Sykes–Picot Agreement in order to create a buffer with Russia, although it retained full economic rights.[161] The fall of Baghdad in February 1917 revived interest, but British commanders focused on blocking a German–Ottoman advance against Persia.[162]

Mesopotamia returned to the forefront of British attention in the summer of 1918 thanks to the efforts of the Cabinet Secretary and Secretary to the Committee of Imperial Defence (CID), Maurice Hankey.[163] On July 30, shortly before the British were to discuss war aims with their Dominion allies, Hankey wrote to the First Lord of the Admiralty, Eric Geddes, concerning Britain's "uncertain" oil position. Considering that the United States might not have an exportable surplus in the future and was soaking up Mexican production, Hankey felt that securing possession of Mesopotamia was imperative.[164] The following day, he wrote to Balfour (now Foreign Secretary) and Lloyd George, enclosing a July 1918 assessment by Slade of Britain's postwar oil position. Hankey insisted that control of Persia and Mesopotamia constituted "a first-class British war aim" and advised resuming operations in Mesopotamia.[165]

Slade had finally left government service in 1917 to become a full-time member of Anglo-Persian's board and its vice-chairman. No one expected him to be impartial, but the aspersions he had cast on Shell's patriotism had harmed his reputation in Whitehall.[166] Geddes had to apologize to the War Cabinet for circulating Slade's 1918 paper without acknowledging his conflict of interest—but he did not revoke his endorsement.[167] Slade contended that Britain's existence as a great power depended on "command" of secure oil supplies. The British Empire had become more dependent upon the United States than ever before, and this was dangerous

[159] De Bunsen et al., "British Desiderata [...]," June 30, 1915, CAB 27/1. See also: Rothwell, "Mesopotamia," 277 and 287; Kent, *Oil*, 121–2.

[160] Rothwell, "Mesopotamia," 279–81.

[161] Rothwell, "Mesopotamia," 288. There is no truth to the claim that Britain ceded Mosul to France under Sykes–Picot and that French Prime Minister Georges Clemenceau relinquished this gain in 1918. Edward Peter Fitzgerald, "France's Middle Eastern Ambitions, the Sykes-Picot Negotiations, and the Oil Fields of Mosul, 1915–1918," *Journal of Modern History* 66: 4 (1994): 697–725; Kent, *Oil*, 137–57. The French only received the northern portion of Mosul; the most promising territories lay within the British sphere. Furthermore, Sykes–Picot guaranteed the prewar rights of the TPC to prospect throughout Mosul. See also: Eric Melby, *Oil and the International System: The Case of France, 1918–1969* (New York: Arno Press, 1981), 26–37; Luigi Scazzieri, "Britain, France, and Mesopotamian Oil," *Diplomacy & Statecraft* 26: 1 (2015): 25–30.

[162] Rothwell, "Mesopotamia," 285–6.

[163] The following discussion draws from: Helmut Mejcher, "Oil and British Policy towards Mesopotamia, 1914–1918," *Middle Eastern Studies* 8: 3 (1972): 382–8.

[164] Hankey to Geddes, July 30, 1918, CAB 21/119.

[165] Hankey to Balfour and Hankey to Lloyd George, both dated August 1, 1918, CAB 21/119. See also: Yergin, *Prize*, 188–9.

[166] Jones, *Industry*, 198–201.

[167] Geddes to Long, September 13, 1918, ADM 1/8537/240; Geddes, "Note [...]," September 17, 1918, G.T. 5710, CAB 24/64.

since, as Slade estimated, "within a measurable period, say 10 years, the amount of the Petroleum that we shall be able to draw from the United States will be greatly diminished if not entirely stopped."[168] The British military concurred wholeheartedly with Slade's assessment. The First Sea Lord, R.E. Wemyss, warned that failure would "wrest from our grasp one of the principal factors on which the maintenance of our naval position depends."[169] To the Chief of the Air Staff, F.H. Sykes, the stakes could not be higher, for "the very existence of the Empire will depend in the first instance upon aerial supremacy."[170]

Balfour dismissed Hankey's suggestion to occupy Mesopotamia as "a purely Imperialist War Aim." Hankey persisted, reminding Balfour that Britain needed to control northern Mesopotamia to secure the headwaters of the Tigris and Euphrates.[171] Hankey succeeded—the following day, on August 13, 1918, Balfour informed the Dominion prime ministers that Britain would maintain either direct or indirect control of Mesopotamia's oil. Lloyd George seconded him by arguing in favor of a renewed military advance toward Mosul, and the assembled ministers gave their assent. The endgame in Mesopotamia came quickly—by October, Ottoman resistance was collapsing, and British forces occupied Mosul several days after the Mudros ceasefire commenced on October 31, 1918.[172] The question of how Britain would exploit such riches was unclear, but Harcourt was developing a strategy.

SECURING SHELL AND THE MIDDLE EAST, 1918–1920

Harcourt used the PIPC to resurrect the idea he had championed in 1916: making Shell "British." Shell was susceptible to British control if Whitehall secured a "redistribution" of TPC shares in its favor.[173] After several months of negotiation, in February 1919 Harcourt and Shell's chairman, Henri Deterding, reached an agreement. Shell guaranteed Whitehall that 75 percent of the directors of Shell Transport and Trading and a majority of several other companies, including the group's Russian, Egyptian, Venezuelan, and Mexican subsidiaries, would be British subjects. Further, there would be no modification of the agreement, sale or lease of Shell's assets, or changes to its Articles of Association without the consent of a "nominee" from the Bank of England, who could also veto prospective British directors. Finally, the group would not interfere with Anglo-Persian's efforts to build a pipeline from Persia to the Mediterranean. British and French consumers would also receive most-favored-nation status in terms of prices, quantities, and

[168] Slade, "Petroleum Situation [...]," July 19, 1918, attached to: Wemyss, "Admiralty Memorandum [...]," July 30, 1918, G.T. 5267, CAB 24/59.

[169] Wemyss, "Admiralty Memorandum [...]," July 30, 1918, G.T. 5264, CAB 24/59.

[170] Sykes to War Cabinet, "Petroleum Situation," August 9, 1918, G.T. 5376, CAB 21/119.

[171] Hankey to Balfour, August 12, 1918, CAB 21/119.

[172] Rothwell, "Mesopotamia," 287–93.

[173] Montague to Harcourt, December 23, 1918, CAB 21/119. Control of the TPC was Deterding's key objective—whether or not Shell was nominally British was irrelevant. Corley, *Burmah*, 257–8.

delivery costs. In exchange, Shell received a share (34 percent) equal to that of Anglo-Persian in the TPC, which would remain under British control, and a promise that Whitehall would not interfere with the group's commercial or financial operations.[174] Harcourt was ecstatic, and Long speculated that it might even render Britain "independent" of U.S. imports within a decade.[175] The board of Royal Dutch granted its approval provided Whitehall assisted Shell in broadening its holdings within British territories, exempted Dutch shareholders from British taxation, and provided Shell with a greater share of the Mesopotamian concession.[176] The Dutch had everything to gain and little to lose, and they indulged Harcourt through "an intricate charade."[177]

Anglo-Persian grasped the danger to its independence, and Slade fired off an attack in September 1918. He questioned the wisdom of leaving Britain's future in the hands of firms whose parent governments might become "members of some 'League of Nations' formed for the purpose (inter alia) of endeavouring to destroy British Naval and Commercial supremacy." Slade was confident that Anglo-Persian would soon "not only be able to supply all of the needs of the Empire but also other countries to a large extent, and thus be in the means of place Great Britain in the position of predominance in regard to Oil she now enjoys in regard to coal."[178]

Anglo-Persian's complaints found little favor in Whitehall, where there was broad support for the Harcourt–Deterding Agreement as part of a wider postwar plan for developing Mesopotamia. A British-controlled Shell, acting in concert with Anglo-Persian, would guarantee British control of the TPC and Mesopotamia, while an intergovernmental agreement with France would allow Mesopotamia's oil to travel via pipeline through Syria and Lebanon to the Mediterranean. Whitehall recognized that the only incentive for Shell in signing the agreement was a larger share of Mesopotamian oil.[179] Meanwhile, Shell saw the agreement as a means to secure control over the TPC. Whitehall intended to divide the TPC's shares under British control equally between Shell and Anglo-Persian—34 percent to each company, with the government holding 2 percent, 20 percent going to France, and a further 10 percent going to the "native" government.[180]

During its negotiations with the PIPC, however, Shell had also been negotiating with Paris to establish a state-sponsored import-quota system. In exchange, a syndicate dominated by Shell would receive any shares in the TPC that Britain transferred from Germany to France, thereby giving Shell majority control of

[174] Harcourt and Deterding, "Memorandum," no date, attached to: Long to War Cabinet, "Acquisition [...]," February 22, 1919, G.T. 6961, CAB 24/76. The complicated structure is described in McBeth, *Policy*, 26ff.

[175] Harcourt to Long, February 7, 1919, and Long to Harcourt, February 7, 1919, both in: POWE 33/64. The Cabinet approved the agreement on May 5, 1919. "British Control [...]," reprinted in: Petroleum Executive, "Negotiations [...] Volume II," p. 17, POWE 33/14 (*PPHM*, ii).

[176] London to Deterding, March 14, 1919, POWE 33/71. For a summary of the negotiations preceding the agreement, see: *PPHM*, i: 11–26; *PPHM*, ii: 3 and 13–20.

[177] Jonker and Zanden, *History of Shell*, 214–16.

[178] Slade, "Views [...]," September 19, 1918, BP 69537. This paper is an updated version of "Threatened Foreign Oil Monopoly," August 21, 1918, ADM 1/8537/240.

[179] Memorandum for Barstow, [?] 1920, T 161/738.

[180] "Note by Foreign Office," I.D.C.E 1022 c. iv, no date (circa December 1919), T 161/738.

the TPC.[181] This plan collapsed after the Standard Oil Company of New Jersey (Jersey) launched an embargo against France, which lasted for several months in 1919–20, and Shell failed to cover the deficit due to the loss of its Russian holdings, damage in Romania, and a shortage of tankers to move oil from overseas. France had learned a valuable lesson—after 1920, it was careful not to antagonize the major oil companies. In 1923, Paris created a new company, the Compagnie Française des Pétroles (CFP) to administer France's shares in the TPC. Much of the CFP's start-up capital came, in fact, from the French subsidiaries and partners of Jersey, Shell, and Anglo-Persian. Paris eventually purchased a 35 percent stake in the CFP in 1929, but the company remained under private control.[182]

In 1915/16, Whitehall had informed Anglo-Persian that it considered the TPC's prewar arrangements to be void and that it would revisit the matter at some unspecified date in future.[183] At war's end, however, the government reversed course and resurrected the TPC to secure control of Mesopotamia.[184] In January 1920, however, the British again reversed course and decided against turning the oilfields over to private firms.[185] The reasons for such zigzagging are unclear.[186] The previous spring, the Civil Commissioner in Mesopotamia, Arnold Wilson, had questioned relinquishing "the whole interests of the Iraq State [*sic*] and of the British Treasury in what might well prove to be the richest oilfields in the world to a multinational consortium." The government should retain control, he argued, because oil revenues were the only collateral Baghdad could offer against the loans it needed from Britain.[187] The Treasury agreed, considering the cost of the British occupation—Wilson estimated this at £25 million over the next five years, and that was before the uprising of 1920.[188] That did not mean state control was a wise move. The head of the Petroleum Department, Frederick Kellaway, complained a year later that this would frustrate negotiations with Shell and argued that private companies were better qualified to handle such a complex industry.[189] If Whitehall

[181] Shell's motivations are summarized in: Jonker and Zanden, *History of Shell*, 213–16, 251, and 256.

[182] Richard Kuisel, *Ernest Mercier: French Technocrat* (Berkeley: University of California Press, 1967), 21–44; Melby, *Oil*, 8–168; Nowell, *Cartel*, 132–41 and 170–83.

[183] Foreign Office, "Memorandum [...]," February 20, 1918, POWE 33/41; Parkes, "[...] Oil Concessions in Mesopotamia," April 27, 1918, 10967, attached to: Alwyn-Parker to the Petroleum Executive, May 3, 1918, POWE 33/41.

[184] Montague to Harcourt, December 23, 1918, CAB 21/119.

[185] "Conclusions of a Conference of Ministers," January 23, 1920, T 1/12544.

[186] Kent describes the decision "to wreck the agreement" as the product of a "sublime misunderstanding of the whole basis of the arrangement." Kent, *Oil*, 152. Jones judged it a response to public antipathy toward the great oil trusts. Jones, *Industry*, 215–16. See also: Scazzieri, "Oil," 38–40.

[187] India Office, "Mesopotamia: Oil Policy," April 10, 1919, Pol. 1858/19, T 1/12544.

[188] Waterfield to Barstow, April 9, 1919, T 1/12544. Wilson's ham-fisted leadership, on the heels of an unpopular occupation, provoked the uprising of 1920. The British response—patronizing Sunni elites over the Shia majority, for instance—laid the foundations for many of Iraq's difficulties following independence. Kristian Coates Ulrichsen, "The British Occupation of Mesopotamia, 1914–1922," *Journal of Strategic Studies* 30: 2 (2007): 349–77. Following his dismissal, Wilson secured a position with Anglo-Persian. Aileen Keating, *Mirage: Power, Politics, and the Hidden History of Arabian Oil* (Amherst: Prometheus Books, 2005), 63–88.

[189] Kellaway, "Mesopotamian Oilfields," April 22, 1920, C.P. 1118, CAB 24/104. For Long and the Treasury's concurrence, see: Long, "Anglo-French Petroleum Agreement [...]," June 29, 1920, C.P. 1554, CAB 24/108; memorandum for Barstow, [?] 1920, T 161/738.

went ahead, Kellaway estimated the cost to taxpayers at £15 million.[190] The new Colonial Secretary, Churchill, shared Kellaway's objections, as did Wilson, who backpedaled from his earlier position.[191]

Shell was now having second thoughts about Harcourt–Deterding. The benefits of closer cooperation with Whitehall seemed outweighed by the disadvantages, since Washington was furious over the exclusion of U.S. firms from Mesopotamia and Palestine. Shell could not afford to alienate the United States, as it was in the process of expanding its Latin American presence and had its eyes on Mexican Eagle.[192] Weetman Pearson (Lord Cowdray after 1910) built a thriving business enterprise in Mexico—initially in construction—after 1889 thanks to the patronage of Mexican dictator Porfirio Díaz. Pearson was also the one of the pioneering figures in the development of the Mexican oil industry after 1901 and founded Mexican Eagle in 1909. The Mexicans favored Pearson with tax breaks and land concessions to prevent U.S. oil companies from dominating the oil industry.[193] Mexican Eagle's discovery of a large oilfield at Potrero del Llano in 1910 turned Mexico into a major oil exporter and made the company the seventh-largest British firm in the world—just as the 1910 Revolution exploded.[194]

As the political situation in Mexico deteriorated, Cowdray became desperate to find partners. Jersey and the British government both declined offers in 1913, and Cowdray even resorted to offering Mexican oil to the Berlin-to-Baghdad Railway.[195] The outbreak of war assured his firm of customers but not of long-term security. Cowdray nearly completed a sale to Jersey in 1917, but Whitehall intervened to block it. Cowdray again tried to sell his interests to Whitehall, but the government found the terms unacceptable and threatened Cowdray with prosecution under

[190] Kellaway, "Anglo-French Petroleum Agreement [...]," June 21, 1920, C.P. 1524, CAB 24/108. See also: Kellaway, "Anglo-French Petroleum Agreement [...]," November 15, 1920, C.P. 2110, CAB 24/115.

[191] Churchill to Hankey, June 20, 1921, attached to: Hankey, "Mesopotamian Oil," June 27, 1921, C.P. 3077, CAB 24/125. Churchill reiterated these points in Churchill, "Mesopotamian Oil [...]," August 29, 1921, C.P. 3271, CAB 24/127. For Wilson's views, see: "Telegram...dated 7th July 1920," attached to: Montague, "Anglo-French Petroleum Agreement [...]," July 12, 1920, C.P. 1607, CAB 24/109.

[192] Brown, Oil, 162ff; Brown, "Fuel Supplies," 178–9.

[193] Before World War I, Mexican Eagle owned concessions covering 2 percent of Mexico's habitable area (1.5 million acres). For background, see: Jonathan Brown, "Domestic Politics and Foreign Investment: British Development of Mexican Petroleum, 1889–1911," *Business History Review* 61 (1987): 389–95 and 415–16; Stephen Haber et al., *Politics of Property Rights: Political Instability, Credible Commitments, and Economic Growth in Mexico* (Cambridge: Cambridge University Press, 2003), 192–8. Díaz wanted to reduce Mexico's dependence upon coal imports, which had reached 4.5 million tons by the end of the nineteenth century. Jonathan Brown and Peter Linder, "Oil," in: *The Second Conquest of Latin America: Coffee, Henequen, and Oil During the Export Boom, 1850–1930*, ed. Steven Topik and Allen Wells (Austin: University of Texas Press, 1998), 157.

[194] For Mexican Eagle's pre-World War I history: Brown, "Domestic Politics," 395ff; Lisa Bud-Frierman et al., "Weetman Pearson in Mexico and the Emergence of a British Oil Major, 1901-1919," Business History Review (2010): 279–87; Jones, *Industry*, 63–77. For its wartime history, see: Brown, *Oil*, 144–53.

[195] The Germans preferred to buy from Anglo-Persian. See: Gwinner, "Oelfeuerung," July 1, 1914, and "Mesopotamisches Petroleum," May 8, 1914, Deutsche Bank Archive (DBA), S-1615.

the Defence of the Realm Act if he went ahead with the sale to Jersey.[196] Whitehall's legal objections evaporated with the war's end, and Cowdray promptly liquidated his Mexican interests, since he had learned in December 1918 that saltwater was seeping into Mexican Eagle's largest oilfield. He eventually sold his stake in both Mexican Eagle and its British subsidiary, Anglo-Mexican, to Shell in April 1919 for £7.7 million.[197]

Shell's purchase of Mexican Eagle appeared to be a coup. By 1919, Mexican Eagle's market value exceeded £62 million, and the Pearson Group as a whole, valued at £79 million, was the most valuable company in Britain. Burmah came in second, but only because of its stake in Anglo-Persian—Mexican Eagle out produced it in terms of oil by a ratio of 4:1 in 1919. It was also worth more than Anglo-Persian and Shell Transport and Trading combined (£47 million).[198] The company was also profitable—it paid dividends of 16 percent in 1915–16 despite the revolution, and in 1920–1, the dividend soared to 45 percent and 60 percent, respectively, while sales profits totaled £20 million in 1921.[199] Soon after, Shell's investment turned sour. Mexican production dropped sharply after 1921 even though the companies continued investment, and Mexican Eagle's share price lost 80 percent of its value between 1921 and 1927.[200]

Fortunately for Shell, Mexico was just one of its investments in the Western Hemisphere. Already the leading producer in Venezuela and a major player within the United States, Shell had much to lose if it became Washington's enemy.[201] Being Whitehall's friend, conversely, counted little for Shell after the government refused to grant a tax exemption for Shell's Dutch shareholders. Once that happened, even Harcourt lost hope, and the agreement soon crumbled.[202]

The failure of the Harcourt–Deterding Agreement was a setback for Whitehall, but it was only one facet of Britain's postwar strategy. Of greater importance was securing the Middle East under British control. With its influence in Persia unchallenged by any external power, Whitehall now set its sights on Mesopotamia. World War I ended with British troops occupying most of what became Iraq after 1922, and Britain's position appeared unassailable. Germany, whose interest in Mesopotamia during the war had concerned Whitehall, was no longer a factor.[203]

196 See the correspondence enclosed with: Pretyman to Hankey, April 26, 1917, G.T. 576, CAB 24/11; Long, "Proposals [...]," July 18, 1917, G.T. 1448, CAB 24/20; letters enclosed with: memorandum, September 10, 1917, no author, G.T. 2000, CAB 24/25. See also: Brown, "Fuel Supplies," 161–2; Jones, *Industry*, 191–2.

197 Shell only bought a 35 percent stake in Mexican Eagle, but it did acquire managerial control. Jones, *Industry*, 217–18. See also: Jonathan Brown, "Why Foreign Oil Companies Shifted Their Production from Mexico to Venezuela during the 1920s," *American Historical Review* 90: 2 (1985): 365–7; Garner, *Lions*, 221–9.

198 Bud-Frierman, "Weetman Pearson," 279 and 285–94.

199 Brown, "Shifted," 371; Hall, *Oil*, 15.

200 Stephen Haber et al., "When the Law Does Not Matter: The Rise and Decline of the Mexican Oil Industry," *Journal of Economic History* 63: 1 (2003): 9–16; Haber, *Politics*, 212–34; Jonker and Zanden, *History of Shell*, 226, 259–61, and 453.

201 Jones, *Industry*, 217–21; Jonker and Zanden, *History of Shell*, 251, 256.

202 Harcourt to Fisher, March 19, 1920, T 172/1387.

203 Admiralty, "Reported Oil Fields [...]," August 2, 1918, G.T. 5313, CAB 24/60.

Nor was there any need to placate potential rivals such as Russia or preserve the Ottoman Empire. Whitehall therefore backtracked on its position that the TPC's claim was invalid. How exactly Britain could reconcile such self-serving behavior with its professed aim of ruling Mesopotamia "in the interests of the inhabitants" was unclear.[204] Moreover, if Britain was to reap its spoils, it required international approval. At the prompting of Long, in late 1919, Cadman briefed Balfour and the British delegation at the Paris Peace Conference on the necessity of securing "British control over additional oil-bearing lands" in the interest of preserving Britain's "naval and mercantile supremacy."[205]

Exploiting Mesopotamia's oil proved harder than expected. A number of parties—British, Dutch, French, U.S., and even Turkish—were clamoring for a share. The British required French cooperation, since oil from Mosul had to travel by pipeline to the Mediterranean. Between 1919 and 1920, Britain and France signed three oil agreements regarding Mesopotamia and other areas of mutual interest: the Long–Bérenger Agreement of April 8, 1919; the Greenwood–Bérenger Agreement of December 21, 1919; and the Cadman–Berthelot Agreement of April 24, 1920, also known as the San Remo Supplemental Oil Agreement. Under the first two, Whitehall promised Paris that France would receive a portion (first 20 percent, then 25 percent) of the shares of the TPC or any successor. The Cadman–Berthelot Agreement dropped any mention of the TPC but left open the question whether a state-run or private company would receive the concession. All of the agreements stipulated that any company that received the concession would remain "under permanent British control." France consented to the construction of pipelines carrying both Mesopotamian and Persian crude oil through Syria and Lebanon. Each of the agreements also gave the Iraqis the option of acquiring up to 20 percent of shares in whatever company worked the concession, but this provision was jettisoned to make room for U.S. firms after 1922.[206]

The Long–Bérenger Agreement was implicitly linked to the Harcourt–Deterding Agreement. According to Hamar Greenwood (Long's successor as Petroleum Minister), the price of securing French support for British ambitions in Mesopotamia was negligible—for a share of its oil, the British would acquire the cheapest transit routes and control of Shell.[207] From his new perch at the Admiralty, Long kept up the pressure in favor of an agreement with Paris, even after the Cabinet temporarily withdrew its support in August 1919.[208] The following March, Long warned his colleagues that failure to reach an agreement with Shell would leave Britain with Mesopotamia and Persia as its only British-controlled sources of oil—only one of which was actually producing at the moment—and, in

[204] Montague to Harcourt, December 23, 1918, CAB 21/119.

[205] "Petroleum Position [...]," December 1918, attached to: Cadman to Cecil, January 6, 1919, Secret 269, POWE 33/60.

[206] The agreements are reprinted in: Kent, *Oil*, 172–8.

[207] Greenwood, "Memorandum [...]," December 6, 1919, C.P. 259, CAB 24/94.

[208] Long, "Memorandum [...]," November 4, 1919, C.P. 59, CAB 24/92, and November 11, 1919, C.P. 115, CAB 24/93.

effect, deepen Britain's reliance on the United States.[209] R.S. Horne, President of the Board of Trade, worried that failure of the Harcourt–Deterding agreement might push Shell "into a definite antagonism, and into the control of some other country."[210] Other officials worried that failure to reach an accord with the French would drive Paris into Washington's arms.[211] As the Foreign Office observed, what truly mattered was control of Shell. Granting a minority stake in the TPC to France was "really a bargain" in the grand scheme.[212]

Whitehall and Paris finally came together in April 1920 to sign the San Remo Agreement. It could not do much to address Britain's immediate shortfall, since Mesopotamian oil would be unavailable until someone built a pipeline. Until then, as British consumption grew, so did Britain's vulnerability. By the summer of 1920, the Royal Navy's requirements would reach 1.11 million tons, with 400,000 tons coming from Persia, 200,000 tons from Mexico, and a further 150,000 tons from Trinidad. This left the navy with a shortfall of 150,000 tons and nowhere near its mandated reserve, which would triple to 4.5 million tons over the decade.[213] Whereas efforts in Mesopotamia were still in their "infancy," Persia was already critical thanks to its low production costs and its location, which spared the Admiralty the need to accumulate larger stocks in Singapore. Replacing the Admiralty's current supplies from Persia from anywhere else would cost "six to seven times" more, which outweighed the costs of protecting Persia's oilfields.[214] Long also kept pushing for the construction of pipelines to the Mediterranean. Persian oil could be both a strategic and a financial boon for Britain, but only if it secured transit rights through the French mandates in the Near East.[215]

Neither Long nor his colleagues appreciated that the primary danger to British ambitions in Persia came from within that country. In a lecture at the Royal Naval College in 1921, Slade argued that Britain's energy future was in the Middle East. Persia and Iraq were Britain's only viable options to reduce its dependence on the United States, and he condemned Whitehall's inadequate defense of British interests in Persia.[216] A visit to Persia in April 1922 confirmed his worst suspicions. Production had exploded between 1911/12 and 1921/2 from 43,000 tons to 2.4 million tons, while annual royalties were £600,000.[217] Slade was still worried—he was convinced that the Americans were seeking to convince Tehran that Anglo-Persian was dragging its heels, and that Persia should modify or cancel the concession for the Americans' benefit.[218]

209 Long, "Oil Supplies," March 18, 1920, C.P. 903, CAB 24/101.

210 Horne, "Mesopotamian Oilfields," April 16, 1920, C.P. 1085, CAB 24/103.

211 Barstow, "Mesopotamian Oil," December 10, year unclear, T 172/1387; *PPHM*, ii: 3–4 and 6–8.

212 "Note by Foreign Office," I.D.C.E 1022 c. iv, no date, T 161/738.

213 Long, "Anglo French Petroleum Agreement [...]," June 29, 1920, C.P. 1554, CAB 24/108.

214 Long, "South Persian Oilfields," December 24, 1920, C.P. 2377, CAB 24/117.

215 Long, "Anglo-French Petroleum Agreement," January 9, 1921, POWE 33/176.

216 "Strategic Aspect [...]," November 14, 1921, BP 69942.

217 Ferrier, *History*, 271 and 370.

218 Slade, "Memorandum [...]," May 23, 1922, BP 72017. The Persians tried vainly between 1921 and 1927 to entice U.S. firms to develop the five provinces not covered by the D'Arcy concession. Michael Rubin, "Stumbling through the 'Open Door': The U.S. in Persia and the Standard-Sinclair Oil Dispute, 1920–1925," *Iranian Studies* 28: 3/4 (1995), 203–29.

Slade's assessment was accurate, albeit for different reasons. Persia was in disarray at war's end. Despite it being a neutral, British, Russian, and Ottoman troops had occupied different portions, and by 1918 the country was suffering from famine and a cholera outbreak. Since 1905, Anglo-Persian had dominated South Persia through treaties with the local Bakhtiari tribes, who received 3 percent of its profits, and Sheik Khaz'al of Khuzestan, which included Abadan. The Russians had consented in 1915 to British control of the "neutral zone" between their spheres of influence under the 1907 Anglo-Russian Convention. After 1918, Britain saw its chance to expand its influence over the entire country. Although Whitehall was reluctant to give its full support to Anglo-Persian's claims to concessions throughout the Persian Gulf and Ottoman Empire, it was determined to protect the Persian concession.[219] The collapse of Tsarist Russia in 1917 left Britain free to make Persia a protectorate and realize the imperialist dream of linking the formal and informal empire from Africa, through the Middle East, to India and the Far East.

Britain's instrument would be the Anglo-Persian Agreement of 1919, which would modernize Persia's army, improve Persia's infrastructure, and provide it with a £2 million loan. None of this would be free—Britain would debit Persia for all expenses, including £131,000 worth of bribes. The Persian government signed the treaty in August 1919, but before it could go into effect, parliament had to approve it and widespread opposition ensured that ratification was impossible. Meanwhile, Tehran started negotiations for a treaty of friendship with Soviet Russia. A few days before the signing of the treaty in February 1921, however, troops under the command of Reza Pahlavi overthrew the government. Although there was some British support for the coup, the new strongman had no interest in resurrecting the Anglo-Persian Agreement and instead signed a treaty with Moscow.[220]

Anglo-Persian's conduct during and after the war stoked Persian hostility. In retaliation for the 1915 sabotage of its pipelines by revolting tribesmen supported by German operatives, the company billed Tehran £615,000 and refused to pay royalties until it received compensation. The Persians disputed the charges and hired an auditor, William McLintock, to vet the company's claims, which turned out to be largely fictitious—actual damages amounted to only £20,000. McLintock also uncovered a host of financial irregularities designed to evade Anglo-Persian's contractual obligation to pay royalties equivalent to 16 percent of the company's profits. Anglo-Persian quickly abandoned its claims for damages, resumed royalty payments, and offered Tehran a £1 million settlement. To avoid a repetition, in 1920, the company enlisted the British expert appointed to manage Persia's finances under the Anglo-Persian Agreement, Sydney Armitage-Smith, to modify its concession.[221] Under the Armitage-Smith Agreement of December 22, 1920, only those operations that took place within Persia would pay the 16 percent

[219] Anglo-Persian wanted Whitehall's support to compel Tehran to accept an extension of the D'Arcy concession to 1986. H.G.N., "Draft Note [...]," February 14, 1918, attached to: Cecil to Petroleum Executive, February 26, 1918, POWE 33/41. The Petroleum Executive agreed. Clarke to Cecil, March 2, 1918, P.E. 0126, POWE 33/41.

[220] Bamberg, *History*, 27–32; Elm, *Oil*, 23–8; Ferrier, *History*, 350–2.

[221] Compare Ferrier, *History*, 358–71, with Elm, *Oil*, 19–22.

royalty. Anglo-Persian would pay royalties for operations beyond Persia only on those profits it "defined and calculated" as stemming from Persia.[222] This revision empowered the company to set the level of compensation it paid Tehran. It also enabled it to build a wall between its operations in Persia and those beyond, much to the consternation of Tehran, which believed it deserved a share of Anglo-Persian's global profits because the company had used Persian oil as its foundation. In the short run, Anglo-Persian had scored a significant victory by extracting favorable terms from a helpless government—but by exacerbating anti-British sentiment in Persia, its success proved counterproductive to the wider aim of British strategy: ending Britain's reliance on the United States for oil.

OIL AND ANGLO-AMERICAN RELATIONS, 1918–1921

In January 1919, the Admiralty informed the Cabinet that the only options to break Britain's dependence on the United States were either synthetic production or the development of the Persian and Mesopotamian oilfields—and there were already doubts about the security of Britain's supply lines to and from the Middle East. Therefore, the Admiralty recommended accumulating a naval fuel reserve in Britain equivalent to one year's wartime consumption—double the prewar standard.[223] Assuming a postwar force 80 percent the size of the prewar fleet, this worked out to 4.5 million tons, which entailed expansion of storage capacity by 2.75 million tons at a cost of £1 per ton.[224] To keep the merchant marine from drawing on naval reserves in wartime, the Admiralty and the Petroleum Executive also recommended creating a civilian stockpile of 1.5 million tons of fuel oil.[225] Until Persia significantly increased output, however, Britain would draw most of its oil from the Western Hemisphere.[226]

Until Middle Eastern oil became plentiful and secure, Britain would have no option but to continue relying upon the United States. Unfortunately, Anglo-American relations in the immediate postwar era were strained, not least because of concerns over the United States' future oil position. British officials fretted over the possible collapse of U.S. oil exports. According to one naval analyst, U.S. oil hegemony was likely transitory because of the insatiable demands of U.S. consumers.[227] Such anxiety echoed official and public sentiment within the United States, which fed on alarmist assessments by the U.S. Geological Service and the American Association of Petroleum Geologists.[228] Consumption in the

222 The agreement is reprinted in Ferrier, *History*, 653–8.

223 Brown, "Fuel Supplies," 102ff and 193–5.

224 Wemyss, "Oil Fuel [...]," January 3, 1919, G.T. 6594, CAB 24/72; Long, "Oil Fuel [...]," January 8, 1919, G.T. 6634, CAB 24/73.

225 Long and Greenwood, "Oil Fuel [...]," February 5, 1920, C.P. 601, CAB 24/98.

226 Wemyss, "Oil Fuel [...]," January 5, 1919, G.T. 6703. CAB 24/74.

227 Miles, "Supply [...]," April 14, 1920, ADM 203/48.

228 Michael Aaron Dennis, "Drilling for Dollars: The Making of U.S. Petroleum Reserve Estimates, 1921–1925," *Social Studies of Science* 15: 2 (1985): 241–65; Roger Stern, "Oil Scarcity Ideology in US National Security Policy, 1908–1997," *Security Studies* 25: 2 (2016): 222–30.

United States was far outpacing new production. Naval consumption increased sevenfold between 1916 and 1923, and a presidential commission worried that one year's naval wartime consumption could consume as much as 10 percent of production.[229] The number of registered automobiles in the United States increased fivefold between 1911 and 1918, and gasoline consumption exceeded production by fifty-eight million barrels (bbl). As a result, U.S. officials and industry executives began calling for the government to promote conservation (making production more efficient) and access to foreign oilfields.[230] U.S. firms soon found that Whitehall had barred access to the British sphere of influence.[231]

Washington determined that Britain was returning to its mercantilist roots, when in fact Whitehall was trying to rebuild the status quo antebellum.[232] To retaliate for British obstruction in the Middle East and Dutch restrictions in the East Indies, in February 1920 the U.S. Congress passed the Mineral Leasing Act, which applied the principle of reciprocity to foreign firms seeking leases on federal lands.[233] Matters came to a head in May, over the San Remo Agreement. The State Department complained that the agreement was an Anglo–French plot to monopolize production within the former Ottoman Empire. Although the United States had neither declared war on the Ottoman Empire nor joined the League of Nations, the State Department claimed that U.S. firms were entitled to "equal treatment" in the mandates. The Foreign Secretary, the Earl Curzon, found U.S. accusations of British perfidy "singularly unintelligible."[234]

British officials took their case against U.S. criticisms public. Cadman lauded "the invaluable assistance which we have received," but argued that "it behooves us to look round to see... from where we are going to be served when the United States finds it difficult to supply us."[235] Such efforts backfired. In May 1920, British Ambassador to the United States, Auckland Geddes, questioned how anyone could claim that Britain was trying to establish an oil monopoly when the British Empire and Persia accounted only for 5 percent of world production whereas the United States controlled more than 80 percent. The State Department's Foreign Trade Adviser, Arthur C. Millspaugh, dismissed this argument as dishonest. He pointed out that the United States and Mexico had only 20 percent of global reserves, and U.S. firms controlled less than 2 percent of those beyond North

[229] Shulman, "Petroleum," 378–81.

[230] Shulman, "Petroleum," 372ff; Yergin, *Prize*, 194–5.

[231] John DeNovo, "The Movement for an Aggressive American Oil Policy Abroad, 1918–1920," *American Historical Review* 61: 4 (1956): 854–76, and American Interests and Politics in the Middle East, 1900–1939 (Minneapolis: University of Minnesota Press, 1963), 176–84.

[232] Compare Davis to Colby, April 28, 1920, National Archives and Records Administration (NARA), Record Group 59: General Records of the Department of State (RG 59), 841.6363/45, with William Stivers, "International Politics and Iraqi Oil, 1918–1928: A Study in Anglo-American Diplomacy," *Business History Review* 55: 4 (1981): 529.

[233] Mira Wilkins, *History of Foreign Investment in the United States, 1914–1945* (Cambridge: Harvard University Press, 2004), 101–2.

[234] The exchanges of May 1920 to February 1921 are reprinted in *Correspondence* [...], Cmd. 1226 (London: HMSO, 1921). See also: memorandum for Chamberlain, "Mesopotamian Oil," September 6, 1921, author unclear, T 161/738; Brown, "Fuel Supplies," 179–80.

[235] John Cadman, "Oil Resources of the British Empire," *Journal of the Royal Society of Arts* 68: 3532 (July 30, 1920): 585.

America. The British and Dutch were in a more favorable position and could end up controlling as much as 75 percent of reserves.[236]

As Geddes explained to Curzon, U.S. complaints about British policy were symptomatic of a deeper anxiety over the country's energy future. The United States had long enjoyed self-sufficiency, but this might soon change. These fears drew nourishment from propaganda disseminated by U.S. oil companies and impolitic remarks by British businessmen.[237] One banker, for example, had boasted that once Britain had a "firmer grip on the world's reserves of oil," it could sit back and collect huge revenues from the United States—as much as $1 billion a year. Such financial transfers would repair Britain's finances and give the empire a new lease of life.[238]

British analysts claimed the divergence in outlook between Britain and the United States stemmed from the fact that, whereas the former believed in "the naval idea that oil suppliers were essential for national defence," the latter maintained the "mercantile idea that oil supplies were essential for money-making."[239] From Whitehall's perspective, U.S. criticism was slanderous.[240] Following the war, Britain became the world's second-largest consumer, but of the 3.37 million tons it imported in 1920, only 2 percent came from British possessions, compared to 61 percent from the United States. To the Petroleum Department, U.S. criticisms were "so unfounded" that they must be a pretext to extort concessions in Mesopotamia.[241] From the sidelines, the German Foreign Office assessed that the Anglo-American "oil conflict" was "considerably greater than any conflict of interest between peoples before the World War."[242]

Luckily, fears of a U.S. petroleum famine proved mistaken. High prices and new technologies soon created an immense surplus.[243] Between 1918 and 1922, U.S. imports (primarily from Mexico) quadrupled, peaking at 21 percent of supplies in 1921, and the United States was a net importer from 1920 to 1922. That said, U.S. production increased continuously between 1918 and 1929, from 355.93 million bbl to 1.01 billion bbl (280 percent), while proven reserves increased every year between 1918 and 1931, from 5.9 billion bbl to 13.6 billion (230 percent). The challenge confronting the U.S. oil industry was not the depletion of reserves, but rather the possibility that consumption would soon overtake production. Consumption increased by 250 percent between 1918 and 1929, from 374.54 million bbl to 941 million bbl (4.3 bbl to 7.7 bbl per capita). During the 1920s and 1930s, new supplies outpaced increases in demand, and the crisis

236 Millspaugh, "British Ambassador's Remarks [...]," May 28, 1920, Sterling Memorial Library, Yale University, Frank L. Polk Papers, Series III, Box 32.

237 Geddes to Curzon, July 29, 1920, A 5494/898/45, CO 323/832/16.

238 Skinner to Colby, "British Quest [...]," April 19, 1920, NARA, RG 59, 841.6363/43; Skinner, "Oil Resources [...]," October 18, 1921, NARA, RG 59, 841.6363/178.

239 Kennedy, "Oil Imperialism," no date (circa 1922), ADM 203/58.

240 Petroleum Department to Harmsworth, July 1, 1920, S.381, CO 323/845/45.

241 Petroleum Department, "Memorandum [...]," February 10, 1921, A 982/44/45, CO 832/865/32.

242 Note to Wirkl and Bücher, January 20, 1921, PAAA, R 118173.

243 Yergin, *Prize*, 218–20.

evaporated. When the ratio of reserves to production began to drop during World War II—from 14.3 bbl to 11.9 bbl between 1942 and 1945—U.S. officials and oil executives again clamored for an aggressive U.S. foreign oil policy (see Illustrations 6 and 7). Unlike after 1918, however, Washington would no longer have to bother with the pretense of Britain being an equal partner in any such global initiative.[244]

Britain's position in 1918 relative to the United States was far superior to that of 1945, but theirs was by no means a relationship of equals. Nothing epitomized the evolving power balance between Britain and the United States better than the waning of British naval predominance. The Royal Navy had already abandoned its "two-power" standard—that the fleet should be as powerful as the next two largest navies combined—before the war. Since the fleet was too small to be dispersed effectively between multiple theaters, the Royal Navy depended upon mobility to concentrate forces wherever a threat arose. Supplying those forces across long distances posed a challenge. The Naval Staff concluded in May 1921 that its reserves of 4.5 million tons were insufficient—an additional five million tons of naval and commercial reserves were required for use beyond the European theater of operations by 1930. During a hypothetical war against the United States in 1930, annual consumption would total 5.11 million tons. For Japan, the figure was only 3.43 million tons, but the navy needed prepositioned reserves, as it lacked the

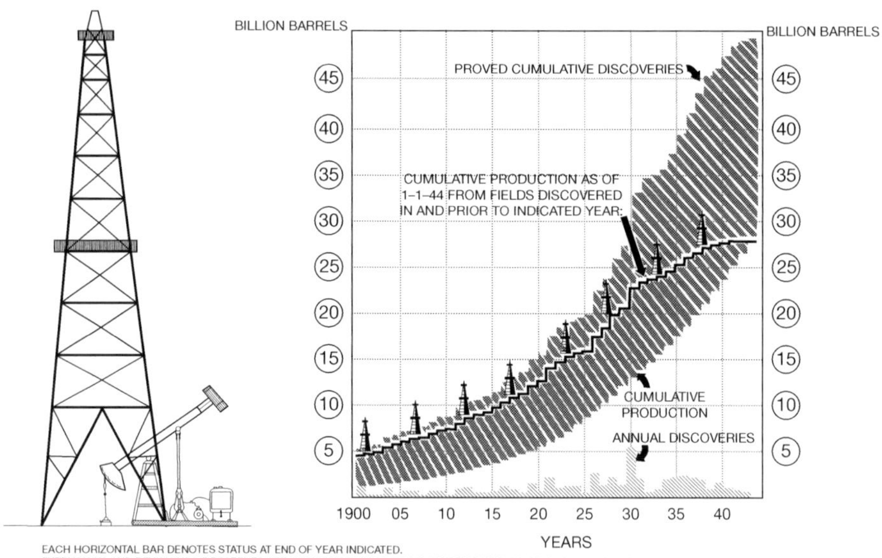

Illustration 6. "Crude Oil Reserves & Production: United States," no date (circa 1945)
Source: NARA, Record Group 59: General Records of the Department of State (RG 59), 800.6363/5-2645.

[244] David Painter, *Oil and the American Century: The Political Economy of U.S. Foreign Oil Policy* (Baltimore: Johns Hopkins University Press, 1986), 9–17, 24–5, and 39–41; Stern, "Ideology," 230–6. The statistics are drawn from: DeGolyer, *Historical Data*, 12, 17, 53, 60–4, 68, and 108.

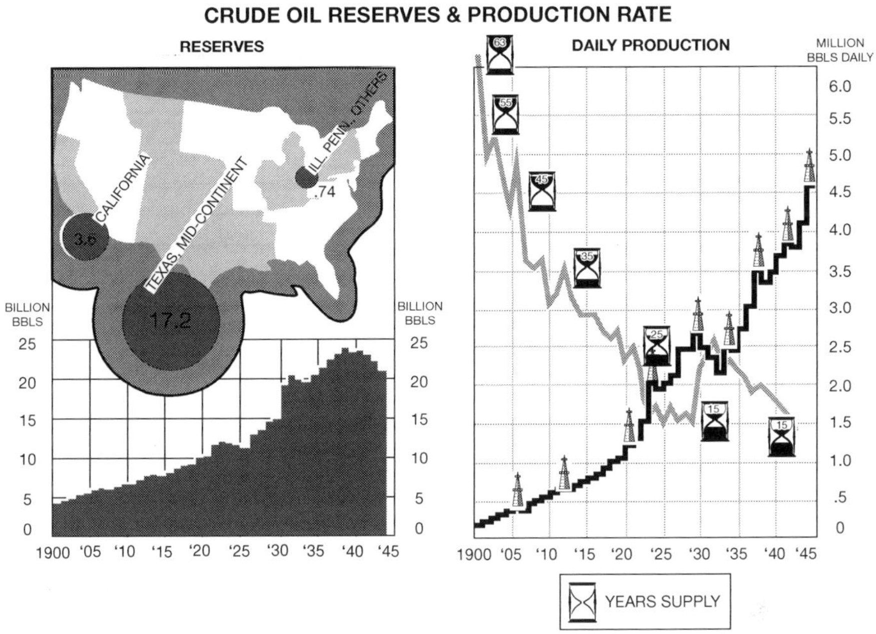

Illustration 7. "Crude Oil Reserves & Production Rate," no date (circa 1945)
Source: NARA, RG 59, 800.6363/5-2645.

resources to transport sufficient quantities to the Pacific—a 10,000-mile journey from Britain.[245]

Even if Britain managed to accumulate nine million tons, the Naval Staff's planning revealed that Britain oil position was inadequate to fight an energy-rich peer competitor. Even if Britain limped through the first year of hostilities, what was it supposed to do thereafter? The empire was incapable of filling any gap, since its total production in 1921, including Persia, was only 3.62 million tons—1.7 percent of global output.[246] Assuming Mexican and U.S. oil was unavailable, the only suppliers left in 1921 were Persia, the East Indies, and the Soviet Union. The East Indies provided 3.7 percent (123,000 tons) of Britain's imports in 1921 (3.31 million tons). As for Russia, in 1920 it provided only 0.7 percent (23,000 tons) of Britain's oil imports. The Soviets' share rose following the signing of the Anglo-Soviet Trade Agreement of March 1921, reaching 5.6 percent in 1925 and its interwar peak of 11.7 percent in 1931.[247] Whitehall nonetheless never overcame its distaste for the Soviet Union. After Moscow halted oil exports to Britain during the General Strike of 1926, Whitehall deemed it "undesirable" for any government department to

[245] Brock, "Oil Fuel Reserves," May 24, 1921, 147-C, ADM 116/3102.
[246] "British Empire [...]," May 10, 1923, attached to: Clarke to Grindle, May 11, 1923, P.D./23/1923, CO 323/903/24.
[247] Appendix I to: CID, "Production [...]," November 29, 1937, 272-A, attached to: Inskip, "Report [...]," February 3, 1938, C.P. 19 (38), CAB 24/274.

depend upon Soviet oil imports.[248] As late as 1934, the government went out of its way to avoid having the Soviet Union satisfy even a trivial quantity of British military consumption.[249]

Arthur Lee (Long's successor as First Lord of Admiralty) took the Naval Staff's case before the CID in June 1921. The Royal Navy, his report explained, was currently incapable of projecting power globally due to the dearth of prepositioned reserves. In the event that Britain had to fight either the United States or Japan, success or failure would depend upon whether adequate supplies of oil were available. Lee requested prepositioning of quantities sufficient for one year of operations near likely areas of operation, and for construction of the necessary infrastructure for sustaining the fleet in transit.[250] At this time, a total of 4.5 million tons was available for operations around Britain but another 3.5 million was required elsewhere.[251] One-third (1.22 million tons) would go to Singapore and Hong Kong, and the remainder to the Indian Ocean (730,000 tons), North America and Caribbean (660,000 tons), Australasia (426,000 tons), and the Mediterranean (300,000 tons). One million tons also had to be available for merchant ships. In all, Britain required a reserve of nine million tons, of which only two million would be available by March 31, 1922.[252]

The Naval Staff and Lee's assessments went to the Standing Defence Sub-Committee of the CID. The Treasury found the Admiralty's strategic assumptions laughable. It never understood the latter's fixation with a one-year reserve, which hinged upon the assumption that Britain would be fighting its major supplier—the United States. If the United States was a neutral, was there any reason to think it would not sell oil to Britain? If not, then the existing six-month reserve should be sufficient. Although the sub-committee agreed with the Admiralty, the costs were prohibitive—stocking the route to Singapore would cost £2 million for the first year of seven. The Admiralty made cosmetic changes, such as dropping its request for a commercial reserve and deleting references to a conflict with the United States, but it did not alter the substance of its proposal.[253] The plan went forward—but the Treasury had not thrown in the towel.[254]

The Great War and its immediate aftermath had numerous implications for Britain's energy future. For Whitehall, the lessons of the conflict appeared to validate the guidelines Churchill had articulated in 1913/14—Britain must seek diversity in its oil supplies, and Whitehall must be prepared to offer energetic leadership.

[248] Cunliffe-Lister, "Purchases [...]," November 22, 1927, C.P. 289 (27), CAB 24/189.

[249] Sassoon, "Purchase [...]," January 23, 1934, C.P. 21 (34), CAB 24/247.

[250] The navy estimated that 2.71 million tons were "required for passage of a fleet to the East and for operations extending over a limited period." Naval Staff, "Naval Oil Fuel Reserves," July 26, 1921, 147-C, CAB 5/4.

[251] The Naval Staff hoped that the Dominions and India would contribute one-quarter of the oil needed for the journey to the Far East. Naval Staff, "Naval Oil Fuel Reserves," July 26, 1921, 147-C, CAB 5/4.

[252] Lee, "Reserves [...]," June 21, 1921, 145-C, ADM 116/3102.

[253] Naval Staff, "Naval Oil Fuel Reserves," July 26, 1921, 147-C, CAB 5/4.

[254] Hankey, "Reserves [...]," October 5, 1921, C.P. 3367, CAB 24/128. The preceding discussion also draws from: Brown, "Fuel Supplies," 200–3 and 223–5.

The matter of safeguarding Britain's future oil supplies took on additional urgency due to fears that the United States might no longer be a reliable supplier thanks to skyrocketing consumption. Furthermore, no self-respecting great power would place its security in the hands of another, not least Britain, whose oil-fueled navy guaranteed access to the foreign markets and resources upon which British prosperity and survival depended.

While Britain's aims were clear, there were differences within Whitehall over the exact strategy Britain should adopt. No single agency was responsible for oil policy, despite the efforts of men such as Long to centralize policymaking. The understaffed Petroleum Department had no executive power and spent most of its time coordinating policy. Among the key players shaping strategy, the Admiralty and Colonial Office took a hard line against cooperation with the United States and preferred that Britain concentrate on imperial development. The Admiralty was also fanatical about defending its special arrangements with Anglo-Persian and unwilling to surrender them, even if it meant allowing Shell to remain under Dutch rather than British control. The Foreign Office and Petroleum Department, by contrast, preferred not to antagonize the United States and sought cooperation with Shell (but not at the expense of the former). As the defender of sound finance, the Treasury sought to minimize the government's interference in private industry and feuded with agencies it believed were scaremongering to justify higher expenditures on defense.

These differences in outlook would define the contours of British oil strategy during the 1920s, but such differences should not obscure the consensus that existed over the broad objectives of British policy. By 1918–19, Britain had developed a two-pronged strategy for the postwar era: promoting British control of Shell (even at the cost of sacrificing the independence of Anglo-Persian or Burmah) and securing development of the only region capable of replacing the United States as Britain's primary supplier—the Middle East. Setbacks such as the failure of Harcourt–Deterding and U.S. displeasure over British policy would convince Whitehall of the benefits of compromise without dampening its ardor for independence.

2

The Years of Complacency
1921–1932

British officials understood that relying on the Middle East entailed certain liabilities. Nationalist sentiment was a growing threat to the integrity of the British Empire. Moving Middle Eastern oil to Europe was also a complex and costly process—firms either had to ship the oil from the Persian Gulf around the Arabian Peninsula and through the Suez Canal or build pipelines to the Mediterranean. Fortunately, the geopolitical environment was relatively benign. At worst, Britain confronted localized threats from disgruntled regional powers. The United States was a partner if not an ally, while British officials were sanguine about the threat posed by Britain's only plausible great power adversary, Japan. This meant that Britain did not have to choose between pursuing an ambitious oil strategy and improving its financial position—although when push came to shove, safeguarding Britain's solvency took a higher priority.

Although this was the ideal time to implement the strategy articulated by Harcourt and championed by Long in 1918–19, Britain's record was disappointing. The dream of creating an "all British" Shell foundered over internal government disputes and a fundamental irreconcilability between Whitehall and Shell's interests. In the Middle East. Britain fended off challengers to the Turkish Petroleum Company (TPC—after 1929 the Iraq Petroleum Company, or IPC) in Iraq but found the rewards meager. The TPC did not begin production in Iraq until 1927 and its Mediterranean pipeline was not complete until 1934. Barring foreigners from imperial oil sources through legislative restrictions only alienated foreign companies and governments, while doing nothing to make Britain energy independent. Despite some progress at increasing British oil supplies, most notably in Venezuela, by the early 1930s, strategic assessments revealed that Britain was still incapable of meeting its wartime requirements from British-controlled sources.

IMPERIAL AND COMMERCIAL DEVELOPMENTS, 1921–1924

Oil was vital to Britain's status as a great power for many reasons besides its importance to the Royal Navy. Oil was also critical to Britain's postwar economic strategy. Contrary to popular mythology, World War I did not bankrupt Britain—its loans to the other Allied powers ($6.7 billion) exceeded its obligations to the

United States ($3.7 billion). Unfortunately, many of its assets, especially loans to Russia and France or reparations from Germany, were of dubious value, unlike its dollar liabilities.[1] Restoring Britain's economic and financial power—what officials in the 1930s called the "fourth arm of defence"—required generating a current-account surplus by reducing imports and/or boosting exports to pay off U.S. debts and restore sterling's value.

Whitehall had two options to achieve this. It could devalue its currency, which would provide a stimulus to Britain's export industries, but this was unacceptable because it would it depreciate the value of British loans to Europe, increase the real value of U.S. loans, harm holders of government bonds, and reduce living standards by increasing the cost of imports. Finally, a weaker sterling would destroy any hope of restoring the gold standard and the City of London as a global financial center. Whitehall therefore opted for deflation by overvaluing its currency and reducing prices through cuts in public expenditure, including defense spending. While there was a risk of unemployment and depressed wages, lower prices would enhance the competitiveness of British exports while preserving the value of sterling.[2]

Although contemporary documents do not argue this explicitly, increasing the supply of Middle Eastern oil sold by British firms and denominated in sterling was indispensable to Britain's broader economic strategy, because it facilitated efforts to control prices by lowering the cost of raw materials inputs. Finally, an oil industry centered in the Middle East, with close links to British banks, shippers, and insurance firms, would facilitate the elite project of transitioning Britain's economy from labor-intensive manufacturing toward capital-intensive services, while restoring a liberalized global financial system with sterling serving as a global reserve currency alongside the dollar.[3] Although these policies foundered consistently for decades, British elites clung tenaciously to the illusion that Middle Eastern oil and sterling could buttress Britain's financial and imperial position until the 1960s.[4]

Britain could only reap the financial advantages of oil if it preserved political and economic control of the oilfields within its orbit. One way of achieving this was through the policy of "control." Pioneered in the colonies before migrating to Whitehall, the policy required that companies operating on many Crown Lands and in protectorates such as Kuwait and Bahrain be registered in Britain or its Dominions, have a majority of British subjects on their board, and not be owned directly or indirectly by foreigners.[5] This last condition was the most important

[1] Zara Steiner, *The Lights that Failed: European International History* (Oxford: Oxford University Press, 2005), 25.

[2] William Stivers, *Supremacy and Oil: Iraq, Turkey, and the Anglo-American World Order, 1918–1930* (Ithaca: Cornell University Press, 1982), 34–41.

[3] Robert Boyce, *British Capitalism at the Crossroads, 1919–1932: A Study in Politics, Economics and International Relations* (Cambridge: Cambridge University Press, 1987).

[4] Steven Galpern, *Money, Oil, and Empire in the Middle East: Sterling and Postwar Imperialism, 1944–1971* (Cambridge: Cambridge University Press, 2009).

[5] Petroleum Department, "Oil Concessions [...]," no date (circa 1929), POWE 34/1. Whitehall developed these regulations to bar Standard Oil from Burma. Payton-Smith, *Oil*, 9–10. These restrictions

one, because it excluded U.S. companies even through their British subsidiaries. The Colonial Office had opposed foreign oil development within the empire since World War I. Its decision to block the sale of an oil property in Trinidad to a Norwegian firm prompted Under Secretary of State for Foreign Affairs Robert Cecil to remark in October 1918 that "control" was self-destructive for a nation that depended on overseas trade.[6] The Admiralty and Board of Trade were content with issuing licenses after "[imposing] conditions which would ensure permanent British control where this appears necessary."[7] Colonial Secretary Long, by contrast, defended the exclusion of foreigners on the grounds of national security.[8]

Once the issue became a source of friction in Anglo-American relations, the Petroleum Department urged Britain to adopt reciprocity with the United States. The passage of the U.S. Mineral Leasing Act in February 1920 was a major setback because it barred British firms from valuable U.S. public lands, and in early 1921, some irate U.S. legislators even threatened an oil embargo to countries that discriminated against U.S. firms.[9] The Petroleum Department questioned not only the wisdom of "control" but also whether excluding foreign firms from the empire would stimulate imperial production—especially since British firms had not distinguished themselves at finding oil.[10] The Colonial Office disagreed—U.S. displeasure was really about Iraq. It also took exception to the Petroleum Department's disparagement of British companies—it was not their fault the empire "contains very little oil." Finally, the alternative of reciprocity was only viable if the entire British Empire abided by it, but Britain could not force the Dominions or India to comply.[11]

The effects of British restrictions on Anglo-American relations were the focus of an Inter-Departmental Committee on Oil during preparations for an Imperial Conference in 1923. Ambassador Geddes advised "removing every possible cause of friction" but cautioned against "opening the door completely."[12] The Petroleum Department agreed—there was no justification for exclusion: "If foreigners like to come and spend money in looking for oil, it seems foolish to prevent their doing so; if they were successful, the oil position of the Empire would be strengthened."[13] Shell was also critical of "control" because it created ill will in the United States.

were not uniform across the empire. Foreigners could acquire concessions in Britain (after 1935), Canada, much of Australia, New Zealand, and South Africa, but not in India, Burma, Trinidad, and Nigeria. "Oil Concessions," Annex to: Board of Trade, "Treatment [...]," October 28, 1932, POWE 33/461; Brown, "Fuel Supplies," 177.

6 Cecil, "Investment [...]," October 14, 1918, E.D.D.C./45, CAB 24/68.

7 Wemyss to War Cabinet, "Investment [...]," December 22, 1918, G.T. 6541; Stanley, "Memorandum [...]," G.T. 6624; both in CAB 24/72.

8 Long, "Investment [...]," October 28, 1918, G.T. 6141, CAB 24/68.

9 McBeth, *Policy*, 61–4.

10 Petroleum Department, "Acquisition [...]," March 23, 1921, attached to: Clarke to Wood, June 7, 1921, R.252, CO 323/879/12.

11 "Copy [...]," no date, attached to: Inter-Department Committee on Petroleum, "Agenda [...] 24th November, 1921," P.D. 1224, CO 323/880/1.

12 Geddes to Curzon, March 23, 1923, attached to: Llewellyn Smith, "Imperial Petroleum Policy," June 7, 1923, Dft. E.C. 24, POWE 33/353.

13 Clarke to Wills et al., "Petroleum Policy," April 21, 1923, P.D. 195; Petroleum Department, "Nationality Restrictions [...]," July 18, 1923, P.D. 267/23; both attached to: Llewellyn Smith, "Imperial Petroleum Policy," June 7, 1923, Dft. E.C. 24, POWE 33/353.

The U.S. Mineral Leasing Act excluded Dutch companies from U.S. public lands to retaliate against Dutch restrictions in the East Indies, and Shell narrowly avoided banishment from Native American lands, as well.[14] A Shell executive tried to convince Whitehall to focus on acquiring oilfields that could supply it in wartime, including fields in the United States. The last thing Britain needed was having other countries emulate its restrictions—British firms would suffer disproportionately since they, unlike their U.S. rivals, had no domestic production upon which to fall back.[15]

The Colonial Office now modified its position to support the so-called "Admiralty Clauses," which stipulated that the Royal Navy should retain the right to call upon any oil produced by an imperial concession during a war or crisis. While it preferred to deal with friendly nationals, the Colonial Office was willing to consider lifting restrictions if nations adopted reciprocity.[16] The Admiralty was not opposed to foreign development of the empire's oil reserves.[17] It did prefer, however, supplies from sources under British control and produced in accordance with the Royal Navy's technical specifications.[18] Something would have to happen soon, however, since "control" was proving an abject failure. British hopes that the end of the war would create a more favorable supply/demand ratio in the British Empire proved unrealistic. Following a drop between 1918 and 1919 from 8.91 million tons to 6.69 million tons, British Empire consumption rebounded in two years to just below the wartime peak (8.83 million tons in 1921). Imperial production between 1918 and 1921 stagnated, however, increasing by only 67,511 tons, and the 1921 figure (1.87 million tons) equaled only 21 percent of consumption.[19]

The "control" clause survived, but the demand for fiscal retrenchment provided a powerful stimulus to reevaluate other policies, particularly the 1919 decision to create a naval stockpile of one year's wartime consumption.[20] In July 1921, the Committee of Imperial Defence (CID) agreed to support a stockpiling plan for naval operations in the Far East that would cost £2 million per annum, but the Treasury balked at spending more than £1.5 million. In August 1921, the government established a Committee on National Expenditure, which concluded in December that "considerable savings" could be extracted from the naval stockpiling plan.[21] Lloyd George thereafter established a cabinet committee chaired by Churchill to review

[14] Wilkins, *Investment*, 103–9 and 263–4.

[15] Cohen to Clarke, May 15, 1923, attached to: Clarke to Ormsby-Gore, May 18, 1923, P.D. 198/23, CO 323/903/25.

[16] Colonial Office, "British Control [...]," June 1923, Dft. E.C. 23, attached to: Llewellyn Smith, "Imperial Petroleum Policy," June 7, 1923, Dft. E.C. 24, POWE 33/353.

[17] Wemyss, "Investment [...]," December 22, 1918, G.T. 6541, CAB 24/72.

[18] "Admiralty Memorandum [...]," no date, POWE 33/353.

[19] "British Empire [...]," May 10, 1923, attached to: Clarke to Grindle, May 11, 1923, P.D./23/1923, CO 323/903/24. For data on British oil imports between 1913 and 1937, see: Appendix I to CID, "Production [...]," November 29, 1937, 272-A, attached to: Inskip, "Report [...]," February 3, 1938, C.P. 19 (38), CAB 24/274.

[20] The following discussion draws from: Brown, "Fuel Supplies," 193–225.

[21] Hankey, "Reserve [...]," February 2, 1924, 479-B, CAB 4/10. For the government's efforts to cut spending, see: Andrew McDonald, "The Geddes Committee and the Formulation of Public Expenditure Policy, 1921–1922," *Historical Journal* 32: 3 (1989): 643–74.

proposed reductions in defense spending. Churchill defended the Admiralty's stockpiling plan by arguing that Britain's position in the Pacific was vulnerable.[22]

The Admiralty and Treasury traded accusations of bad faith. Finally, in July 1922, the CID sided with the Treasury. With great reluctance, Churchill and the Admiralty bowed to austerity and reduced the amount of oil to be stockpiled to 8.14 million tons—4.5 million tons in Britain, one million tons of commercial reserves, and the remainder around the world. The Far Eastern stockpile would be ready by 1931, and the home reserve six years later.[23] As of March 1923, fuel oil reserves in Britain had reached 2.5 million tons—more than six months of wartime naval consumption. The Treasury now questioned the need to collect a one-year reserve within Britain when the current figure would suffice.[24] The Chancellor, Neville Chamberlain, tried to negotiate with First Lord of the Admiralty Leo Amery, and the Treasury solicited the opinion of Cadman, now a member of Anglo-Persian's board. He was convinced there was no need for a one-year reserve since any conflict with Japan was unlikely to be protracted, while ample supplies would be available from the United States and Persia.[25] Amery dismissed the Treasury's arguments, but before the CID could adjudicate the matter, the Conservative government fell in January 1924.[26]

The new Labour ministers carried on their Conservative predecessors' work—unsurprisingly, considering that the First Lord of the Admiralty, Lord Chelmsford, was a Conservative peer and only accepted the job to preserve the navy's existing force structure. He argued that the 4.5 million ton reserve within Britain was necessary if the Royal Navy expected to conduct operations beyond the North Sea and Mediterranean. He also disagreed with the claim that there was no need to worry about acquiring supplies from the United States and worried that Britain lacked sufficient tankers to both supply its fleet and import oil from the United States.[27] The Treasury did not dispute the need for reserves, but it argued that the size of the stockpile, its rate of accumulation, and its location deserved scrutiny. The Treasury pointed to the inconsistency at the heart of the Admiralty's position. In 1919, the Admiralty had claimed that 4.5 million tons equaled one year's wartime consumption. The request in 1921 to double the amount to be accumulated implied one of two scenarios: a war against the United States or one against Japan. The latter was no longer under consideration, and the Treasury believed the Admiralty was exaggerating the threat. Why then was Britain accumulating a two-year reserve

[22] Churchill, "Report [. . .]," February 4, 1922, C.P. 3692, CAB 24/132.

[23] Lee, "Reserves [. . .]," July 7, 1922, 175-C, CAB 5/4; Treasury, "Reserves [. . .]," November 19, 1927, CA, CHAR 22/174. Britain would gather these reserves in secret to avoid arousing Japanese suspicions. Secretary of the Admiralty to the Secretary of the CID, "Reserves [. . .]," October 3, 1922, 180-C, CAB 5/4.

[24] Joynson-Hicks, "Home Reserve [. . .]," August 14, 1923, C.P. 384/23, CAB 24/161.

[25] Brown, "Fuel Supplies," 204–5.

[26] Treasury, "Reserves [. . .]," November 19, 1927, CA, CHAR 22/174.

[27] Chelmsford, "Reserve [. . .]," January 30, 1924, 476-B, CAB 4/10. At the time, Abadan could only refine 1.5 million tons of fuel oil. British and Australian refineries could double the amount available depending on the supply of tankers. Beatty, "Oil Fuel," February 14, 1924, 482-B, CAB 4/11.

to fight Japan? Five million tons of reserves, divided between Britain and overseas depots, seemed a more reasonable figure.[28]

Despite the Treasury's objections, the Admiralty prevailed in February 1924, when the CID upheld the original decision to establish a one-year naval reserve by 1937.[29] The Treasury had, however, breached the wall and was on its way to establishing itself as the primus inter pares within Whitehall. While many date this development to the post-World War II period, in fact it began much earlier.[30] Since 1919, the Treasury had challenged the Admiralty's judgment, positing that Britain's situation did not justify such expenditure.[31] In future, the Admiralty would find itself unable to defend its oil requirements against calls for retrenchment, which soon forced a reassessment of British strategy in the Middle East.

The key to Britain's energy future was controlling Iraq and Persia. British troops had managed to occupy the most promising area in the former, Mosul, at the end of the war but numerous obstacles delayed exploitation of its oil riches. By the end of 1920, two of the most pressing difficulties had been resolved: in April, Whitehall and Paris reached an understanding at San Remo, and by October, British forces had quelled the uprisings that had convulsed Iraq since the summer. That left the United States, which opposed Anglo-French efforts to monopolize Iraq's oil wealth.

Britain's position rested upon the contention that the TPC possessed a valid concession. In fact, the TPC had only received the promise of a concession in 1914 from a government that no longer existed. To admit U.S. oil companies into Iraq after 1918 without any preconditions would have defeated the objective of British policy—to gather as much oil under British control as possible. The United States was naturally opposed to this. At the end of World War I, when fears of an oil shortage abounded, U.S. companies hounded Washington to facilitate access to overseas reserves. The U.S. government's support for an "Open Door"—equal opportunity for U.S. businesses—in the Middle East continued even after new U.S. production had wiped out fears of depletion, except now Washington was assisting U.S. companies in finding foreign oilfields to supply their overseas markets.[32] By December 1921, the Petroleum Department did not put much stock in speculation that the United States was running out of oil. It characterized Washington's fealty to the "Open Door" as a ploy to ensure that "[the] door is to be open wide enough to let Americans in, even if others stay outside."[33] Whitehall soon realized, however, that it needed U.S. help if it was to retain control of Iraq.[34]

In the wake of the Iraqi uprisings of July–October 1920, which claimed 1,500 British casualties and cost £600,000 a week to suppress, Britain had to sustain its presence in Iraq despite massive cutbacks in government expenditures. The British

[28] Snowden, "Reserve [...]," January 31, 1924, 477-B, CAB 4/10.

[29] Birkenhead, "Report [...]," February 20, 1928, C.P. 47 (28), CAB 24/192; Treasury, "Reserves [...]," November 19, 1927, CA, CHAR 22/174; Brown, "Fuel Supplies," 206–8.

[30] Galpern, *Money*, 8–9.

[31] Brown, "Fuel Supplies," 264–5.

[32] Stivers, *Supremacy*, 194–9.

[33] See notes attached to: Clarke, December 5, 1921, POWE 33/93. See also Yergin, *Prize*, 196–6.

[34] Stivers, *Supremacy*, 76–9.

were caught in a trap, for they could not acknowledge publicly their reason for being in Iraq—control of its oil.[35] Officials soon devised a new way to preserve British control at a fraction of the current cost. Under the "Cairo Strategy" developed at the Middle East Conference of March 1921, Britain would gradually withdraw its garrison of regular and imperial troops and use Iraq's oil royalties to establish an Iraqi army and stabilize the fledgling Hashemite dynasty. These would be supported by Royal Air Force (RAF) squadrons that would terrorize intransigent natives and deter external threats such as the Turks or the Saudi Ikhwan tribesmen; British advisors scattered throughout the Iraqi government; and a High Commissioner who would effectively rule the country as a proconsul. Beyond its relatively low cost, the strategy also promised major political dividends for key stakeholders. Churchill, for one, needed a major success to save his floundering career. The prominent role of air power suited Hugh Trenchard, the Chief of the Air Staff, who wanted to preserve the independence of the RAF. British officials and oil executives were obviously pleased to retain access to Iraqi oil. Finally, the strategy would empower local Sunni elites, who were determined to cement their authority in a post-Ottoman Iraq.[36]

For the strategy to be viable, however, Iraq needed oil revenues, which meant starting production as soon as possible and dealing with the Americans. According to Churchill, Britain had three options by March 1922. It could accept U.S. demands to nullify the TPC's claims, submit the issue to international arbitration, or change U.S. opinion. Churchill rejected the first option on the grounds of prestige. The second was undesirable because it risked exposing the shaky legal foundation of the British position. He therefore backed a proposal suggested by the Foreign Office—and supported by both the Petroleum Department and Anglo-Persian—that the TPC's partners offer U.S. companies an equal portion of shares as the French in exchange for Washington's assurance that it would recognize the TPC's position in Iraq.[37]

The breakthrough came the following month, when Greenway, Cadman, and A.C. Bedford (Jersey's chairman) reached an understanding that U.S. companies would join the TPC. Both sides wished to remove the dispute from "political channels" and handle it as a "commercial transaction," which was also Whitehall's position.[38] The only difficulty was getting the three governments involved to agree. The French might complain, since the inclusion of U.S. oil companies into the TPC would compel them to yield a portion of their shares. They were not obligated to do so under the San Remo Agreement unless the Iraqis received shares.[39] Negotiations in the summer of 1922 produced an agreement in principle between

[35] Sluglett, *Iraq*, 34.

[36] For the genesis of the "Cairo Strategy," see: Sluglett, *Iraq*, 39–41; Stivers, *Supremacy*, 75–109. For the conference report, see: Colonial Office, "Report [...]," June 1921, CO 935/1. For an official statement of British policy, see: Amery, "Situation in Iraq," May 11, 1925, C.P. 235 (25), CAB 24/173. For the role of airpower, see: David Omissi, *Air Power and Colonial Control: The Royal Air Force, 1919–1939* (Manchester: Manchester University Press, 1990), 18–38; Sluglett, *Iraq*, 184–92.

[37] Churchill, "Iraq Oil," March 13, 1922, C.P. 3832, CAB 24/134; Venn, "Foreign Office," 76–8.

[38] Cadman, "Memorandum... 9th April, 1922 [...]," attached to: letter to Harmsworth, April 21, 1922, author unclear, S. 542, POWE 33/95; Oliphant to the Secretary to the Petroleum Department, December 28, 1922, E 14034/132/65, POWE 33/95.

[39] Letter to Harmsworth, April 21, 1922, author unclear, S. 542, POWE 33/95.

the TPC and a consortium of U.S. companies. The denouement came in December 1922, when Anglo-Persian and Shell signed a memorandum of understanding whereby the former would yield half of its shares in the TPC to the Americans for a 10 percent overriding royalty on any oil the TPC produced (reduced to 7.5 percent in 1931).[40]

The resolution of the dispute over Iraq's oil came as a tremendous relief in both London and Washington.[41] The Americans were unwilling to accept any British fait accompli. The State Department's Foreign Trade Adviser, Millspaugh, had argued that securing U.S. access to Iraq's oil was a diplomatic rather than a commercial issue, and he opposed British efforts to monopolize Iraq's oil.[42] Luckily, he was also committed to a collaborative relationship with Britain. Millspaugh considered the resolution of the Anglo-American oil dispute a prerequisite to building a stable postwar world order. He argued that the best way to satisfy U.S. and British oil needs was through an international agreement based on the "Open Door."[43] Following consultations with other government departments and private industry, he completed a draft Anglo-American oil agreement in May 1921.[44] Millspaugh believed that the Washington Naval Conference was the ideal venue to broach the issue, since a successful arms-control treaty would reduce naval fuel consumption, thereby easing competition for oil.[45] Whitehall was not interested. Once news leaked that the United States would push for an agreement at the Washington conference, Lloyd George made it clear that Britain's oil assets were "much too important" and promised to fight any attempt "to give them away."[46]

Millspaugh's model for reorganizing Anglo-American oil relations collapsed—at least until 1944, when Washington tried again to reach a multilateral oil agreement with Britain—but the threat to Anglo-American comity had passed.[47] Even at the height of the tensions, Cadman reassured U.S. officials "that the policy of the British Government was one of the open door" and lamented the "folly in our opposing each other beyond the point where healthy and natural competition

40 "Memorandum [...]," December 12, 1922, attached to: Clarke to McNeill, December 13, 1922, S. 610, FO 839/10. The Foreign Office did not intend to have France ratify the new agreement, lest the French use it as an excuse to renegotiate the San Remo Agreement. Oliphant to the Secretary to the Board of Trade, December 28, 1922, E 14034/132/65, POWE 33/95.

41 "Anglo-American Oil Relations," October 16, 1922, attached to: Clarke to Grigg, October 17, 1922, T 172/1268; Rubin, "Stumbling," 211ff.

42 Millspaugh, "Memorandum [...]," April 9, 1920, NARA, RG 59, 800.6363/166, and "Memorandum," November 29, 1920, NARA, RG 59, 867.602 OT81/170.

43 Millspaugh, "Informal and Provisional Memorandum [...]," February 19, 1921, attached to: Merle-Smith to Hughes, March 11, 1921, NARA, RG 59, 800.6363/325; Millspaugh to Hughes, "Petroleum Negotiations [...]," March 29, 1921, NARA, RG 59, 890G.6363/69.

44 "Draft," May 12, 1921, attached to: memorandum for Millspaugh, May 10 (*sic*) 1921, NARA, RG 59, 811.6363.

45 Millspaugh, "Petroleum Situation [...]," 1921, Hoover Institution, Papers of Stanley Hornbeck, Box 436.

46 Lloyd George to Pretyman, November 25, 1921, POWE 33/93; Brown, "Fuel Supplies," 182; Ferrier, *History*, 376–7.

47 The fate of the 1921 agreement is detailed in Anand Toprani, "An Anglo-American 'Petroleum Entente'? The First Attempt to Reach an Anglo-American Oil Agreement, 1921," *The Historian* 79: 1 (2017): 56–79.

should prevail."[48] Washington considered the British presence in the Middle East useful and was happy to let Britain carry the political and military burdens if U.S. firms could reap the commercial benefits. For its part, Whitehall believed that harnessing U.S. capital and expertise would solidify Britain's position by promoting Baghdad's solvency.[49] Like the United States in 2003, Britain hoped that oil revenues would fund both the occupation of Iraq and its development.

The United States turned out to be the least of Britain's problems in the Middle East. A thornier issue was the Anglo-Turkish dispute over Mosul.[50] From the start, the Colonial Office was convinced that the French and the Americans were aiding the Turks.[51] During the Lausanne Conference of 1922–3, Britain tried to negotiate a new peace treaty with Turkey that would force Ankara to abandon its claims to Mosul. Losing Mosul would have nipped Britain's plans for Iraq in the bud. Turkey would thereafter pose a threat not only to Iraq but also to other British interests in the Persian Gulf.[52] Britain demanded that Turkey confirm prewar concession agreements, but the United States supported Turkey in blocking such terms from the treaty.[53] Britain and Turkey eventually submitted their dispute for arbitration by the League of Nations. When the latter decided in 1925 that Iraq should receive Mosul, Turkey protested but, under the threat of sanctions, it relented the following year in exchange for 10 percent of the royalties from Mosul for the next twenty-five years. Turkey also had the option to sell its rights to revenues for a lump-sum payment of £500,000, which it quickly exercised. The British were delighted and made the Iraqis pay.[54]

The Petroleum Department was pleased with the course of developments.[55] Amery, Colonial Secretary since February 1924, took a cynical view. He was contemptuous of the "Open Door" and surmised that once the Americans received their cut, the British were "likely to hear nothing more" about the matter. He also admitted that the companies were uninterested in making the TPC profitable, since they wanted to avoid paying corporate taxes in Britain.[56] The Iraqis were the only losers. Although the San Remo Agreement left open the possibility of Baghdad receiving a 20 percent stake in the TPC, none of the companies was willing to

[48] Dearing, memorandum of conversation, December 23, 1921, NARA, RG 59, 841.6363/188; Millspaugh, "Memorandum [...]," January 17, 1922, NARA, RG 59, 841.6363/203.

[49] Michael Hogan, *Informal Entente: The Private Structure of Cooperation in Anglo-American Economic Diplomacy, 1918–1928* (Columbia: University of Missouri Press, 1977), 159–85; Stivers, "Iraqi Oil," 519–26, and *Supremacy*, 87–91 and 192–3.

[50] The following discussion draws from: Sluglett, *Iraq*, 71–86.

[51] Colonial Office, "Foreign Incitement [...]," December 13, 1921, C.P. 3566, CAB 24/131.

[52] Peter Beck, "'A Tedious and Perilous Controversy': Britain and the Settlement of the Mosul Dispute, 1918–1926," *Middle East Studies* 17: 2 (1981): 256–8.

[53] DeNovo, *American Interests*, 191–6; Fiona Venn, "Oleaginous Diplomacy: Oil, Anglo-American Relations and the Lausanne Conference, 1922-23." *Diplomacy and Statecraft* 20 (2009): 414–33.

[54] Beck, "Mosul Dispute," 270–1; Geoff Berridge, *British Diplomacy in Turkey, 1583 to the Present: A Study in the Evolution of the Resident Embassy* (Leiden: Martinus Nijhoff Publishers, 2009), 145–51.

[55] Clarke, "Turkish Petroleum Company [...]," February 4, 1924, P.D. 13, POWE 33/183.

[56] Amery, "Turkish Petroleum Company," February 23, 1925, C.P. 108 (25), CAB 24/172. See also: Amery, "Turkish Petroleum Company," March 18, 1925, C.P. 171 (25), CAB 24/172.

share ownership with Iraq.[57] The TPC finally acquired a seventy-five-year concession from the Iraqi government in 1925—originally for 192 square miles but expanding in 1931 to cover most of Iraq east of the Tigris—but had yet to find large quantities of oil. As the British financial mission to Iraq warned, Baghdad was unlikely to "receive any appreciable revenue" from the TPC for several years.[58] Britain had nevertheless established the political preconditions for Iraq to become a major oil supplier. Air power and the façade of sovereignty allowed Whitehall to slash expenditures without yielding its political influence.[59] The British would not be so fortunate when it came to the other pillar of their postwar strategy.

SHELL GAMES, 1922–1924

Besides cementing British control of Middle Eastern oil, Whitehall's primary objective after 1918 was creating an "all British" major oil company to break the United States' dominance of the global oil market. Efforts with Shell had stalled following the failure of the Harcourt–Deterding Agreement in 1919, but elements within Whitehall, Shell, and Burmah were determined to try again. The most ambitious attempt took place in 1922, at the behest of Shell, Burmah, and certain departments in Whitehall.[60] It entailed a merger of Shell, Anglo-Persian, and Burmah, with Whitehall relinquishing its shares in Anglo-Persian. Shell and Burmah worked out a formula whereby the latter would purchase the government's shares in Anglo-Persian and combine with Shell Transport and Trading to "extinguish" the Dutch majority within the Shell Group while exempting subsidiaries operating on Dutch territory. Of the shares, 10.68 percent would go to Burmah, 33.24 percent to Shell, and 6.25 percent to Anglo-Persian—thus, 50.17 percent would go to British interests. The merger would have created a true British competitor to the major U.S. companies.[61] Shell, Anglo-Persian, and Burmah annually produced 13.1 million tons of oil—12 percent of global production—compared to the 15.9 million tons produced by the various Standard Oil companies (14 percent).[62]

The government began studying the matter in January 1922. Petroleum Minister Philip Lloyd-Greame conceded that the Shell Group was a tempting prize—a global enterprise comprising more than 100 subsidiaries and capitalized at £300 million.[63] It produced three times more oil than Anglo-Persian and Burmah combined—9.7 million tons versus 3.44 million tons. Its marketing subsidiary,

[57] Edith Penrose and E.F. Penrose, *Iraq: International Relations and National Development* (London: Ernest Benn, 1978), 60–6.

[58] After running a budget deficit in its first two years, Iraq had a surplus of £670,320 in 1923/4. This would not last due to spending obligations imposed by Whitehall. Colonial Office, "Report [...]," May 1925, Middle East No. 6, CAB 24/173; Sluglett, *Iraq*, 61–5 and 86–91.

[59] Sluglett, *Iraq*, 182–4 and 210–18.

[60] Ferrier, *History*, 371–85; Jones, *Industry*, 222–9; McBeth, *Policy*, 36–42 and 44–6.

[61] Lloyd-Greame, "Proposed Amalgamation [...]," January 6, 1922, C.P. 3637, CAB 24/132.

[62] Petroleum Department, "Production of Petroleum [...]," March 20, 1922, O.S.C. 7, ADM 116/3452.

[63] Lloyd-Greame, "Proposed Amalgamation [...]," January 6, 1922, C.P. 3637, CAB 24/132.

Shell-Mex, was Britain's largest importer, at 1.24 million tons (28 percent)—twice as much as Anglo-Persian's 653,000 tons (15 percent).[64] The pros and cons were roughly equal. British consumers would benefit, but the Admiralty's prospects were unclear. A merger could antagonize the United States, but eliminating Whitehall's stake in Anglo-Persian would also remove a source of friction with Washington.[65] Unless the government was willing to fund Anglo-Persian's development, the company would have to issue additional stock, which would dilute the government's stake.[66] The Admiralty promptly dismissed the merger.[67] The Royal Navy received 40 percent of its annual requirements through its supply contract with Anglo-Persian and purchased much of the rest on the open market. The Admiralty feared that by reducing the number of sellers, the merger would result in higher prices. The only way to win the Admiralty's support was to convince Shell to sign a new supply contract. The Admiralty demanded a "perpetual" agreement for 250,000 tons with an option to purchase an additional 150,000 tons at 20/- per ton. This was hardly a serious proposal—the Admiralty even admitted that it had drafted the terms with the intention of provoking Shell's rejection.[68]

In March 1922, the government established a cabinet committee chaired by the President of the Board of Trade, Stanley Baldwin, to study the matter.[69] The merger drew fire from across Whitehall.[70] One critic claimed that Shell's oilfields in the Western Hemisphere (Mexico) and Egypt did not yield oil of suitable quality for the Admiralty. The government could also not afford to ignore the political consequences—creating a gigantic conglomerate contradicted the rationale for bailing out Anglo-Persian in 1914.[71] During a meeting on March 2, 1922, the Admiralty Board concluded that the merger would provide only "shadow" British control of Shell.[72] The merger did not even increase the amount of British-controlled oil owned significantly, since many of Shell's assets were located in countries without formal ties to Britain.[73] Another consideration was that Whitehall's purchase of shares in Anglo-Persian turned out to be a windfall. One 1922 valuation concluded that the government had already earned £17,410,000 through higher

[64] Committee on Oil Companies Amalgamation, "Memorandum from Petroleum Department," March 20, 1922, O.S.C. 7, CAB 27/180.

[65] A point reiterated by Geddes in Washington on January 18, 1922 and May 4, 1922, both in T 161/142.

[66] Lloyd-Greame, "Proposed Amalgamation [...]," January 6, 1922, C.P. 3637, CAB 24/132.

[67] Brown, "Fuel Supplies," 180–2.

[68] "Note [...]," no date, Appendix IV to: Lloyd-Greame, "Proposed Amalgamation [...]," January 6, 1922, C.P. 3637, CAB 24/132.

[69] St. Quintin Hill, "Committee [...]," March 6, 1922, O.S.C. 1, CAB 27/180. See also: "Heads for Discussion," March 10, 1922, O.S.C. 3, CAB 27/180.

[70] The Admiralty was the first to take up cudgels, but the Treasury, Board of Trade, Colonial Office, India Office, and Foreign Office eventually joined it. Director of Contracts to the Fourth Sea Lord, November 28, 1922, ADM 116/3452.

[71] C.W., "Proposed Amalgamation [...]," February 8, 1922, ADM 116/3452.

[72] "Extract from Board Minutes," March 2, 1922, ADM 116/3452.

[73] Committee on Oil Companies Amalgamation, "[...] Statement of Admiralty Views," March 11, 1922, O.S.C. 5, CAB 27/180.

tax receipts, share appreciation, savings through the Admiralty's contract with Anglo-Persian, and indirect savings from other suppliers thanks to competition from Anglo-Persian.[74] Even the Treasury, which had long considered the government's ownership of shares in Anglo-Persian to be "embarrassing," was lukewarm.[75]

The committee rendered its decision in June 1922. It concluded that the merger did not offer "effective and permanent control" of the new company. The committee also agreed with the Admiralty that the merger would jeopardize the navy's lucrative contract with Anglo-Persian. Further, Shell would have little incentive to develop high-cost fields within the empire. Finally, the merger would harm Anglo-American relations by creating a new rival to U.S. firms. The committee therefore recommended that the government retain its shares in Anglo-Persian and oppose the merger.[76]

While this verdict was the effective death knell of the six-year project to bring Shell under British control, the government did consider selling its shares in Anglo-Persian to Shell in late 1923. The Treasury had never been comfortable with the government's shareholding, and Anglo-Persian's poor financial performance validated the belief of Conservative leaders such as Baldwin (now Prime Minister) that government interference in private enterprise was suboptimal. The latest scheme had the backing of Churchill, who was temporarily out of a job. He had opposed the merger the year before but changed his mind due to the company's financial difficulties. Shell considered Churchill's support essential since he was probably the only person capable of winning over the Admiralty.[77] Churchill was desperate for cash while searching for a new parliamentary seat, and he asked Shell and Burmah for £10,000 if his services proved unsuccessful and £50,000 if the merger was successful. He then brokered a draft agreement between Shell, Anglo-Persian, and Burmah in October 1923. The government was also willing to tax only profits from Shell's foreign subsidiaries paid to British shareholders.[78]

In November, Churchill withdrew his services and returned his £5,000 advance in order to run for Parliament in the upcoming General Election.[79] Even before Churchill had recused himself, the Admiralty had already judged his arguments as "entirely unsatisfactory." Creating a new, nominally "British" company meant little unless the Admiralty could protect its interests by nominating directors.[80] Otherwise, there was no reason to alter the arrangement with

74 This figure did not include Whitehall's investments after 1914. "Estimate [...]," January 25, 1922, no author, O.S.C. 4, ADM 116/3452. By contrast, Iran's royalties between 1912 and 1924 totaled £3.7 million. Elm, *Oil*, 17–22.

75 Horne, "Amalgamation [...]," no date, T 161/142.

76 Committee on the Proposed Amalgamation, "Report," June 12, 1922, attached to: Baldwin to Cabinet, June 23, 1922, C.P. 4050, CAB 24/137.

77 Cohen to Churchill, September 14, 1923, CA, CHAR 2/128/5–10.

78 Lloyd-Greame to Churchill, October 18, 1923, CA, CHAR 2/128/24–25. A draft of the agreement along with supplementary memoranda on the deal's benefits may be found in: CHAR 2/129/1–52.

79 Cohen to Churchill, November 15, 1923, CA, CHAR 2/128/58. See also: Brown, "Fuel Supplies," 183–4; Corley, *Burmah*, 298–304; Yergin, *Prize*, 192–4.

80 Note to Amery, "Mr. Churchill's Replies [...]," November 14, 1923, ADM 1/8658/55.

Anglo-Persian.[81] The fate of the Admiralty's rebate was a major bone of contention. The Admiralty demanded that Shell not only fulfill the supply contract with Anglo-Persian but at a lower price if there was no rebate.[82] Churchill replied that it was unreasonable to expect Shell to pay above market prices for Anglo-Persian shares and still pay the rebate, which was of minor significance anyway since the Treasury was footing much of the bill as Anglo-Persian's majority shareholder. Nor could it expect Shell to double the quantity it sold the Admiralty unless the navy was willing to pay market prices.[83] Churchill urged Amery, still First Lord of the Admiralty, to sell the government's stake in Anglo-Persian before its share price, which had declined by over half over the past year, fell any further.[84]

Sticking with Anglo-Persian required confidence in the productivity of the Persian oilfields, and supporters of a merger pointed to the example of Mexico, where production had plummeted from 193.4 million bbl to 149.59 million bbl (22 percent) between 1921 and 1923.[85] The Admiralty discounted such a possibility in Persia since Anglo-Persian was the sole concessionaire. If anything threatened the health of the Persian oilfields, it was Shell, which would certainly boost cheaper Persian production.[86] Despite its reservations, the Admiralty again consulted with Shell about a new supply contract that would lock in its existing price and prevent the company from recouping its losses through higher prices to meet the Royal Navy's outstanding demand. The terms included 1.2 million tons per year—more if the navy needed to increase stockpiles—from any Shell oilfield the Admiralty specified at a cost of 20/- per ton.[87] Shell again refused, and Amery concluded that, since "our interests and those of the Group are not in harmony," a merger would be "entirely unsatisfactory."[88]

While it was deliberating the matter, on January 16, 1924 the Conservative government fell and was replaced by the first Labour government. This change did not have any effect on the prospects for the merger, as opposition was widespread. The Admiralty now had the support of the Treasury. The latter had always been skeptical of Whitehall owning shares in Anglo-Persian but was determined to make the best of the situation. Although the value of Anglo-Persian's stock was plummeting (from £4 to £2½), the Treasury advised against dumping the government's shares. It also placed little credence in fears that Persian production would soon collapse—a wise judgment since, in 1927, Anglo-Persian implemented unit

[81] Amery, "Admiralty Views [...]," January 10, 1924, C.P. 20 (24), CAB 24/164. Anglo-Persian's annual sales of fuel oil to the Admiralty between 1921/2 and 1932 averaged more than 500,000 tons (peaking at 621,118 tons in 1930), although the latter's share of the former's total fuel oil sales declined from one-quarter to one-seventh. Ferrier, *History*, 668.

[82] "Questions [...]," no author or date, ADM 1/8658/55.

[83] Churchill, "Proposed New Oil Scheme [...]," 1923, CA, CHAR 2/129/68–155.

[84] Churchill to Amery, October 29, 1923, and November 1923; CA, CHAR 2/128/29–35 and 44–6.

[85] DeGolyer, *Historical Data*, 5.

[86] Amery, "Admiralty Views [...]," January 10, 1924, C.P. 20 (24), CAB 24/164.

[87] Note to Amery, "Mr. Churchill's Replies [...]," November 14, 1923, ADM 1/8658/55.

[88] Amery, "Admiralty Views [...]," January 10, 1924, C.P. 20 (24), CAB 24/164.

operation of its Persian fields, which extended their lifespan.[89] While the political situation in Persia was unsettled, the Treasury was still confident that "the obvious interest of the Persian Government in the maintenance of a reliable source of revenue" was "the best protection against political disturbance."[90]

On a pure profit–loss basis, the government's decision continued to pay dividends. In 1923, Churchill estimated that the government's initial £2.2 million investment was worth £40 million.[91] Between 1922 and 1928, savings through the Admiralty's supply contract—deliveries averaged over 500,000 tons between 1922/23 and 1932—and rebate totaled £2,935,000, plus £4,272,000 in dividends and interest payments. In 1926–7 alone, Whitehall realized a profit of £8,125,000 when it received half of the five million bonus shares distributed by the company, not including £11,250,000 of appreciation of existing shares.[92] Whitehall did pay a price—its refusal to part with its shares in Anglo-Persian doomed efforts to secure British control of the Shell Group.

By 1924, one of the key aims of Britain's postwar strategy—creating an "all British" major oil company capable of satisfying the requirements of empire and challenging the U.S. giants—was out of reach. Whitehall's preoccupation with eliminating "foreign" control of Shell seems peculiar today. Shell had rendered yeoman service to the Allied cause between 1914 and 1918, and there was no reason to doubt that in the future. This fact eliminate the need for British control of Shell, especially if it doing so jeopardized the Admiralty's contract with Anglo-Persian and promoted further cartelization of the oil industry. By 1923, Whitehall was realizing that the security of oil-rich territories was more important than the exact composition of the firms that produced the oil.[93] By that standard, British progress since 1918 had been unimpressive, as a new interagency committee tasked with evaluating wartime requirements—the Oil Board—soon discovered.

MOVING BRITAIN'S OIL, 1925–1931

By the mid-1920s, it seemed that the whole world was awash with oil except the British Empire, whose share of global production was negligible—1.8 percent of total output and 3.8 percent of Britain's imports in 1926 (1.73 million bbl of 46 million bbl). This figure would contract to 3.2 percent in 1927—1.56 million bbl of 49 million bbl.[94] Elsewhere around the world, however, discoveries as well as

89 Ferrier, "Iranian Oil Industry," 646–7.

90 Snowden, "Anglo-Persian Oil Company," January 26, 1924, C.P. 32 (24), CAB 24/164.

91 Churchill, *World Crisis*, 139–40.

92 Cadman to Maund, March 14, 1928, BP 71044; Ferrier, *History*, 474.

93 Sluglett, *Iraq*, 71.

94 Petroleum Department, "Oil Supplies [...]," July 1926, O.B. 19, CAB 50/3. The Oil Board was especially pleased that Anglo-Persian was completing a third pipeline from its Persian oilfields to Abadan, increasing total production from 4.75 million tons to seven million tons per annum. Hermon-Hodge, "Persian Oil Field," February 26, 1927, O.B. 22, CAB 50/3.

improved extraction and refining techniques had created a glut of oil. The companies reacted by colluding to stabilize prices. In 1928 they began a multiyear project under the so-called Achnacarry (or "As-Is") Agreement, with Whitehall's approval.[95] Signatories agreed to preserve existing market shares, pool resources, and supply markets from the closest sources. To compensate for discrepancies in costs of production and freight, they also used the price of oil at the Gulf of Mexico as the global benchmark.[96] Companies sold their oil as if it was priced and shipped from the Gulf—a system known as "Gulf-Plus" pricing. Even outsiders such as the Soviet Union abided by Achnacarry's provisions, since they too benefited from an artificially high price.[97]

The fact that Britain still depended on foreign-controlled supplies was not yet a liability. Only two great powers, the United States and the Soviet Union, possessed significant internal production, and there was no expectation that Britain would be fighting either. Britain's likely adversaries—Japan, Italy, and Germany—lacked its easy access to overseas supplies or close relationship to the United States. The latter was still a major supplier for Britain—40 percent of U.S. oil exports in 1927 (6.9 million tons) went to the empire, and 40 percent of Britain's imports in 1926 came from the United States.[98] A number of British firms were also significant U.S. oil producers.[99] Even if the United States banned exports to the British Empire, the British could still deal with U.S. oil companies in Latin America and import U.S. oil indirectly through third parties.[100]

Scrutiny of British planning documents reveals, however, a more alarming picture. In 1924, the Admiralty sought control over the allocation of fuel oil/diesel and tanker tonnage in wartime. Other departments opposed granting control over a vital resource to the military. Since oil was becoming more important to the civilian economy, as well as the military, it made sense to create an interagency committee to assess Britain's wartime requirements and allocate between civilian and military consumers.[101] Nothing happened until 1925, when the Principal Supply Officer's Committee (the logistical arm of the CID) completed an alarming assessment. It estimated Britain's requirements in wartime during the first year of operations at 11.17 million tons of fuel oil, 355,000 tons of aviation fuel, 1.68 million tons of gasoline, 416,000 tons of kerosene, and 166,000 tons of lubricants—quantities suitable for maritime or expeditionary operations rather than a major conventional land war.[102]

[95] Petroleum Department, "Oil Production [...]," no date (circa 1929), O.B. 32, CAB 50/3; Yergin, *Prize*, 262–3.

[96] The agreement is reprinted in Bamberg, *History*, 528–34. See also: Wilkins, Maturing, 207ff; Yergin, *Prize*, 260–8.

[97] Bamberg, *History*, 107–17; John Blair, *Control of Oil* (New York: Pantheon Books, 1976), 54–63, 113–15; Nowell, *Cartel*, 191–200 and 221–2.

[98] Llewellin et al., "Thirteenth Annual Report," January 24, 1939, O.B. 294, CAB 50/7.

[99] Wilkins, *Investment*, 226–7 and 263–6.

[100] "Effect of a Prohibition [...]," no date or author (circa 1928), POWE 33/765.

[101] See the notes by Chelmsford, Webb, and Thomson (500-B, 501-B, 506-B, 507-B, and 509-B) dated between 3 and 25 July 1924, in: CAB 4/11.

[102] Birch et al., "Control of Oil [...]," February 28, 1925, P.S.O. 37, attached to: Hermon-Hodge to the CID, May 1, 1925, O.B. 2, CAB 50/3.

Most troubling was the tanker shortfall. The Admiralty estimated that it would need 1.6 million tons of tonnage to move its annual requirements of fuel oil, plus a further 24,000 tons for other petroleum products. The Admiralty's tanker fleet only totaled 260,000 tons, while the combined tonnage of every British tanker larger than 3,000 tons was only 2.3 million. In other words, the Royal Navy alone would tie up 70 percent of Britain's available tonnage, leaving the other services and the civilian economy to divide the remainder and whatever neutral tonnage was available.[103] Overall, Britain would need 470 tankers. In 1923, however, it could rely upon 312 tankers with a gross capacity of 1.69 million tons, which sufficed to move only half of its requirements.[104] As a result of this troubling assessment, the CID established the Oil Fuel Board, renamed the Oil Board in 1927, chaired by a cabinet minister and staffed by representatives from the services and various departments to gauge Britain's oil and tanker requirements in wartime, determine sources of supply, and offer recommendations to resolve any problems.[105]

In 1926, the Admiralty advised the Oil Fuel Board that, in the first year of any hostilities in the Far East—presumably against Japan—48 percent of the Admiralty's requirements (3.56 million tons) would come from reserves, while the remainder (3.9 million tons) would be imported. Since existing stocks in January 1926 were 3.2 million tons short of the necessary figure of one year's wartime consumption (7.5 million tons), the Admiralty would have to import an additional 3.77 million tons on top of the aforementioned 3.9 million tons to make up its shortfall. This doubling of the Royal Navy's import requirements meant that a further eighty-three tankers were necessary either to carry the oil or to serve as floating storage tanks, which worked out to 295 tankers moving 7.66 million tons. Virtually all of the excess oil would be imported from Mexico, Venezuela, and most importantly the United States, which would have to send 2.01 million tons to the Far East to compensate for British stocks being 2.04 million tons short.[106]

In 1927, the CID assigned the Oil Board the task of evaluating the requirements during a Far Eastern war (against Japan) assuming no "European complications."[107] For the next eight years, the Oil Board confined itself to examining, "at a rather leisurely pace," Britain's needs during conflicts with either Japan or (after 1929) the Soviet Union in Afghanistan. Between 1926 and 1930, the board only met three times and it took four years to draft its first plan for a war against Japan.[108] During

103 "Departmental Requirements [...]," Appendix A to Birch et al., "Control of Oil [...]," February 28, 1925, P.S.O. 37, CAB 50/3.

104 Birch et al., "Control of Oil [...]," February 28, 1925, P.S.O. 37, attached to: Hermon-Hodge to the CID, May 1, 1925, O.B. 2, CAB 50/3.

105 Hermon-Hodge, "Oil Fuel Board," May 1, 1925, O.B. 1, CAB 50/3; Brown, "Fuel Supplies," 212–13. The key works on the Oil Board are: Marian Kent, *Moguls and Mandarins: Oil, Imperialism, and the Middle East in British Foreign Policy, 1900–1940* (London: Frank Cass, 1993), 147–69; Orest Babij, "The Royal Navy and Inter-War Plans for War against Japan: The Problem of Oil Supply," in: *Merchant Marine in International Affairs, 1850–1950*, ed. Greg Kennedy (London: Frank Cass, 2000), 84–106.

106 Egerton and Oliver, "Admiralty Estimate [...]," March 22 and 25, 1926, O.B. 16, CAB 50/3.

107 Hermon-Hodge, "First Annual Report," February 25, 1927, O.B. 21, CAB 50/3.

108 Payton-Smith, *Oil*, 39–41.

its entire existence—from 1926 to the outbreak of World War II, when the Petroleum Department assumed its functions—the Oil Board produced thirteen annual reports. In the first (1926), it estimated consumption in a Far Eastern war in 1937 at 7.5 million tons of fuel oil for the Admiralty, 227,000 tons of gasoline for the army, and 181,000 tons of various fuels for the RAF. Civilian consumption would rise from 4.88 million tons to 7.46 million tons between 1925 and 1937. Britain would require 467 tankers of more than 3,000 tons—221 for military needs, 136 for the civilian economy, and 110 to meet imperial requirements. By 1937, 482 tankers of British, Admiralty, and neutral registry would be available, leaving a surplus of fifteen tankers. In 1926, however, there was a shortage of sixty-two tankers—459 required versus 397 available. Finally, as of 1925, only seven out of the eighteen million tons of oil consumed within the empire came from imperial or British-controlled sources. While the Persian oilfields were "vital," so were those of the United States.[109]

The shortage of tankers persisted throughout the interwar period.[110] Between 1926 and 1935, only in 1930 did Britain's requirements after civilian rationing not exceed supply, as can be seen in Table 2.1.

Ironically, Whitehall's solution to its oil supply problem—replacing the United States with the Middle East—was the worst choice in terms of logistics. Yet officials made this choice despite the possibility, based on their experience during World War I, that they could neither guarantee safe transit through the Mediterranean nor provide sufficient tankers to reroute supplies. That also assumed Britain could get the oil to the tankers in the first place. While Persian oil enjoyed relatively easy access to the Persian Gulf, Iraqi oil did not, and even a decade after World War I, the British were still nowhere near getting a pipeline to move it to the Mediterranean.

Table 2.1. Britain's estimated tanker deficit, 1926–1935[111]

Oil Board Annual Report (Date)	Deficit (w/civilian rationing)
First Annual Report (1926)	62 (35)
Second Annual Report (1928)	60 (14)
Third Annual Report (1929)	49 (3)
Fourth Annual Report (1929)	64 (10)
Fifth Annual Report (1930)	56 (0)
Sixth Annual Report (1931)	85 (16)
Seventh Annual Report (1932)	74 (9)
Eighth Annual Report (1933)	83 (28)
Ninth Annual Report (1934)	69 (19)
Tenth Annual Report (1935)	98 (46)

[109] Peel et al., "First Annual Report," December 31, 1926, O.B. 20, CAB 50/3.
[110] Brown, "Fuel Supplies," 239ff.
[111] These figures are drawn from the Oil Board's annual reports. See: CAB 50/1–7.

Britain, Iraq, and Turkey reached a settlement on the Mosul question in 1926, and in October 1927, the TPC struck oil at Baba Gurgur. It seemed like the British investment in Iraq might finally yield tangible gains. Whereas Britain had been spending £20 million per annum on Iraq in the period 1920–1, by 1929 the cost would be only £500,000, leaving aside any oil revenues.[112] Before Britain could reap its rewards, however, a pipeline was necessary. Its route became a point of contention. Whitehall wanted the pipeline to pass through mandates of Transjordan and Palestine before terminating at Haifa but wanted to avoid any insinuation that it was exploiting its privileges as a mandatory power for commercial gain.[113] Paris preferred the northern route via its mandates in Syria and Lebanon, with a terminus at Tripoli. As an added incentive for Britain, the Admiralty planned to use Haifa as an alternative source of fuel in Eastern Mediterranean, assuming the TPC built a refinery there.[114] The Iraqis also insisted on a Haifa terminus to maximize their share of the pipeline's economic benefits.[115] King Faisal of Iraq doubtless wished to support his brother, King Abdullah, in Transjordan and relished the prospect of repaying the French for ousting him from Damascus in 1920 after he was crowned the King of Syria.

The pipeline was a godsend because of the potential savings in tanker tonnage—a reduction in sailing time of twenty-seven days versus oil shipped from Abadan to Britain via Suez.[116] Whitehall's preferences clashed with those of the IPC. The Tripoli line was shorter than the Haifa route (529 miles versus 640 miles) and would cost less (£8.5 million versus £11 million)—an average cost of £16,070 per mile of pipeline versus £17,200.[117]

Due to the 100-mile differential, the cost to transport one ton of oil worked out to 7/6d for Tripoli and 10/6d for Haifa. No matter where the pipeline terminated, it would improve the cost competitiveness of Middle Eastern oil. The issue was not the price, since the oil companies had used the price at the Gulf of Mexico as the global benchmark since 1921, but rather the cost of freight. Oil going from the Persian Gulf to Britain via the Suez Canal had to travel 1,500 miles farther than oil from the Gulf of Mexico (6,500 miles versus 5,000 miles). Canal tolls also added 5/- per ton to the final cost. While the cost of transportation per 100 miles was identical (7/-), the longer journey from the Persian Gulf meant that Middle Eastern oil cost 37/6d per ton to import into Britain compared to 28/6d from the Gulf of Mexico. An IPC pipeline completely changed the commercial calculus. For Shell, oil from the IPC pipeline traveled 1,000 fewer miles compared to Venezuelan or Mexican oil, 2,000

112 Webb, "Our Position [...]," July 18, 1929, C.P. 214 (29), CAB 24/205.

113 Chamberlain, "The Oil Position [...]," May 7, 1929, C.P. 145 (29), CAB 24/203.

114 Amery, "Oil Position [...]," March 11, 1929, C.P. 73 (29), CAB 24/202; "[...] Possible Alignments [...]," January 20, 1930, no author, BP 68387.

115 Humphrys to Passfield, June 18, 1930, I.O.C. (30) 10, Appendix I to: Henderson, "Interim Report," June 20, 1930, C.P. 209 (30), CAB 24/213.

116 Brown, "Fuel Supplies," 188.

117 "Memorandum [...]," no date, attached to: Cadman to Henderson, March 20, 1930, B.H. (30) 2, Appendix I to: Committee on the Baghdad-Haifa Railway and Pipeline, "Report," May 1, 1930, C.P. 136 (30), CAB 24/211.

fewer compared to Texan oil, and 5,000 fewer miles compared to Californian oil via the Panama Canal. For Anglo-Persian, the savings reached 3,500 miles compared to oil shipped from Abadan via Suez. Moreover, the company did not have to worry about the additional influx of supplies—while Iraqi oil would displace Persian production from Europe, Anglo-Persian could redirect it to its growing markets in Africa and India thanks to marketing agreements with Shell and Burmah.[118]

The TPC/IPC's discovery of oil in 1927 set off a scramble to finalize the pipeline's route. The TPC used the prospect of a pipeline to induce Baghdad to grant it an extension before it had to begin operations under its existing concession.[119] The TPC made no mention of the pipeline's route, but after the company let slip it was considering a northern terminus, officials in Whitehall became furious. The government tried to convince Cadman and Deterding of the benefits of an "all British" route. Both were sympathetic provided Whitehall offered additional incentives—construction of a complementary railroad would sweeten the deal, but the oilmen were unwilling to build one without a subsidy (£300,000 per annum over fifty years).[120]

In March 1928, the Cabinet delegated responsibility for drafting the government's position to the CID. Cadman testified that Anglo-Persian was interested in building a supplementary line to transport Persian crude oil to the Mediterranean to save money on Suez Canal tolls, which cost the company roughly £900,000 each year. Although a Tripoli terminus would cost £3 million less to construct and have lower maintenance fees, Whitehall was determined to have a Haifa terminus. France presented numerous headaches. Besides the possibility that the Royal Navy might not be able to take supplies from Tripoli during a war in which France was neutral, there was also the inconvenience of the San Remo Agreement, which guaranteed France 25 percent of any oil piped from Persia through the French mandates. The subcommittee recommended that the government wait until the TPC was willing to meet certain preconditions before offering assistance.[121] No one disputed the need for pipeline—exports from the Levant were 1,000 miles closer to Britain than any other British-controlled source, on hand during a Mediterranean conflict, and located along Britain's lines of communication.[122]

Whitehall's decision to side with the Iraqis in supporting a pipeline to Haifa infuriated Paris.[123] The two sides traded accusations of bad faith, with Paris

[118] Iraq Oil Committee, "Some Notes [...]," June 1930, I.O.C. (30) 5, AIR 9/43. In 1927/8, Anglo-Persian, Shell, and Burmah divided the Indian market. Anglo-Persian and Shell also established a joint subsidiary that would operate in Egypt, East and South Africa, and the Indian Ocean. Bamberg, *History*, 106–41; Yergin, *Prize*, 262–3.

[119] "Historical Note," March 16, 1929, enclosed with: Amery, "Oil Position [...]," March 16, 1929, C.P. 80 (29), CAB 24/202.

[120] Amery, "Trans-Desert Pipe Line," February 29, 1928, B.H.R. (28) 2, and Committee on the Proposed Baghdad-Haifa Railway and Pipeline, "Minutes [...]," March 7, 1928, B.H.R. (28)—1st Minutes, both in: CA, CHAR 22/220.

[121] CID, "Report," June 13, 1928, 886-B, attached to: Hankey to Cabinet, "Proposed [...]," March 7, 1929, C.P. 68 (29), CAB 24/202. Cadman's report is included as Appendix I.

[122] Trenchard et al., "Baghdad-Haifa Pipeline [...]," May 24, 1928, C.I.D. Paper No. 884-B, Appendix II to: CID, "Report," June 13, 1928, 886-B, CAB 24/202.

[123] Cadman claimed that Anglo-Persian supported the Haifa route but was "handicapped... by the necessity of carrying his French colleagues with him." Amery, "Oil Position [...]," C.P. 73 (29),

complaining of British meddling in a commercial matter.[124] Whitehall pleaded ignorance, claiming that the initiative for a pipeline to Haifa had come from a rival of the TPC, the British Oil Development Corporation (BOD), which had offered to construct a railway to Haifa if Baghdad granted it a concession.[125] Anglo-French friction stemmed in part from the two parties' conflicting needs. The French did not want to import refined petroleum from Iraq. Starting in 1928, Paris imposed import quotas to stimulate the development of both a domestic refining industry and a French tanker fleet. It wanted France's share of Iraqi production piped to Tripoli, whence it would travel on French tankers to French refineries.[126]

Whitehall, by contrast, did not support the creation of a domestic refining industry until after 1947.[127] It preferred to construct a refinery in Haifa to refine Iraqi crude oil prior to its export to Britain, which Shell and Anglo-Iranian finally completed in 1939.[128] Although Whitehall was unanimous in its support for a Haifa terminus, divisions appeared over the diplomatic repercussions—the Foreign Office counseled compromise while the Colonial Office fulminated against appeasement. The Cabinet finally determined on March 20, 1929 that the government should dispense with the charade of disinterest and inform Paris of its preference for Haifa.[129]

The 1929 General Election, however, brought down the Conservatives and returned Labour to power. In September, a new committee led by Colonial Secretary Lord Passfield informed the Cabinet that it was unable to reach a unanimous decision about whether to support the pipeline. Supporters argued that government assistance would not entail any new responsibilities. Rather, Britain would reap considerable benefits by not only bringing Iraqi production online, but also increasing production in Persia if Anglo-Persian also built a supplementary pipeline to Haifa. A pipeline would slash the distance Persian oil needed to travel

March 11, 1929, CAB 24/202. Whatever the exact reasons, Anglo-Persian had yet to commit officially to either a Haifa or a Tripoli terminus as of January 1930. "[...] Possible Alignments [...]," January 20, 1930, no author, BP 68387.

124 Amery, "Oil Position [...]," April 24, 1929, C.P. 125 (29), CAB 24/203.

125 Established in 1928, the BOD expanded to include Italian, German, and French partners, and it received a concession in 1932 for most of Iraq west of the Tigris River. Amery, "Conflicting [...]," C.P. 164 (28), May 22, 1928, CAB 24/195; Petroleum Department, "Mosul Oilfields Limited [...]," June 1, 1933, attached to: Allen, "Mosul Oil Fields [...]," June 14, 1933, O.B. 116, CAB 50/5. See also: Helmut Mejcher, *Die Politik und das Öl im Nahen Osten: I. Der Kampf der Mächte und Konzerne vor dem Zweiten Weltkrieg* (Stuttgart: Klett-Cotta, 1980), 102–210. The IPC bought out the BOD in 1937 after the latter ran into financial difficulties. Bamberg, *History*, 166–71; Penrose, *Iraq*, 138–41.

126 "Frankreichs Ölversorgung," *Deutsche Wehr* (November 29, 1934); Mohamed Sassi, "The Emergence of the French Oil Industry between the Two Wars," *Business and Economic History On-line* 1 (2003), accessed August 15, 2018, <http://www.thebhc.org/sites/default/files/Sassi.pdf>, and "Evolution of the Structure of the French Oil Economy between the Wars: Toward a French Holding Company," *Business and Economic History On-line* 2 (2004), accessed August 15, 2018, <http://www.thebhc.org/sites/default/files/Sassi_1.pdf>.

127 Corley, "Oil Companies," 167–8.

128 Bamberg, *History*, 165

129 Amery, "Oil Position [...]," April 30, 1929, C.P. 134 (29), CAB 24/203. The debate, which consumed the attention of the Conservative government until its collapse in May 1929, is detailed in: C.P. 73, 78, 80, 125, and 134 (29), CAB 24/202–203.

by sea to reach Europe by 6,400 miles (round trip) and eliminate Anglo-Persian's expenditures on canal tolls (£900,000 in 1927).[130] Opponents doubted that Paris was planning to construct a rival railway, or that the project would stimulate employment in Britain.[131] The Royal Navy was still opposed to a Tripoli terminus, and its hostility was understandable considering that only 14 percent of the navy's wartime requirements came from sources under British control. Increasing the output of Iraqi oil without having to share any with France, however, would allow Whitehall to reduce the size of its naval reserves. An Iraqi pipeline would also create momentum for a Persian one, which would shave sixteen days off the journey from Persia to Britain and free fifty tankers for service in other theaters.[132]

To break the impasse, the government formed yet another committee, now chaired by Foreign Secretary Arthur Henderson. With the country reeling from the Great Depression, the Treasury was unwilling to offer financial assistance to make the Haifa terminus an attractive alternative to Tripoli. Although the Iraqis were opposed to a Tripoli pipeline, it was in their interest to cooperate, since they needed the oil revenues. The Labour government was also confident it could make "satisfactory arrangements" with France to ensure oil supplies during wartime.[133] The Colonial Office remained intransigent and challenged Cadman's argument that the Tripoli pipeline was more economical than Haifa, hinting that he was putting his company's interests above those of Britain. If the IPC went ahead with a Tripoli terminus, the Colonial Office advised, the company could "no longer be considered as a British interest." It even questioned whether to entrust Iraq's oil to the IPC, as a "little healthy competition" would benefit Baghdad without harming British security.[134]

Tempers soon cooled—Whitehall continued to back the IPC, which reconsidered angering Baghdad over pipelines as the consequences would probably "outweigh the economic advantages."[135] In 1931, the IPC partners reached a compromise brokered by Jersey, whereby they would build a bifurcated pipeline terminating at Tripoli and Haifa by 1935.[136] Their combined cost exceeded £10 million, and throughput by 1938 reached 4.3 million tons.[137] The pipelines reduced the transit

[130] Passfield et al., "Report," September 11, 1929, B.H.R. (29) 11, attached to: Passfield, "Baghdad-Haifa Railway [...]," September 19, 1929, C.P. 247 (29), CAB 24/205. See also: Colonial Office, "Berlin-Haifa Railway [...]" and "Historical," both dated July 12, 1929 and attached to: Passfield, "Berlin-Haifa Railway and Trans-Desert Pipeline," July 18, 1929, C.P. 215 (29), CAB 24/205.

[131] See annexures 1–3 to: Passfield et al., "Report," September 11, 1929, B.H.R. (29) 11, attached to: Passfield, "Baghdad-Haifa Railway [...]," C.P. 247 (29), CAB 24/205.

[132] Madden, "Baghdad-Haifa Railway [...]," April 5, 1930, attached to: Committee on the Baghdad-Haifa Railway and Pipeline, "Report," May 1, 1930, C.P. 136 (30), CAB 24/211.

[133] Committee on the Baghdad-Haifa Railway and Pipeline, "Report," May 1, 1930, C.P. 136 (30), CAB 24/211.

[134] Colonial Office, "Oil Policy in Iraq," May 5, 1930, Appendix B to: Passfield, "Oil Policy in Iraq," May 16, 1930, C.P. 164 (30), CAB 24/212. See also: Baghdad to RAF Hinaidi, "Oil Interests," August 2, 1930, AIR 23/393.

[135] Foreign Office, "Oil Policy in Iraq," May 29, 1930, Annex to: Iraq Oil Committee, "Note [...]," no date, I.O.C. (30) 6, CAB 27/436.

[136] Jersey wanted France to cease importing Soviet oil and grant it a preferential import quota. Edward Peter Fitzgerald, "Business Diplomacy: Walter Teagle, Jersey Standard, and the Anglo-French Pipeline Conflict in the Middle East, 1930–1931," *Business History Review* 67: 2 (1993): 207–45; Melby, *Oil*, 113–41; Nowell, *Cartel*, 268–76.

[137] Jung, "Ölleitungen [...]," *Vierjahresplan*, 1942/II.

distance from Iraq to Europe by 1,500 miles and made Iraqi oil cost-competitive with Iranian oil. The main beneficiary was France, which virtually ceased to import oil through Suez, where oil shipments declined by two-thirds in 1935 (although they had rebounded by 1937).[138]

Whitehall understood that there was little demand for Iraqi oil because of the global surplus.[139] The British and U.S. partners in the IPC would probably stifle production, even if the Iraqis and the French were pushing for more.[140] They could do so thanks to the "Red Line" Agreement of 1928, which retained the "self-denying" ordinance from the "Foreign Office" Agreement of 1914 forbidding signatories from seeking concessions anywhere within the Arabian Peninsula (except Kuwait), the Levant, Iraq, and Anatolia without the consent of all partners.[141] To prevent the oil-starved Compagnie Française des Pétroles from boosting production in Iraq, the Americans, Shell, and Anglo-Persian insisted that any resolution before the board could only pass with the approval of three of the four voting members.[142]

BRITAIN'S POSITION BEFORE 1933

A viable IPC was good for its partners' bottom lines, but it did not materially reduce Britain's dependence on the United States. Nonetheless, the absence of geopolitical tension emboldened those who advocated for reducing defense expenditures. The Admiralty had fended off the Treasury's attempts in 1923–4 to slash naval oil reserves. The Treasury—led by the converted fiscal hawk, Churchill—kept chipping away. In 1925, due to the repercussions of the return to the gold standard, the Treasury forced the Admiralty to scale back its procurement target for 1929, and the CID—with the Cabinet's blessing—added that the odds of war with Japan within the following decade were negligible. To add insult to injury, a parliamentary committee criticized the Admiralty's interpretation of the Ten-Year Rule, positing that it should assume "at least" a decade of peace in its planning. When the Admiralty revised its reserve requirements upward in 1926 and 1927, its opponents pounced.[143]

138 D.A. Farnie, *East and West of Suez, 1854–1956* (Oxford: Clarendon Press, 1967), 601 and 614.

139 Petroleum Department, "International Control [...]," October 24, 1932, POWE 33/461.

140 Iraq Oil Committee, "Some Notes [...]," no date (circa May 1930), I.O.C. (30) 5, CAB 27/436.

141 Partially reprinted in Payton-Smith, *Oil*, 32–7. For the negotiations leading to the agreement, see: William Stivers, "A Note on the Red Line Agreement," *Diplomatic History* 7: 1 (1983): 23–34. The "self-denying" ordinance was repealed in 1948. Anand Toprani, "The French Connection: A New Perspective on the End of the Red Line Agreement, 1945–1948," *Diplomatic History* 36: 2 (2012): 261–99.

142 Theodore Moran, "Managing an Oligopoly of Would-Be Sovereigns: The Dynamics of Joint Control and Self-Control in the International Oil Industry Past, Present, and Future," *International Organization* 41: 4 (1987): 580–3.

143 Developments between 1924 and 1927 are summarized in Treasury, "Reserves [...]," November 19, 1927, CA, CHAR 22/174; Brown, "Fuel Supplies," 208–13.

Under its interpretation of the Ten-Year Rule, the Admiralty estimated that during a war with Japan in 1937 it would require 7.15 million tons of fuel oil annually. This meant that 3.3 million tons had to be added to existing reserves over the following decade at a cost of £13 million.[144] The Treasury was unconvinced—the seven-million-ton "requirement" was unreasonable considering the geopolitical situation.[145] Churchill reminded the Admiralty that 4.5 million tons had been sufficient in 1919. First Lord of the Admiralty William Bridgeman retorted that the Admiralty used a conflict against Japan as its baseline not because it considered war likely, but because Japan was a plausible rival for planning purposes. To break the deadlock, in October 1927 the Cabinet tasked the Naval Programme Committee to consider Britain's requirements during a Far Eastern conflict.[146] The committee sided with Churchill and recommended that reserves should increase by only 100,000 tons for the coming year.[147] Bridgeman did not take defeat graciously. In late 1928, he worked out an agreement with Churchill that an additional 100,000 tons be added in 1929, while deferring orders for "the normal instalment of 323,000 tons." He acceded to this arrangement against his better judgment and warned that British security was "for financial reasons, being retarded much beyond the limit which the Board of Admiralty think safe."[148]

Even this compromise proved short-lived. In 1929, the new Labour government suspended all further purchases of fuel oil for naval reserves, and the following year, it demanded cuts to the existing reserve due to the London Naval Treaty.[149] The Admiralty was furious, since naval reserves totaled only 4 million tons against an estimated annual consumption of 7.5 million tons in a war against Japan. Britain's position (a six-month reserve) compared unfavorably, the Sea Lords complained, with that of the other great powers.[150] Technological developments also threatened the navy's oil supplies. By the late 1920s, thanks to catalytic cracking, it was possible to convert heavier petroleum distillates into gasoline. Since the price for gasoline was higher than that of fuel oil (£12/10 versus £2/10 in 1928), refiners preferred to produce the former. This was unhelpful from the Admiralty's perspective—besides reducing the amount of fuel oil each barrel of crude oil yielded, cracking altered the chemical composition of remaining heavy oils, which might no longer satisfy Royal Navy specifications.[151]

[144] Bridgeman, "Admiralty Memorandum [...]," January 27, 1928, N.P. (27) 10, CA, CHAR 22/174.

[145] "Reserve [...]," no date or author, and Fisher, "Oil," February 2, 1928, both in: CA, CHAR 22/174.

[146] Howarth, "Naval Programme Committee," February 2, 1928, N.P. (27) 6th Cons., CA, CHAR, 22/174.

[147] Birkenhead, "Report [...]," February 20, 1928, C.P. 47 (28), CAB 24/192. The cabinet concurred: "Oil Fuel [...]," February 22, 1928, CA, CHAR 22/174.

[148] Bridgeman, "Oil Fuel [...]," December 18, 1928, C.P. 410 (28), CAB 24/109. See also: Brown, "Fuel Supplies," 213–15.

[149] Babij, "Oil Supply," 87–90; Brown, "Fuel Supplies," 215–17.

[150] "Oil Fuel Reserve," no date, attached to: Alexander, "Oil Fuel Reserve," February 17, 1931, C.P. 51 (31), CAB 24/220. See also: Thomson, et al., "Fifth Annual Report," July 1, 1930, O.B. 51, CAB 50/3.

[151] Brown, "Fuel Supplies," 216; Kent, *Moguls*, 151.

In April 1929, a special Sub-Committee on Oil Supply in Time of War laid bare the weakness of Britain's position. In a global war, the empire's annual wartime requirement in 1927 would be 19.34 million tons (11.89 million tons west of Suez), or 17.39 million tons after rationing. Assuming that Persian and Iraqi production continued and that the IPC pipeline was operational, 9.14 million tons would come from the Middle East. Since it was impractical to rely on Far Eastern oil west of Suez because of its transportation costs and vulnerability to Japanese attack, most of the remainder would have to come from the Western Hemisphere. The United States had exported 6.89 million tons to the British Empire in 1927—2.6 million tons of which went to Canada—and if it was "unfriendly," rationing would be necessary. Venezuela would have to pick up the slack to an "alarming" extent—5.5 million tons.[152] Overall, with a "friendly" United States, Britain would manage, but without that, Britain's position would be "precarious."[153]

During the 1920s, the Japanese navy was a "useful" adversary because it allowed the Admiralty to justify its existing force structure and doctrine.[154] The Japanese were also a convenient foe because they could not threaten British access to Persian oil. The same was not true of the Soviet Union. During an Anglo-Soviet conflict in Afghanistan (an "Asiatic" war), the greatest challenge would be the loss of Persian and Soviet oil exports. Their absence would have a ripple effect, since other nations would have to find alternatives to the missing 2.14 million tons of Persian and 2.84 million tons of Soviet exports. Combined with the eight million tons of new oil Britain would require, an Anglo-Soviet war would case a global disruption totaling thirteen million tons. The tanker situation would not be dire, as Britain would have to charter only eighteen neutral tankers. Unfortunately, everything depended on Japan's neutrality, as most of the British Empire east of Suez depended upon East Indies oil if Persian supplies were unavailable. In other words, Britain could not defend its empire in Asia without U.S. assistance when fighting conventional wars in multiple theaters.[155]

In 1932, the Oil Board made no changes to its previous estimate of Britain's requirements in a war against the Soviet Union but stipulated it would review the situation annually.[156] The following year, it determined Britain would need to import 21 million tons—16 million tons of which were for civilian requirements, three-quarters of which went west of Suez—on 359 tankers, which required the

[152] Stanhope et al., "[...] Oil Supply in Time of War," March 20, 1929, O.B. 27, CAB 50/3.

[153] Peel et al., "Third Annual Report," April 26, 1929, 937-B, attached to: Hankey, "Oil Fuel," May 6, 1929, C.P. 142 (29), CAB 24/203. By "unfriendly," the Oil Board assumed that Washington would only limit exports from the United States and not interfere with third party suppliers or the chartering of tankers. The CID and the Cabinet quickly approved the report. The CID also directed "the Oil Board to revise the scheme for meeting Empire oil requirements in time of war triennially," and to determine "how the oil supplies of the British Empire could best be met in the event of a war with Russia in Afghanistan." Maund, "Third Annual Report—Approval Of," May 10, 1929, O.B. 35, CAB 50/3.

[154] Joseph Maiolo, "Did the Royal Navy Decline between the Two World Wars?" *RUSI Journal* 159: 4 (2014): 20.

[155] Memorandum, September 22, 1930, attached to: Allen, "Proposed Scheme [...]," June 23, 1931, O.B. 60 (Revise), CAB 50/4.

[156] Londonderry et al., "Seventh Annual Report," October 26, 1932, O.B. 83, CAB 50/4.

chartering of thirty-five neutral vessels. The loss of Persia and Iraq would not seriously impair Britain's war effort, because an Asiatic war did not threaten British access to Western Hemispheric oil. The Admiralty's requirements for imported fuel oil (4.5 million tons) would come from five sources: Venezuela (1.6 million tons), Mexico (947,000 tons), Texas (850,000 tons), Trinidad (746,000 tons), and the East Indies (300,000 tons).[157]

The situation would not be so rosy when fighting an enemy that could threaten British oil supplies. Since it had proven impossible to end Britain's reliance on the Americas, Whitehall finally resolved to end any lingering disputes with Washington over oil. Starting in 1930, the British re-evaluated the "control" clause when a U.S. firm, the Standard Oil Company of California (SOCAL), sought a concession in Bahrain.[158] Back in 1913, the ruler of Bahrain had agreed never to grant a concession to a foreign company without British approval. In the early 1920s, however, Frank Holmes, a British prospector, began purchasing concessions throughout the Persian Gulf, including Bahrain. He planned to sell them to British companies such as Anglo-Persian but found no takers. Anglo-Persian had an appalling track record at finding oil, and after World War I, its geologists were convinced that none existed on the Arabian side of the Gulf.[159] Holmes also bought an option for the al-Hasa province from Ibn Saud in 1923, with the proviso that he could not sell the concession to Anglo-Persian, but it expired in 1925, before Holmes could find a buyer. He eventually sold the Bahrain concession to Gulf Oil in 1927; Gulf Oil relinquished it a year later when it joined the TPC and signed the "Red Line" Agreement.[160]

The Petroleum Department again led the charge to modify policy. The exclusion of foreign capital from imperial oil development through "control" had yielded no discernible benefits and probably retarded development. The department urged the government to adopt reciprocity if foreign companies offered guarantees concerning the sale of concessions, registered their companies in British territories, ensured that a majority of their employees and "some" directors were British subjects, and refined some of the oil on British territory at Admiralty specifications while giving Britain preferential supplies in wartime.[161] Whitehall eventually allowed

[157] "Empire Requirements [...]," no date or author, Annexure A to: Londonderry et al., "Eighth Annual Report," July 31, 1933, O.B. 122, CAB 50/5. The following year's evaluation is appended as Annex A to: Londonderry et al., "Ninth Annual Report," October 31, 1934, O.B. 134, CAB 50/5.

[158] Petroleum Department, "Oil Concessions [...]," no date (circa 1929), POWE 34/1.

[159] Leatherdale, *Arabia*, 192. The Treasury lamented that U.S. companies secured concessions in Bahrain and Saudi Arabia "because the Anglo-Iranian was technically incompetent to find the oil." Eady to Padmore, "Oil Conference," February 17, 1944, T 161/1195. Shell's record was hardly better. Jonker and Zanden, *History of Shell*, 283.

[160] The story of Holmes, known to the Arabs as "Abu al-Naft" ("the Father of the Oil"), is recounted in Keating, *Mirage*, 207–315; Yergin, *Prize*, 280–302.

[161] Petroleum Department, "Enclosure," July 25, 1930, enclosed with: Hankey, "Policy [...]," September 16, 1930, 1017-B, attached to: Hankey, "Policy [...]," October 7, 1930, C.P. 328 (30), CAB 24/215. The CID and Cabinet both concurred with the Petroleum Department: CID, "Policy [...]," September 29, 1930, attached to: Hankey, "Policy [...]," October 7, 1930, C.P. 328 (30), CAB 24/215; Passfield, "Policy [...]," November 24, 1930, C.P. 393 (30), CAB 24/216. See also: McBeth, *Policy*, 122–8.

SOCAL to take up the concession after it agreed to use a subsidiary registered in Canada and under British management.[162]

A higher-profile dispute arose in 1932–3 over Kuwait, where Anglo-Persian was competing with Gulf Oil. Anglo-Persian had shown little interest in Kuwait before 1932 and passed on Holmes' concessions. SOCAL's 1932 strike in Bahrain set off a scramble for the Saudi concession to al-Hasa, but Anglo-Persian and the IPC lost out to SOCAL in May 1933.[163] Combined with the shock of Persia's cancellation of the 1901 concession in December 1933 (see Chapter 3), Anglo-Persian was desperate to secure Kuwait. Rather than compete, it chose to cooperate with Gulf and formed the Kuwait Oil Company (KOC) in December 1933, which received a seventy-five-year concession the following year.[164] Although ownership of the KOC was split 50/50, Anglo-Persian exercised de facto control.[165]

The KOC discovered the massive Burgan oilfield in 1938, but Kuwait was of marginal consequence before World War II, and Cadman did not even mention Anglo-Iranian's activities there during the company's annual meeting in 1939.[166] Of more immediate significance was whether the "control" clause would apply in Kuwait. As in Bahrain, the ruler of Kuwait had agreed never to grant a concession to a foreign company without British consent. Following the setback in Saudi Arabia, Whitehall suddenly awoke to the potential danger posed by U.S. penetration of the Middle East. The Admiralty had originally wanted a British firm to hold this concession, which they considered strategically important in the case that supplies from Persia ever became unreliable. The Foreign Office, by contrast, welcomed an Anglo-American partnership, and in April 1932, the government agreed, in no small part to win U.S. debt relief.[167] The Admiralty did not give up—it was furious with the Bahrain settlement, which did not require SOCAL to construct a nearby refinery to Admiralty specifications or offer Britain "pre-emption" in wartime. Moreover, Anglo-Persian wanted Whitehall to waive the "control" clause entirely to facilitate an agreement with Gulf in Kuwait.[168] The Admiralty had to settle for half a loaf in Kuwait—under the terms of a 1934 "political agreement" with KOC, the company granted Britain "pre-emption" but was not obligated to build a refinery.[169]

[162] Wilkins, *Maturing*, 214 and 218; Yergin, *Prize*, 282–3.

[163] Irvine Anderson, *ARAMCO, the United States, and Saudi Arabia: A Study in the Dynamics of Foreign Oil Policy* (Princeton: Princeton University Press, 1981), 21–34; Edward Peter Fitzgerald, "The Iraq Petroleum Company, Standard Oil of California, and the Contest for Eastern Arabia," *International History Review* 13: 3 (1991): 441–60; Leatherdale, *Arabia*, 193–208.

[164] "History [...]," Appendix I (F) to: Ministerial Oil Committee, "Oil Policy," no date, M.O.C. (44) 5, POWE 33/1399; Bamberg, *History*, 146–55; DeNovo, *American Interests*, 202–9; Benjamin Shwadran, *Oil and the Great Powers* (New York: Council for Middle Eastern Affairs Press, 1959), 384–8.

[165] Blair, *Control*, 42–3.

[166] Wilkins, *Maturing*, 213–14.

[167] Foreign Office, "American Wish [...]," March 30, 1932, attached to: Simon, "Koweit [...]," April 2, 1932, C.P. 120 (32), CAB 24/229; Fiona Venn, "'A Futile Paper Chase': Anglo-American Relations and Middle East Oil, 1918–1934," *Diplomacy & Statecraft* 1 (1990): 165–84.

[168] Admiralty, "Oil [...]," no date, and "Admiralty Policy," no date, both attached to: Admiralty, "Oil Situation," January 8, 1934, ADM 1/8773/57.

[169] "Political Agreement [...]," Appendix VII (B) to: Ministerial Oil Committee, "Oil Policy," no date, M.O.C. (44) 5, POWE 33/1399.

British appeasement in Bahrain and Kuwait was not simply the handiwork of a Foreign Office eager "to sacrifice Britain's global oil interests" to buy U.S. goodwill, as the Admiralty complained.[170] Rather, the Foreign Office was keen to eliminate sources of friction, which explains why it opposed allowing any firms—British or U.S.—to seek concessions within the disputed Saudi–Kuwaiti Neutral Zone.[171] Nor was it willing to allow access to oil to jeopardize overarching British foreign policy aims. Compared to repairing the damage to Anglo-American relations, particularly after Britain's abandonment of the gold standard and free trade, and its de facto repudiation of its war debt to the United States, oil and even British predominance in the Persian Gulf were secondary concerns.[172] Other agencies in Whitehall may have griped, but German analysts recognized that the entrance of U.S. firms into the Gulf strengthened Britain by giving the United States a stake in the region's security.[173] In view of Britain's strategic overextension and economic weakness, there was no alternative to cooperation with the United States. By the early 1930s, energy independence through the Middle East was still a dream, and until that changed, Britain would still have to import much of its wartime needs from the Western Hemisphere.

Although the Cabinet accepted the recommendations of the Petroleum Department concerning the "control" clause in October 1930, it made no official announcement until July 1936 and did not pass relevant legislation until 1938. The most important change was that foreign-controlled companies could now purchase previously restricted concessions if their parent governments reciprocated to British firms. The new policy also satisfied the Admiralty by requiring concessionaires to refine at least 50 percent of their production within the empire and allowing the government to "preempt" production during an emergency.[174] The reasons for the delay in announcing the shift were twofold.[175] First, Britain had to pass model legislation in 1934 and 1935 eliminating the "control" clause within Britain. Second, Whitehall needed to convince the Dominions to cooperate to ensure compliance with the U.S. Mineral Leasing Act. Officials argued that Britain's case for reciprocity would be stronger if the entire empire abandoned the "control" clause.[176] There was also little to lose—by 1936, Whitehall had concluded that the empire had little oil. As one Treasury official observed, "we stand to gain more than we offer by this reciprocity business."[177] Whitehall's flexibility over Bahrain and Kuwait, and the gradual elimination of the "control" clause, ensured that Washington raised no objections in 1935 when Whitehall queried whether British firms were eligible to prospect on federal lands.[178]

170 Corley, "Oil Companies," 166.

171 Leatherdale, *Arabia*, 208–11.

172 Venn, "Foreign Office," 73 and 79ff.

173 "Ölwirtschaft [...]," *Vierjahresplan*, 1938/X.

174 Thomas, "Policy [...]," April 27, 1936, C.P. 119 (36), CAB 24/262.

175 Developments between 1930 and 1938 are summarized in: Petroleum Department, "Position [...]," September 1938, POWE 34/1. See also: Charles More, *Black Gold: Britain and Oil in the Twentieth Century* (London: Continuum, 2009), 65–6.

176 Board of Trade, "Treatment [...]," October 28, 1932, POWE 33/461.

177 Bridges, "Oil Development [...]," May 6, 1936, T 161/172.

178 See: NARA, RG 59, 841.6363/412–425; Wilkins, *Investment*, 362.

Britain's reversal on colonial policy was a belated recognition of the fact that its energy supplies in wartime depended more on the Western than the Eastern Hemisphere. Contrary to popular belief, the Middle East was not Britain's largest regional supplier of oil before World War II.[179] That honor belonged to South America. Before World War I, the British economic presence in Latin America dwarfed that of the United States. In 1913, the region received 10 percent of Britain's exports and 25 percent of its foreign investment, which increased from £80 million to £1.1 billion between 1890 and 1913. Foreign investment from the United States also grew, from $308 million to $1.6 billion between 1897 and 1913, but it still lagged behind Britain.[180] The Great War shifted the balance toward the United States, as Britain liquidated its foreign investments (one-third of its national wealth in 1905) to pay for the war.[181] During the conflict, Washington and U.S. oil companies sought to expand their Latin American gains at Britain's expense.[182] Even after the dispute over Iraq ended, Washington still opposed Anglo-Persian acquiring a concession in Venezuela.[183] By 1929, U.S. foreign investment in Latin America had increased to $3.46 billion, which was probably the year that the United States surpassed Britain in this category.[184]

As production in Mexico dwindled, Venezuela rose to take its place (see Illustration 8). Shell was among the pioneers of Venezuela's oil industry, discovering one of the world's richest oilfields (Barroso No. 2) along the eastern shore of Lake Maracaibo in December 1922. Six years later, it had 125,000 bpd of production in Venezuela, compared to nil for Jersey, and 148,000 bpd across Latin America, compared to 89,300 bpd for Jersey.[185] Shell's production share nevertheless gradually shrank,[186] as firms in the United States reduced its position from 99 percent of Venezuelan output in 1922 to one-third by 1939.[187] Between 1927 and 1939, Jersey's production went from nil to 273,000 bpd—40 percent of its global output.[188] United States investment in Venezuela more than doubled

[179] The following discussion draws from: Brown and Linder, "Oil," 125–87; B.S. McBeth, *British Oil Policy, 1919–1939* (London: Frank Cass, 1985), 86–120; Stephen Rabe, *Road to OPEC: United States Relations with Venezuela, 1919–1976* (Austin: University of Texas Press, 1982), 3–93.

[180] Garner, *Lions*, 16–18. [181] Bud-Frierman, "Weetman Pearson," 277–8.

[182] In 1919, U.S. oil companies maintained 28 percent of their capital in North America (Mexico and Canada) and 17 percent in South America, whereas in 1939 the figures were 12 percent and 42 percent. Brian McBeth, "Venezuela's Nascent Oil Industry and the U.S. Tariff on Crude Oil Imports, 1927–1935," *Journal of Iberian and Latin American History* 27: 3 (2009): 435.

[183] Rabe, *OPEC*, 22–33. Anglo-Persian had expressed interest in purchasing concessions in Venezuela but relented after Venezuela passed legislation prohibiting the sale of concessions to companies owned by a foreign government. Bamberg, *History*, 144–6; Ferrier, *History*, 551–61.

[184] Rabe, *OPEC*, 195; Wilkins, *Maturing*, 29–32 and 155.

[185] The imbalance was even greater in refining: 318,000 bpd for Shell compared to 47,000 for Jersey. Brown and Linder, "Oil," 139.

[186] Brown and Linder, "Oil," 148.

[187] Jonker and Zanden, *History of Shell*, 249–51; McBeth, "Industry," 430, and Juan Vicente Gómez and the Oil Companies in Venezuela, 1908–1935 (Cambridge: Cambridge University Press, 1983), 64; Rabe, *OPEC*, 33–42.

[188] Henrietta Larson et al., *New Horizons, 1927–1950*, vol. 3 of *History of Standard Oil Company (New Jersey)* (New York: Harper & Row, 1971), 41–2 and 144.

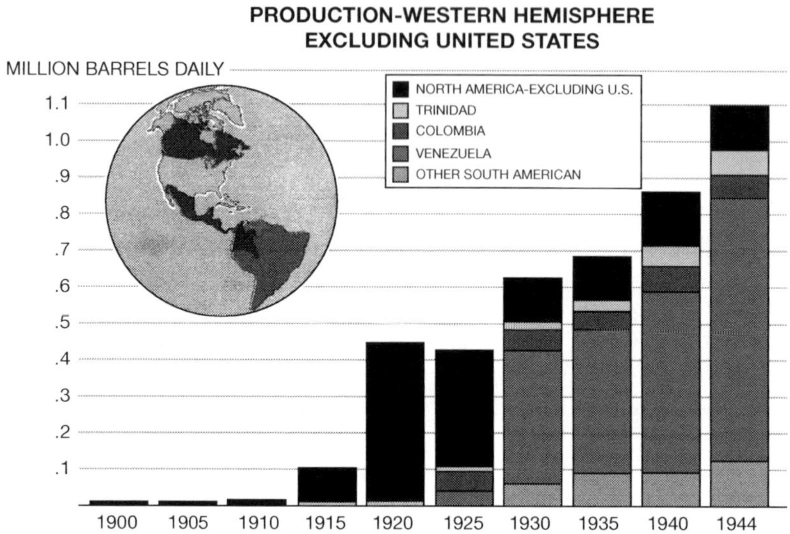

Illustration 8. "Production—Western Hemisphere," February 22, 1945
Source: LOC, Harold L. Ickes Papers, Box 221.

from $75 million to $161.6 million in just five years (1924–9).[189] From 1929 until nationalization in 1976, U.S. companies accounted for more than half of Venezuela's production.[190]

Luckily, there was plenty of oil to go around. Between 1921 and 1935, Venezuelan output rose from 240,000 tons to 15.1 million tons, and it was the world's second-largest producer until the Soviet Union overtook it in 1931.[191] In 1928, oil wells in Venezuela were three times more productive than wells in the most prolific U.S. state, California (1.25 million bbl per year versus 454,000 bbl), and production costs were less than one-third those of the United States ($0.65 versus $1.98 in 1931).[192] Mexico was not even close—daily production from new wells drilled in 1928 was 268 bbl, compared to 1,376 in Venezuela. The oilfields in Venezuela were also closer to New York and Britain than the major Mexican oil port (Tampico), and only 644 miles from the Panama Canal.[193] Shell and Pan American Petroleum and Transport (a subsidiary of the Standard Oil Company of Indiana) built refineries on Curaçao and Aruba to process Venezuelan crude oil and export it to Europe.[194]

[189] Brown and Linder, "Oil," 148. U.S. oil investment in Venezuela averaged 95 percent of total U.S. investment in the country between 1929 and 1940. McBeth, "Industry," 435–6.

[190] Brown and Linder, "Oil," 169–71; McBeth, *Policy*, 92–3; Rabe, *OPEC*, 34.

[191] Jonker and Zanden, *History of Shell*, 249; McBeth, *Venezuela*, 63; Rabe, *OPEC*, 196–7.

[192] McBeth, *Venezuela*, 1–2.

[193] Brown, "Shifted," 381–2; McBeth, "Industry," 441, Policy, 93, and Venezuela, 1–2; Rabe, *OPEC*, 59.

[194] By the end of the 1920s, the Curaçao refinery handled 80 percent of Shell's Venezuelan production (140,000 bpd). Brown and Linder, "Oil," 147–8. Jersey and Shell's refineries had a capacity of seventeen million tons per year by 1935. "Brief Survey [. . .]," no date, attached to: Norfolk, note, June 26, 1936, O.B. (P.R.) 14, CAB 50/14.

In 1928, Venezuela surpassed the United States as the world's largest oil exporter, and by 1936, it was exporting almost as much oil as the next seven exporters combined—the United States, Peru, Iran, Romania, the Dutch East Indies, Iraq, and the Soviet Union (43.05 million tons versus 44.39 million tons).[195] Venezuela's exports in 1930 (twenty million tons) were also double Britain's imports (nine million tons).[196]

Although the death of the dictator Juan Vicente Gómez in 1935 and the passage of new legislation authorizing workers to unionize portended a less friendly environment, conditions in Venezuela were still better than in either Mexico or the Middle East.[197] Mexican output collapsed during the 1920s. Its government also received twice as much in revenues per barrel as Venezuela's between 1922 and 1930, and even after 1931, the latter's per-barrel revenues were still less than 80 percent of Mexico's.[198] Venezuela did receive higher revenues than Middle Eastern producers ($0.23 per barrel versus $0.21 per barrel), but it was also more hospitable to new firms seeking a foothold than the Middle East, where large firms or joint ventures already controlled most of the region. The Venezuelan per-barrel cost of production was higher than in the Middle East ($0.52 versus $0.44), which meant that the companies earned a higher per-barrel profit in the latter ($0.88 versus $0.66).[199] Since Venezuela's oilfields lay close to the Atlantic Ocean, they did not require expensive pipelines like those in Iraq.[200] Finally, as in the Middle East, the absence of antitrust regulations in Venezuela allowed companies to pool resources and implement unit operation of shared oilfields.[201]

Venezuela also benefited temporarily from a larger share of the U.S. market. By 1928, it had overtaken Mexico as the largest oil exporter to the United States (31.47 million bbl including exports from the Dutch West Indies versus 26.02 million bbl) and contributed 70.6 percent of U.S. imports by 1929–30.[202] In June 1932, however, U.S. independent producers convinced the U.S. Congress to impose a tariff on imported oil. The tariff varied from $0.21 per barrel for crude oil to $1.05 for gasoline and $1.68 for lubricants.[203] To give a sense of proportion, U.S. oil prices in 1931 varied from $0.39 per barrel to $0.83 per barrel.[204] On the bright side, U.S. companies could still freely import crude oil for re-export to Europe.[205] After taking into considering the varying transportation costs and the value of the refined products—U.S. crude oil cost more to produce and transport but was more valuable after refining—Venezuelan crude oil enjoyed a price advantage over U.S. crude oil of $0.45 per barrel before the 1932 tariff and $0.25 afterwards.[206]

[195] Reichs-Kredit-Gesellschaft, "Treibstoffwirtschaft [...]," April 1938, NARA, Record Group 242: Foreign Records Seized, T-84/51 (EAP 66-c-2-10/22).

[196] Petroleum Department, "International Control [...]," October 24, 1932, POWE 33/461.

[197] Wilkins, *Maturing*, 224.

[198] Brown, "Shifted," 381–2; McBeth, *Venezuela*, 66.

[199] McBeth, *Venezuela*, 65–6.

[200] Jung, "Ölleitungen [...]," *Vierjahresplan*, 1942/II.

[201] Larson, *New Horizons*, 138.

[202] Brown, "Shifted," 384; McBeth, "Industry," 432.

[203] McBeth, "Industry," 452–4.

[204] DeGolyer, *Historical Data*, 107.

[205] McBeth, "Industry," 455. The tariff was reduced in 1939 and 1942. See: Appendix II (B) to Ministerial Oil Committee, "Oil Policy," no date, M.O.C. (44) 5, POWE 33/1399.

[206] McBeth, "Industry," 442 and 455.

One consequence of the tariff was that Indiana sold its Latin American operations to Jersey in April 1932. This sale helped Jersey to overtake Shell as the major producer in Venezuela. More importantly, by closing access to U.S. markets, the tariff redirected Venezuelan oil to Europe, and Venezuela replaced the United States as Europe's primary source of imports until World War II.[207] Venezuela's share of European crude oil imports rose from 13.4 percent in 1928 to 21.4 percent in 1933, while the U.S. share fell from 38.8 percent to 18.6 percent. This trend was most noticeable in Britain, where Venezuela's share rose from 17 percent to 49 percent during the 1930s.[208] Britain was now awash in oil from a source besides the United States and Persia. The only problem was that this source lay within the U.S. political and economic orbit.

Britain's record during the 1920s was unimpressive. Exerting British control over Shell proved a dead end—even when Whitehall had a willing suitor in Shell, it could not present a united front, since Anglo-Persian and the Admiralty both had reasons to torpedo any merger. Equally futile was the policy of promoting British "control" of oil resources within the empire, whose output remained trivial during the interwar period, fluctuating between 2 percent to 4 percent of global production. British imports, meanwhile, increased from 5.3 million tons in 1918 to 10.5 million tons by 1937.[209] Britain's greatest success, Venezuela, did not contribute to British energy independence as it lay within the U.S. sphere of influence.

On the plus side, Persia was finally meeting expectations. The productivity of its oilfields was unmatched—up to four times greater than the U.S. average (12,000–18,000 bpd versus 4,000 bpd). Finally justifying Churchill's confidence in 1914, Anglo-Persian's production rose from 1.74 million tons in 1920–1 to 5.94 million tons in 1930. The company had also become more efficient, reducing production costs by 5/- per ton between 1920/21 and 1925/26. The Abadan refinery showed particular improvement, almost tripling throughput between 1919–20 and 1932 (from 1.2 million tons to five million tons). Finally, by 1933, the company had completed surveys of more than 80 percent of its 1901 concession area.[210] This happy situation was not replicated in Iraq. Whitehall had secured the IPC's predominance by allowing French and U.S. participation, but Iraqi exports did not arrive in Europe until 1934—and then only in limited quantities due to limited throughput of the IPC's pipelines and low oil prices, which discouraged the companies from exploiting Iraq's full capacity.

There was little urgency in Whitehall about the underwhelming results of Britain's postwar strategy thanks to the geopolitical tranquility of the 1920s. The British were living on borrowed time. Not only would Britain remain dependent on U.S. oil supplies in wartime, but also a shortage of tankers might prove crippling even if sufficient oil was available. Starting in 1932, events would take an ominous turn, exposing the gap between Britain's ambitions and the reality of its position.

[207] Larson, *New Horizons*, 47–50; Wilkins, *Maturing*, 209–11. See also: Rabe, *OPEC*, 55.

[208] McBeth, *Venezuela*, 204–5, *Policy*, 93–4, and "Industry," 453–4.

[209] These figures are drawn from various Oil Board reports located in: CAB 50/3–7. See also: Bamberg, *History*, 175

[210] Ferrier, *History*, 370, 397–9, 415, 428ff, and 601.

3

The Reality of Dependence

1932–1939

Many officers in and around Whitehall were concerned about the security of British oilfields in the Middle East, but they were fixated on the internal threat. Slade consistently spoke out about the risks in Persia, urging Britain to station forces in the Middle East to defend the Persian oilfields and warning that Britain might have to "submit to an ignominious peace" if wartime supplies from Persia were disrupted.[1] Senior British officers shared Slade's anxiety—one admiral warned the First Sea Lord in 1933 against leaving Britain's wartime oil supplies to the "goodwill" of countries such as Persia, whose government "could upset our supplies... if circumstances made it worth their while."[2]

As it turned out, Britain navigated the 1932–3 cancelation crisis in Persia, and then another following Mexico's nationalization of its oil industry five years later, with a minimum of disruption to its oil supplies. British officials were pleased with their success, but what they failed to consider was that these incidents also revealed the inadequacy of force to compel even weak governments to respect the rights of British firms. Whitehall would learn this lesson at great cost following Iran's nationalization of its oil industry in the period 1951–4 and during the Suez crisis of 1956–7.

The immediate threat to Britain's energy supplies, however, came from Britain's fascist rivals, starting with Italy following the Abyssinian crisis of 1935–6. British oil strategy assumed safe passage through the Mediterranean, but Whitehall's halfhearted support for League of Nations sanctions to punish Italian aggression only succeeded in enraging Rome without doing much to hinder its genocidal war against Abyssinia. Although the Royal Navy maintained its qualitative superiority over the Italians, it could not guarantee the safety of Britain's Mediterranean lifeline, which was the most efficient route for Middle Eastern oil exports sailing west of Suez. Compounding this was the United States' lurch toward isolationism under the Neutrality Acts of 1935–7, which complicated British access to U.S. oil supplies by requiring immediate payment in hard currency and transport on British tankers. Supplies of both would prove inadequate to meet the challenge of what British planners euphemistically called the "dual contingency" of simultaneous conflicts in the European and Pacific theaters.

[1] Slade, "Oil Supplies in War," *RUSI* 71: 481 (1926): 119–35, and "Strategic Aspect [...]," no date (circa 1926), BP 68940.

[2] Dunbar-Nasmith to Chatfield, December 11, 1933, ADM 1/8773/57.

(ANOTHER) CRISIS IN PERSIA, 1932–1933

In 1930, the Oil Board had warned that Britain could not rely on Persia if it was fighting the Soviet Union. This was an unlikely prospect, and Whitehall still rated Persia a dependable supplier in wartime.[3] Relations between Britain and Persia had not improved after the failure of the Anglo-Persian Agreement of 1919. They reached a new low in 1928, when the Persians demanded the islands of Abu Musa and the Greater and Lesser Tunbs, which are located at the eastern end of the Persian Gulf and sit athwart the Strait of Hormuz. The Persians also claimed sovereignty over Bahrain, Muscat, and the Trucial States themselves. The Secretary of State for India, the Earl of Birkenhead, warned his Cabinet colleagues in 1928 that Persia aimed at nothing less than the "total elimination" of Britain's presence in the Gulf.[4]

Britain had been committed to maintaining its predominance in the Persian Gulf since 1903, when Foreign Secretary Lord Lansdowne announced that Britain would consider the expansion of any foreign military presence in the area "as a very grave threat."[5] A generation later, the Persian Gulf Sub-Committee of the Committee of Imperial Defence (CID) described the Gulf's significance in terms of its role as a trade and communications link as well as a source of energy.[6] Colonial Secretary Amery suggested that Britain might "adopt the attitude that the Persian Gulf is a special interest, comparable to the Monroe Doctrine" and exclude all potential rivals. Although the subcommittee recommended maintaining the status quo, Amery preferred a radical approach. In terms that presaged the Carter Doctrine of 1980, he wished the British to stipulate that "we regard the Persian Gulf as an area where interference with our supremacy or with our established rights will be resisted as a direct act of aggression upon ourselves."[7] Birkenhead believed that Britain should not be deterred by Persian threats to its oil supplies, and he advised his colleagues to call Tehran's bluff if it canceled the Anglo-Persian concession. Not only did the Persians depend on oil revenues, but international opinion would also turn against Persia if it violated British rights.[8] Other departments were less eager. In 1932, the Foreign Office pressed for a treaty to reduce "friction," blunt Soviet influence in Persia, and protect Britain's "important vested interests."[9]

[3] Kent, *Moguls*, 153–4.

[4] Birkenhead, "British Policy [...][...]," October 11, 1928, C.P. 299 (28), CAB 24/197.

[5] Yergin, *Prize*, 140.

[6] CID, "Persian Gulf," October 29, 1928, 169-D, attached to: Hankey, "Persian Gulf," October 30, 1928, C.P. 321 (28), CAB 24/198.

[7] Amery "Persian Gulf," October 29, 1928, C.P. 322 (28), CAB 24/198. Compare Amery's words with President Jimmy Carter's warning that any "attempt by any outside force to gain control of the Persian Gulf region will be regarded as an assault on the vital interests of the United States of America, and such an assault will be repelled by any means necessary, including military force." "State of the Union [...]," January 23, 1980, The American Presidency Project, accessed August 15, 2018, <http://www.presidency.ucsb.edu/ws/index.php?pid=33079>.

[8] Birkenhead, "British Policy [...]," October 11, 1928, C.P. 299 (28), CAB 24/197.

[9] Simon, "Anglo-Persian Relations," October 24, 1932, C.P. 358 (32), E 5481/208/34, CAB 24/234. The India Office also took a softer line, except on Bahrain. Hoare, "Anglo-Persian Relations," October 31, 1932, C.P. 371 (32), CAB 24/234.

Perhaps the greatest source of "friction" was Anglo-Persian's 1901 concession—it was certainly Britain's most significant foreign investment besides the Suez Canal. Oil revenues accounted for more than half of Persia's export earnings by the end of World War I, but fluctuations in royalties were a problem.[10] After becoming Shah in 1926, Reza Khan pressured the company to renegotiate. His government required at least £1 million annually to fund its modernization program. Between 1918 and 1927, royalties increased from £419,000 to £1.4 million. In 1928, however, they cratered to £502,000, and the Shah denounced the Armitage-Smith Agreement of 1920 and began negotiations for a new concession. Unhappy that Whitehall owned Anglo-Persian shares but they did not, the Persians wanted a stake in the company (25 percent), board representation, and a lump sum payment of £500,000 plus 2/- per ton of oil produced in Persia to guarantee revenues of £1 million per year—less than the £1.4 million they had received in 1927.[11]

Cadman blanched at these terms, but he also wanted a settlement that reconciled Persia to the company's presence. The Persians complained that they had no control over how Anglo-Persian calculated its royalty payments or any share of its global profits. Cadman sympathized and wanted to establish an "identify of interest" that preserved Anglo-Persian's future in Persia, ideally by extending its concession.[12] The company had tried once before in 1920 to placate Persia by offering non-voting shares in the company to settle outstanding differences, but Whitehall vetoed the idea.[13] Eight years later, however, Cadman sold Whitehall and his board on sharing ownership of the company with Persia, provided Tehran agreed never to sell its shares.[14] He then traveled to Persia in February 1929 and offered Tehran a 20 percent stake, royalties of 2/- per ton, and a reduction of the company's concession area to 100,000 miles in exchange for a thirty-year extension to the 1901 concession. The ground, however, had shifted. There was intense political unrest under the Pahlavi regime through 1928, and Persia's financial position had deteriorated due to the decline in the price of silver, which backed its currency until 1932.[15]

Accordingly, Tehran had less room to compromise and refused to budge on its initial demand for a 25 percent stake and a guaranteed annual minimum payment, while offering a twenty-year extension of the concession. Anglo-Persian's board rejected Tehran's counterproposal and negotiations collapsed. The company had

[10] Ferrier, *History*, 629.

[11] Elm, Oil, 7ff and 28ff. Besides Elm, the following discussion draws from Peter Beck, "The Anglo-Persian Oil Dispute, 1932–1933," *Journal of Contemporary History* 9: 4 (1974): 123–51; Gregory Brew, "In Search of 'Equitability': Sir John Cadman, Rezā Shah and the Cancellation of the D'Arcy Concession," *Iranian Studies* 50: 1 (2017): 125–48; Ferrier, *History*, 588–631.

[12] "Memorandum by the Government Directors," October 19, 1928, T 273/357; Brew, "Equitability," 130–2.

[13] See: T 161/2, especially Greenway to Treasury, June 9, 1920, and Barstow to Greenway, July 5, 1920.

[14] Note to Barstow, November 23, 1928, author unclear, T 273/357. The Treasury was skeptical but willing to hold its tongue so long as the Admiralty's supply contract was unaffected and Tehran extended the concession. See Fisher to Lindsay, October 15, 1928, and Oliphant to Fisher, October 19, 1927, T 273/356; the memoranda of October 24–27, 1928, enclosed with: T 273/357.

[15] Brew, "Equitability," 132–3.

Table 3.1. Oil production, royalties, and profits in Persia, 1929–1931[16]

Year	Oil production in Persia	Royalties (£)	Royalty per ton	Company profits (£)
1929	5,460,955 tons	1,436,764	5/3d	4,274,000
1930	5,939,302 tons	1,288,312	4/4d	3,786,000
1931	5,750,498 tons	306,382	1/1d	2,413,000

not been entirely forthcoming—it was disingenuous to deny Tehran's special "guarantees" when the company was selling fuel oil to the Admiralty at a steep discount. The Persians, however, had overplayed their hand—the 1929 spike in royalties to £1,436,764 was an aberration.[17] By the time negotiations restarted in 1931, the world was in the throes of the Great Depression, and Persia's finances were crumbling. The Persians' objectives during the new round of negotiations were modest—20 percent of the company's profits and a guaranteed minimum payment similar to the one Iraq received from the Iraq Petroleum Company (IPC). Anglo-Persian agreed to 20 percent profit sharing and offered a £1,000,000 pay off, and the two sides signed a provisional agreement in May 1932, but it unraveled in June after the latest royalty figures were released. Royalties in 1931 had slumped by three-quarters from those of 1929 while Anglo-Persian's profits sagged by 44 percent. See Table 3.1.

Incensed, the Shah abrogated the 1901 concession on November 27, 1932. He did not nationalize the industry, which now comprised 68 percent of Persia's exports. Rather, he informed Anglo-Persian's local manager that he wished to renegotiate the concession, and production continued without interruption throughout the crisis.[18] Outsiders surmised, however, that the Persians were after more than just money—cutting Anglo-Persian down to size was a matter of "prestige" and a way for the Pahlavi dynasty to distinguish itself from the Qajar one it overthrew.[19]

Whitehall warned Tehran that it would "take all legitimate measures to protect their just and indisputable interests."[20] Cadman had already warned Foreign Secretary John Simon in October that the company might require government assistance.[21] Although Britain was not obligated to defend Anglo-Persian's concession any more than Shell's in Mexico, the Cabinet instructed the Chiefs of Staff to draft contingency plans to secure the oilfields.[22] The British were fortunate that they never had to follow through since their forces could accomplish little.[23] Whitehall began studying how to secure the Persian oilfields in 1921, and more than once over the intervening decade opted against stationing troops—their

[16] Ferrier, *History*, 601. [17] Elm, *Oil*, 30.

[18] Brew, "Equitability," 137; Ferrier, "Iranian Oil Industry," 645–6.

[19] Dieckhoff to Neurath, November 30, 1932, PAAA, R 118175.

[20] "Persian Oil Concessions," December 8, 1932, no author, BP 69287; Brew, "Equitability," 131–7; Bamberg, *History*, 33–7; Elm, *Oil*, 18, and 23–31.

[21] Brew, "Equitability," 136.

[22] Churchill, "Proposed Arrangement [...]," May 11, 1914, CA, CHAR 13/31/24; note to Fisher, "Defence [...]," February 23, author and year unclear, T 161/805.

[23] Beck, "Anglo-Persian Dispute," 133, 135–6, and 146–9.

presence would be too costly and provoke anti-British sentiment—or creating a private militia of Anglo-Persian employees. If Persia occupied the oilfields, Britain should embargo any exports and bankrupt Tehran.[24]

Little changed over the intervening decade. Just a few weeks before the Shah canceled Anglo-Persian's concession, Whitehall agreed to deploy a brigade from India to Persia during any civil disturbance. Only a single battalion would go directly to the oilfields, and there was no expectation that it should "deal with an attack in force by organized troops equipped on modern lines [...]."[25] The Chiefs of Staff and Anglo-Persian were unanimous in their desire to avoid any "provocation."[26] In the event of conflict, existing forces in the region could only evacuate British nationals from Abadan. Nothing could protect the oilfields and retaking them would require several months and considerable expenditure of effort. The Persians could also create problems in Iraq by working with the Shia or Kurdish populations. The most dangerous contingency, however, was intervention by the Soviet Union under its 1921 and 1927 treaties with Persia.[27]

An armed response was also impractical because the British public opposed any unilateral application of force. Whitehall's best option was to refer the matter to the League of Nations or the Permanent Court of International Justice.[28] Involving the League might also repair some of the damage it suffered following its ineffectual response to Japan's invasion of Manchuria in 1931.[29] In December 1932, the Foreign Office advised the Cabinet against any escalation until all diplomatic options had been exhausted.[30] By the end of December, the prospect of armed action had subsided, allowing the British to redeploy two destroyers from the Persian Gulf to the Bay of Bengal.[31]

After the matter went to the League of Nations in January 1933, the two sides agreed to resume negotiations. The Persians again sought a 25 percent share, and the talks nearly collapsed again. In April, Tehran, by now desperate for revenue, and Anglo-Persian agreed on a new concession after Cadman bypassed the Persian negotiators and dealt personally with the Shah. Anglo-Persian's concession area shrank by 80 percent, from 500,000 square miles to 100,000 square miles—although it retained its most promising territories—and the company agreed to pay Tehran a fixed sum of 4/- per ton of oil exported or sold in Persia, but not including

[24] Wilson, "Protection [...]," October 27, 1921, 157-C, CAB 5/4. See also: 168-C, 172-C, 173-C, 178-C, and 179-C, all in: CAB 5/4; Foreign Office, "Memorandum [...]," January 7, 1931, enclosed with: Baxter to CID, January 7, 1931, and note to Upcott, "South Persian Oilfields," no date or author (circa January 1931), both in: T 161/805.

[25] Hankey, "South Persian Oilfields [...]," November 14, 1932, C.P. 390 (32), CAB 24/234.

[26] Milne et al., "Protection [...]," December 2, 1932, C.O.S. 300, attached to: Hankey, "Protection [...]," December 3, 1932, C.P. 419 (32), CAB 24/235.

[27] Haining et al., "Protection [...]," December 10, 1932, C.O.S. 300 (J.P.), attached to: Hankey, "Protection [...]," December 13, 1932, C.P. 430 (32), CAB 24/235.

[28] Beck, "Mosul Dispute," 259 and 271–2.

[29] Beck, "Anglo-Persian Dispute," 129–34 and 149–50.

[30] Vansittart, "Cancellation [...]," December 6, 1932, C.P. 421 (32), and Vansittart, "Memorandum [...]," C.P. 428 (32), December 13, 1932, both in: CAB 24/235.

[31] Beck, "Anglo-Persian Dispute," 139.

any used for the company's Persian operations or its tankers. The company also forfeited its exclusive right to own and operate pipelines within the concession area and agreed to pay a sum equivalent to 20 percent of its dividend payments beyond the first £671,250. It also guaranteed an annual minimum payment of £750,000. As a result, Persia's royalties for 1931 more than quadrupled, from £306,872 to £1,339,132, and nearly tripled between 1931 and 1937—from £1,339,132 to £3,545,313. In exchange, Tehran forfeited its right to 16 percent of Anglo-Persian's annual profits, much to the delight of the company's leadership.[32] Tehran also extended the life of the concession from 1961 to 1993 and did not receive any shares in the company.[33]

The surge in royalties following the settlement stemmed from a 3.5 million ton increase in production between 1933 and 1938 (from 6.45 million tons to 10.2 million tons) but did not buy peace. "Such a spectacular rate of growth could not possibly be sustained," the company later concluded, but "excited hopes and aroused expectations which stood not the remotest chance of being fulfilled." The British had only boosted production to gather wartime stockpiles, and Tehran's royalties shrank by £200,000 in 1938. Only Cadman's intervention and the promise of British export credits dissuaded the Shah from canceling Anglo-Iranian's concession again in the spring of 1939.[34]

This crisis of 1932–3 was evidence that Britain's hold over Middle Eastern oil was tenuous, but it prompted no major reassessment of Britain's oil position.[35] The Oil Board merely lauded the new concession's favorable terms.[36] Far from being "plucked," as he stated in public, in private Cadman bragged to Simon.[37] Anglo-Persian's general counsel praised Cadman's handiwork, opining "that it was the best Concession he had ever seen and the best that could be devised." The company's board unanimously approved the new agreement.[38] That year, the company announced a 7.5 percent dividend, in the midst of the Great Depression.[39] Anglo-Persian had won a great victory, as the new deal was superior to the D'Arcy concession in almost every respect, but it "was too shrewd to last."[40] The Persians were resentful—how could they not be, when Whitehall received more in taxes from Anglo-Persian than did Tehran?[41] Britain's most-valuable overseas oil asset remained in jeopardy—and not just because of the animosity simmering in Tehran.

[32] Greenway Speech, January 25, 1927, CA, CHAR, 2/151/16–24.

[33] Bamberg, *History*, 33–51; Brew, "Equitability," 137–40; Elm, *Oil*, 31–43. Shwadran, *Oil*, 41–56, is good on the legal issues.

[34] Bamberg, *History*, 51–62 and 69.

[35] In retrospect, the British conceded that the 1932–3 crisis in Persia, followed six years later by Mexican nationalization, proved that Whitehall's policy "was built on shaky foundations." Payton-Smith, *Oil*, 24–5.

[36] Londonderry et al., "Eighth Annual Report," July 31, 1933, O.B. 122, CAB 50/5.

[37] Contra Yergin, *Prize*, 271, see: "Memorandum," no date, attached to: Simon, "Anglo-Persian Relations," May 19, 1933, C.P. 135 (33), CAB 24/241. Cadman also assured Simon that the moment was ripe to seek a broader diplomatic settlement. Simon, "Persia," December 8, 1933, C.P. 297 (33), CAB 24/245. See also: Simon, "Anglo-Persian Relations," January 23, 1934, C.P. 18 (34), CAB 24/247.

[38] "Board Meeting, 15th May, 1933," BP 88373.

[39] Beck, "Anglo-Persian Dispute," 144.

[40] Elm, *Oil*, 36.

[41] £173.1 million versus £100.5 between 1933 and 1951. Brew, "Equitability," 140.

Britain's entire strategy of seeking energy independence by developing the Middle East assumed safe passage through the Mediterranean. The elimination of any rival naval force there after 1918 had afforded Britain a welcome degree of security, and even the rise of the Third Reich in 1933 did not pose any immediate danger. The only plausible risk came from Britain's former wartime ally, Italy. Any souring of Anglo-Italian relations would undermine the entire basis of Britain's oil strategy.

WAR PLANNING, 1935–1937

After 1933, Britain found itself in an increasingly difficult position. Its most likely military adversary in the 1920s had been Japan, which could not project power beyond the Pacific or threaten British oil supplies other than those in the East Indies or Burma. Germany's abrogation of the Versailles Treaty and the fracturing of Britain's relationship with Italy over its invasion of Abyssinia in 1935 changed the strategic calculus. Germany and Italy, although inferior to Britain at sea, could sever access to the Middle East through the Mediterranean—through which 34 percent of Britain's oil imports traveled in 1934—and possibly even invade the region.[42]

Ironically, the man who had put oil at the center of British strategy, Admiral Fisher, was also the same man who had argued that Britain needed to reduce its naval commitment to the Mediterranean. Since Britain's main threat lay across the North Sea, Fisher believed that Britain should leave the Mediterranean to allies such as France. In addition, the advent of submarines and aircraft made the inland sea too dangerous for British capital ships. Herein lay a strategic paradox—as its consumption of Middle Eastern oil grew, Britain's energy supply depended upon safe transit through the Mediterranean, but British planners increasingly doubted Britain's ability to operate there safely. During World War I, German submarines wrought havoc in the Mediterranean and forced a diversion of shipping around the Cape in 1916 until the adoption of convoying the following year, and losses remained high through.[43] While the British were confident they could prevail in any struggle for control of the Mediterranean, many doubted whether they could succeed at an acceptable cost.[44]

Prior to 1935, Italy had every incentive to contain Germany to thwart its ambitions in Austria. Anglo-French-Italian cooperation flourished briefly in April 1935 with the formation of the Stresa Front after Germany abrogated the military restrictions of the Versailles Treaty. Britain then spent the rest of year undermining the Stresa Front, initially by signing a naval arms-limitation agreement with Germany in June without bothering to consult with either France or Italy. Any lingering

[42] Kent, *Moguls*, 148–9.

[43] Paul Halpern, *Naval History of World War I* (Annapolis: Naval Institute Press, 2012), 381–402.

[44] Michael Simpson, "Superhighway to the World Wide Web: The Mediterranean in British Imperial Strategy, 1900–1945," in: *Naval Policy and Strategy in the Mediterranean: Past, Present and Future*, ed. John Hattendorf (London: Frank Cass, 2000), 51–60.

hopes for Anglo-Italian cooperation against Germany vanished after Britain joined the League of Nations in imposing economic sanctions (not including oil) against Italy following its invasion of Abyssinia in October 1935.[45]

The potential efficacy of a League oil embargo against Italy remains the subject of debate. The prevailing opinion is that an oil embargo, combined with a pledge by U.S. producers to maintain exports at peacetime levels, would have derailed the Italian war effort.[46] Moreover, a closure of the Suez Canal would have been doubly devastating for Italy by cutting off its army in Africa. There is also disagreement over whether an embargo was politically feasible. Critics of London and Paris claim that sanctions might have been an effective tool but point out that neither government wanted to risk a conflict.[47] Defenders respond that the League of Nations could not guarantee the compliance of the United States upon whom the support of other oil exporters depended.[48] In other words, both sides argue that the success of an oil embargo hinged upon U.S. and League support.[49]

Both the Petroleum Department and representatives from Shell and Anglo-Iranian agreed that an oil embargo would cripple Italy within a few months—a finding affirmed by a League of Nations committee that studied the feasibility of sanctions in January 1936. Neither sanctions nor a "voluntary" restriction of exports to Italy to prewar levels would have a decisive effect without the cooperation of its three largest suppliers: Romania, the Soviet Union, and the United States. In 1934, those countries accounted for two-thirds of Italy's oil imports, and between August and November 13, 1935 for 77.6 percent.[50] Unfortunately, the United States, unlike Romania and the Soviet Union, was not a member of the League. Moreover, even if the major oil companies restricted exports, U.S. independents, with 80,000 tons of daily production, could replace them, as Italy's daily consumption was only 8,000 tons.[51] Even though many of the major companies abided by Washington's "moral embargo," U.S. exports to Italy in November 1935 were 50 percent higher than the previous month and were three times larger than the monthly average.[52]

[45] Zara Steiner, *Triumph of the Dark: European International History, 1933–1939* (Oxford: Oxford University Press, 2011), 100–36.

[46] Cristiano Andrea Ristuccia, "The 1935 Sanctions against Italy: Would Coal and Oil Have Made a Difference?" *European Review of Economic History* 4 (2000): 85–110.

[47] George Baer, *Test Case: Italy, Ethiopia, and the League of Nations* (Stanford: Hoover Institution Press, 1976), 99–100, 103–4, and 301–4. The chiefs advised that French support and two months of preparation were required before any conflict. Chiefs of Staff, "Italo-Abyssinian Dispute," August 9, 1935, C.O.S. 392, attached to: Hemming, "Italo-Abyssinian Dispute," C.P. 166 (35), August 19, 1935, CAB 24/256.

[48] CID, "Economic Pressure on Italy," no date (circa July 1934), A.T.B. (E.P.) 29 (4th Revise), attached to: Hemming, "Economic Pressure [. . .]," August 21, 1935, C.P. 169 (35), CAB 24/256. See also: Bruce Strang, "'The Worst of All Worlds': Oil Sanctions and Italy's Invasion of Abyssinia, 1935–1936," *Diplomacy and Statecraft* 19 (2008): 210–35.

[49] Elliot, "[. . .] Sanctions," October 3, 1935, A.T.B. 131, attached to: Hankey, "Italo-Abyssinian Dispute," October 4, 1935, C.P. 186 (35), CAB 24/257.

[50] "Oil Sanctions and Italy," December 1935, Annex I to: Runciman, "Oil Supplies for Italy," December 9, 1935, C.P. 236 (35), CAB 24/257.

[51] Both Jersey and Shell limited deliveries to their Italian subsidiaries to prewar levels. Larson, *New Horizons*, 336; Jonker and van Zanden, *History of Shell*, 461.

[52] Baer, *Test Case*, 199–201; Robert Divine, *Illusion of Neutrality* (Chicago: University of Chicago Press, 1962), 129–30.

A diversion of tankers would only have a negligible effect without the support of the Scandinavian nations, which controlled the third-largest bloc of tonnage behind Britain and the United States. Foreign Secretary Anthony Eden still believed Britain should join an oil embargo, if only for symbolic reasons, but the French refused to participate, since they did not wish to antagonize Italy needlessly.[53]

The Cabinet accepted Eden's recommendation, but it was too late, as Addis Ababa fell in May. Ultimately, neither Britain nor France was prepared to risk a war—although not because of any fear of the Italian Navy and air force. The Royal Navy was confident that it could smash the Italians even if it began a war on the defensive. Why then were the British so reluctant to provoke Italy? British planners worried that a war would undermine Britain's global position by siphoning resources away from the containment of Japan and Germany, eliminate the Mediterranean as a transit route for operations in the Far East, and undermine the Royal Navy's readiness if losses were high. The British and French were also anxious not to lose Italy as an ally against Germany, and Paris both promised Rome it would oppose closing the Suez Canal and refused to place its fleet at Whitehall's disposal except under impossible preconditions.[54] Nor could Britain take the support of the Dominions for granted. In Canada, Prime Minister William Lyon Mackenzie King had resisted Britain's strategic leadership since the Chanak Crisis of 1922.[55] His Liberal Party depended on votes from Francophone Quebec, where anti-British sentiment ran high. King also wanted to devote his political capital to fighting the Great Depression.[56] Australian Prime Minister Joseph Lyons reluctantly agreed to support sanctions but expressed his preference that Britain not antagonize Italy when the British Empire already had Japan and Germany to deal with.[57]

Fighting Italy may have been unpalatable, but appeasing it was probably fruitless. Self-sufficiency was at the heart of Benito Mussolini's conception of fascism, and he believed that Italy's future depended upon control of the Mediterranean and access to the high seas.[58] In any event, Britain abandoned any

[53] Hoare and Eden, "Embargo [...]," November 27, 1935, C.P. 212 (35), CAB 24/257; Runciman, "Oil Supplies [...]," December 9, 1935, C.P. 236 (35), CAB 24/257; Eden, "Dispute [...]," February 22, 1936, C.P. 53 (36), CAB 24/260. See also: BP 68621; Baer, *Test Case*, 205–13 and 221–7.

[54] Farnie, *Suez*, 594–603; Arthur Marder, "The Royal Navy and the Ethiopian Crisis of 1935–36," *Journal of American History* 75: 5 (1970): 1341ff; R.A.C. Parker, "Great Britain, France and the Ethiopian Crisis, 1935–1936," *English Historical Review* 89: 174 (1974): 293–332.

[55] For King's sabotage (abetted by the South Africans) of British attempts to promote closer defense cooperation with the Dominions after 1921, see: Correlli Barnett, *Collapse of British Power* (London: Eyre Methuen, 1972), 166–233.

[56] B.J.C. McKercher, "From Isolation to Intervention: Anglo-Canadian Defense Relations from the Canadian Perspective, 1935–1939," in: *Multinational Operations, Alliances, and International Cooperation Past and Future*, ed. Robert Rush and William Epley (Washington, DC: Center for Military History, 2006), 85–7.

[57] Christopher Waters, *Australia and Appeasement: Imperial Foreign Policy and the Origins of Appeasement* (New York: Palgrave, 2012), 11–15.

[58] Knox, *Mussolini Unleashed*, 3–44, "Conquest, Foreign and Domestic, in Fascist Italy and Nazi Germany," Journal of Modern History 56: 1 (1984): 1–57, and Hitler's Italian Allies: Royal Armed Forces, Fascist Regime, and the War of 1940–43 (Cambridge: Cambridge University Press, 2000), 5–21.

effort at appeasement following the collapse of the Hoare–Laval Pact Agreement. The course Britain did choose, however, was doubly disappointing. The League of Nations—with British support—did impose sanctions against Italy, which included weapons, rubber, and metals, in February, by which time it was too late to help the Abyssinians. Britain's unwillingness to approach the Abyssinian crisis ruthlessly—a reflection of British leaders' ignorance of strategic affairs and their unwillingness to defy public opinion—alienated Italy, without doing much for Abyssinia or the League's credibility. Whitehall's dithering continued until the outbreak of World War II, as evidenced by its decision to forego a preemptive attack again Italy, which left the initiative to Mussolini, who grasped it in June 1940.[59]

With regard to oil, what is most remarkable about the Abyssinian Crisis is the fact that British officials never considered the consequences for Britain's energy security—a remarkable oversight considering that Whitehall understood that the most efficient transit route for Middle Eastern oil was through the Mediterranean.[60] As Italy prepared to invade Abyssinia, in August 1935 an interdepartmental committee surmised that this might "constitute a more serious threat than at present to our strategical position in the Red Sea."[61] The Foreign Office concurred.[62] The Dominions Secretary was the only official who mentioned that Britain might have "to contemplate diverting British shipping form the Mediterranean to the Cape Route," but only in the short run.[63] Luckily, Italy proved incapable of severing British lines of communication in the Mediterranean without German assistance. Although it consumed an increasing share of the prewar defense budget, the Italian Navy suffered from a crippling disadvantage as against the Royal Navy due to its failure to embrace radar. The Italian air force, meanwhile, was wedded to the theories of Giulio Douhet and refused to establish a naval-air arm, which meant that Italy entered the war without aircraft carriers or torpedo-bomber capability.[64] Still, before the war and during its early stages, there was every indication that Italy,

[59] Williamson Murray, "The Role of Italy in British Strategy, 1938–1939," *RUSI Journal* 124: 3 (1979): 43–9, "Appeasement and Intelligence," Intelligence and National Security 2: 4 (1987): 47–66, and "The Collapse of Empire: British Strategy, 1919–1945," in: Making of Strategy, ed. Williamson Murray et al. (Cambridge: Cambridge University Press, 1994), 414.

[60] There was concern about Britain's supply lines to the Far East. Murray, *Balance*, 51–2; Brian Sullivan, "More than Meets the Eye: The Ethiopian War and the Origins of the Second World War," in: Origins of the Second World War, ed. Gordon Martel (London: Routledge, 1999), 191.

[61] Maffey et al., "[…] British Interests in Ethiopia," June 18, 1935, attached to: Hoare, "Italo-Ethiopian Dispute," August 16, 1935, C.P. 161 (35), CAB 24/256.

[62] Foreign Office, "Points […]," August 9, 1935, attached to: Hoare, "Italo-Ethiopian Dispute," August 16, 1935, C.P. 162 (35), CAB 24/256. The department was more concerned about the diplomatic repercussions than the risk to British security. Foreign Office, "Memorandum […]," August 13, 1935, attached to: Hoare, "Italo-Ethiopian Dispute," August 20, 1935, C.P. 167 (35), CAB 24/256.

[63] Thomas to Hoare, August 15, 1935, attached to: Thomas, "Italo-Abyssinian Dispute," August 20, 1935, C.P. 168 (35), CAB 24/256.

[64] Brian Sullivan, "The Italian Armed Forces, 1918–40," in: *The Interwar Period*, vol. 2 of *Military Effectiveness*, ed. Allan Millet and Williamson Murray (Boston: Allen & Unwin, 1988), 169–217; MacGregor Knox, "The Italian Armed Forces, 1940-3," in: *The Second World War*, vol. 3 of *Military Effectiveness*, ed. Allan Millet and Williamson Murray (Boston: Allen & Unwin, 1988), 136–79, and Mussolini Unleashed, 1939–1941: Politics and Strategy in Fascist Italy's Last War (Cambridge: Cambridge University Press, 1982), 14–33.

singly or with German assistance, posed a credible threat to the Royal Navy and Britain's global interests, which makes Britain's handling of the Abyssinian crisis even more baffling.[65]

Experts soon grasped the repercussions. British bases near the Suez Canal were now vulnerable to Italian bombers based out of the Dodecanese, and Whitehall would have to repair its relations with the Arab world to strengthen its bases in Haifa, Aden, and Basra. With the Mediterranean route now questionable, the British also revived the Cape route for mail service.[66] This was not a practical option for oil shipments. As one speaker explained to the Royal United Service Institution in 1936, assuming that tankers carrying Iranian oil made it out of the Persian Gulf and through the Suez Canal, they "would have to run the gauntlet of the Mediterranean." Rerouting Iranian and East Indies oil around the Cape was "a very unhappy prospect" considering the fourteen-week transit. Iraqi oil was unlikely even to make it to the Mediterranean, considering the vulnerability of the IPC pipelines and the coastal ports. "It would appear," he concluded, "that the fate of the British Empire will probably lie in the hands of the United States, Mexico, Venezuela, and the Dutch East Indies."[67]

This must have come as a shock to British officials. Since 1925, the Oil Board's evaluations had given Britain a false sense of security because its most plausible adversary, Japan, could not interfere with British supplies beyond East Asia. The changing strategic picture after 1930 forced Britain to reconsider its position even before the break with Italy. Following the Manchurian crisis (1931–3) and the failure of the World Disarmament Conference in Geneva (October 1933), in March 1934 the Chiefs of Staff secured an increase in defense spending.[68] Finally, in November 1934, the chiefs advised the CID to begin preparations for a war against Germany within five years. In February 1935, the Oil Board's President suggested that, with the deteriorating situation in Europe and the jettisoning of the Ten-Year Rule in 1932, the board should study requirements during either a Far Eastern or a European war.[69]

By 1936, the Royal Navy was only 500,000 tons short of its objective of accumulating within Britain reserves equal to six months of wartime consumption (3.5 million tons). The army and air force were also trying to gather a similarly sized reserve until 1939, when the Air Ministry raised its requirements to one year's wartime consumption.[70] Also in 1936, CID suddenly backtracked and

[65] Richard Hammond, "An Enduring Influence on Imperial Defence and Grand Strategy: British Perceptions of the Italian Navy, 1935–1943," *International History Review* 39: 5 (2017): 810–35.

[66] Farnie, *Suez*, 606–7.

[67] Bristow, "Oil from Coal [...]," RUSI 81: 521 (1936): 40–54.

[68] The British national security apparatus often functioned at an excruciatingly slow place. There was also a lack of coordination between the services and the Foreign Office. Paul Kennedy, "British 'Net Assessment' and the Coming of the Second World War," in: *Calculations: Net Assessment and the Coming of World War II*, ed. Alan Millet and Williamson Murray (New York: Free Press, 1992), 19–59; Murray, *Balance*, 55–7, 62–4, and "Collapse," 398–9.

[69] Londonderry, "Proposals [...]," February 22, 1935, O.B. 140, CAB 50/5; Londonderry et al., "Tenth Annual Report," November 26, 1935, O.B. 147, CAB 50/5.

[70] Payton-Smith, *Oil*, 61–4.

decided to reestablish a naval reserve of twelve months' wartime consumption. By the outbreak of World War II, the Admiralty had accumulated 6.3 million tons, against an estimated annual requirement of 8.73 million tons during a great power war.[71] Finally, in February 1936, the Oil Board established a new Sub-Committee on Petroleum Products Reserves (PPR) to evaluate Britain's requirements in a war against Germany.[72]

As the PPR's vice-chairman explained in March, the task of fulfilling Britain's needs in a war against Japan "was a comparatively simple one" compared to one with Germany. During the former, there would be no need to build up stocks since army and air force requirements were low. Britain's oil supplies were not in danger since Japan could not attack most of its major suppliers. Britain would enjoy none of these advantages against Germany.[73] In June 1936, the Petroleum Department prepared a preliminary assessment of requirements in either a European or Far Eastern war. Britain's annual consumption without drawing from stockpiles would be as shown in Table 3.2.

If one included the requirements of the empire as a whole, the figures rose to 23.75 million tons in a European war and 25.25 million tons in a Far Eastern one.[74]

The following month, the PPR considered how to satisfy those requirements by 1940.[75] It estimated the British Empire's consumption during the first year of hostilities at 28.5 million tons in a European war and 30.5 million tons in a Far Eastern conflict—or 23.75 million tons and 25.75 million tons of "new supplies," respectively. The primary variable was the United States. The PPR still believed the United States would supply Canada, and that U.S. companies would sell any oil they produced beyond the United States to Britain. Although the situation would become "difficult" if the United States was "unfriendly," enough oil would be available if Latin America was exporting. Britain's largest supplier, however, was Iran, which would provide ten million tons of oil per year—40 percent of the empire's requirements. The second and third would be a "friendly" United States and Venezuela, at four million tons each. The East Indies, Iraq, and Trinidad would provide a further 3.5 million tons. The tanker situation was less favorable. By 1940,

Table 3.2. Annual consumption without drawing from stockpiles

Total Requirements		East of Suez
European War	17,952,210 tons	1,346,680 tons
Far Eastern War	19,536,000 tons	7,253,933 tons

[71] De La Warr et al., "Twelfth Annual Report," December 28, 1937, O.B. 230, CAB 50/6; Brown, "Fuel Supplies," 220ff.

[72] Clarke and Norfolk, "Sub-Committee [...]," February 21, 1936, O.B. (P.R.) 1, CAB 50/14. See also: Payton-Smith, *Oil*, 40–1.

[73] Lindsay, "Sub-Committee [...]," March 24, 1936, O.B. (P.R.) 2, CAB 50/14.

[74] Petroleum Department, "Summary [...]," no date, attached to: Norfolk, note, June 26, 1936, O.B. (P.R.) 13, CAB 50/14.

[75] Except where noted, the following discussion draws from: Crookshank, "First Report," July 21, 1936, O.B. (P.R.) 21, attached to: Ormsby-Gore, "Reserves [...]," July 27, 1936, O.B. 162, CAB 50/5.

Table 3.3. Journey duration

Port of origin	Distance to London (nautical miles)	Length of journey, traveling at 8 knots (days)
Haifa via Strait of Gibraltar	3,320	17
Curaçao	4,206	22
Houston	5,011	26
Abadan via Suez Canal	6,559	34
Long Beach via Panama Canal	7,711	40
Singapore	8,247	43
Abadan via Cape of Good Hope	11,292	59

Britain would require 420 tankers against the 354 currently owned by either British firms or the Admiralty. The only way to close the shortfall was for Britain to commandeer neutral tankers. Another source of concern was a "closed" Mediterranean. Diverting tankers around the Cape increased the length of their journey by 70 percent and almost doubled the duration, as shown in Table 3.3.[76]

Forcing tankers to take the longer route would increase the deficit from 66 to 230, which was more than twice the number of neutral tankers in service. The only solution the PPR could suggest was to adopt convoying.

In his covering letter, President of the Oil Board, William Ormsby-Gore, advised accumulating a six-month military reserve by 1939/40 and stressed that the PPR's favorable assessment of Britain's tanker position assumed the Mediterranean was open.[77] Ormsby-Gore was no stranger to the Middle East, having spent much of his career working in or on the region, and he recognized the many internal or external threats to British interests there. The CID approved the PPR's report on July 30, 1936 and instructed that a six-month military reserve be gathered as soon as possible. It also acknowledged the possibility of the Mediterranean's closure and the need to enlarge Britain's tanker fleet.[78]

The PPR had revealed the weakest link of Britain's oil strategy—its Mediterranean supply line.[79] Before World War II, twelve million tons of oil traveled through the Mediterranean annually, 75 percent from east to west.[80] Britain and France were especially vulnerable to any disruption. During the first ten months of 1936, 31 percent of Britain's gasoline imports and 45 percent of its oil imports came via the Mediterranean.[81] Yet few officials had questioned the wisdom of basing

76 The following data comes from: <https://sea-distances.org/>, accessed August 15, 2018. The average speed of convoys during World War II was 7.5 knots for slow convoys and 9 knots for fast ones, although in practice both traveled much slower.

77 Ormsby-Gore, "Reserves [...]," July 27, 1936, O.B. 162, CAB 50/5.

78 Clarke, "Reserves [...]," August 12, 1936, O.B. 166 (O.B. (P.R.) 24), CAB 50/5.

79 Crookshank, "First Report," July 21, 1936, O.B. (P.R.) 21, attached to: Ormsby-Gore, "Reserves [...]," July 27, 1936, O.B. 162, CAB 50/5.

80 WRo V to WWi VI, "Die Erdöltransporte [...]," January 23, 1939, NARA, RG 242, T-77/649 (Wi/VI. 38).

81 The situation was worse in France, where 58 percent of its oil imports came via the Mediterranean. Griffiss, "Fuel Situation [...]," December 2, 1936, University Publications of America, *U.S. Military Intelligence Reports: France, 1919–1941* (Frederick: University Publications of America, 1986), Reel 10.

Britain's energy future upon the assumption of secure lines of communication across the Mediterranean. Planning for a "closed" Mediterranean only began in 1936, after the Italo-Abyssinian War. The year before, the Oil Board had been sanguine about Britain's oil supply, possibly because it was only considering British requirements during a war against Italy alone.[82]

Fighting Germany was another matter. In response to the PPR's troubling first report, the CID instructed the Oil Board to find alternative suppliers and additional tonnage if the Mediterranean was "closed."[83] Britain had access to ample oil in the Middle East but lacked the tanker capacity to transport enough of it around the Cape. As the Director of Sea Transport concluded, if Britain replaced six million tons of oil from Iran and Iraq traveling around the Cape with an equal amount from the Western Hemisphere, it would save 218 tankers. He asked the Petroleum Department if Britain could increase its imports from the Western Hemisphere if the Mediterranean were "closed" since the tanker shortfall was too large.[84] The Petroleum Department replied that replacing Iranian oil was more complicated than finding alternative supplies. Iran was, for instance, the primary source of petroleum for the Admiralty, providing 70 percent of its wartime requirements.[85] Neither department mentioned the foreign exchange benefits to Britain of purchasing oil price in sterling rather than dollars—this issue would only become paramount in 1939, by which point it was too late to take corrective action.

In the autumn of 1936, officials from the Petroleum Department, the Mercantile Marine Department, and the Admiralty demonstrated that Britain was incapable of replacing Middle Eastern oil if the Mediterranean was "closed" and the United States "unfriendly." The problem was Iran—at best, two million tons of Iranian oil could be replaced by imports from Latin America, but this came at the expense of Britain's allies while still leaving 7.25 million tons to be imported from Iran and Iraq. This oil was irreplaceable, but there were not enough tankers to transport it around the Cape. Every one million tons of Iranian and Iraqi oil that went through the Mediterranean, however, would save twenty-seven tankers.[86]

The PPR presented its evaluation of a "closed" Mediterranean in December 1936. Under such circumstances, Iran and Iraq could not fulfill their primary function west of Suez (supplying the Royal Navy), while the tanker shortage ruled out transporting Middle Eastern oil to Britain around the Cape. The shortfall west of Suez presented "no insuperable difficulty" if the United States was "friendly" and unused Middle Eastern oil was reallocated east of Suez and replaced by U.S. supplies. If the United States was "unfriendly," the two million tons of Iranian oil previously allocated to Britain could be replaced by additional imports from Latin America. Demand for Iranian and Iraqi oil would drop to 7.5 million tons,

[82] Londonderry et al., "Tenth Annual Report," November 26, 1935, O.B. 147, CAB 50/5.

[83] Clarke, "Reserves [...]," August 12, 1936, O.B. 166 (O.B. (P.R.) 24), CAB 50/5.

[84] Hynard to Starling, September 7, 1936, attached to: Norfolk, note, no date, O.B. (P.R.) 26, CAB 50/14.

[85] "Note by Petroleum Department [...]," attached to: Norfolk, note, no date, O.B. (P.R.) 26, CAB 50/14.

[86] PPR, Report, no date, O.B. (P.R.) 34, CAB 50/14.

with four millions tons going to Britain. These supplies would have to travel through the Mediterranean at a rate of twenty tankers per month (ten each way) for every one million tons per annum. On the bright side, the tanker situation had improved—the number of neutral tankers that would be available by 1940 had increased to 150. This lowered the deficit to only sixteen, and Britain could meet the remaining shortfall by returning obsolete vessels to service and seizing enemy vessels.[87] A Board of Trade study of Britain's tanker requirements the following year actually forecast an increase of 200 tankers by 1940. Britain would thereafter have a surplus of nineteen, but this could turn into a deficit if the Mediterranean were "closed."[88] In March 1938, the Board of Trade forecast a surplus of twenty-two tankers, although the Oil Board—heeding the advice of its Tanker Tonnage Committee—suggested that the government might want to accumulate an Admiralty reserve of as many as twenty-five ships.[89]

Whitehall could not afford to ignore events in the Far East, where Japan was not only pursuing an aggressive foreign policy against China but also moving closer to Germany with the signing of the Anti-Comintern Pact in November 1936. In May 1937, the PPR reassessed Britain's oil position during a Far Eastern war. The British Empire would have to import 25.69 million tons—15.44 million tons west of Suez and 10.25 million tons east of it. Britain's key suppliers would be Iran, Venezuela, the United States, Mexico, Peru, Trinidad, Iraq, Bahrain, and Romania. The latter two were replacements for the Dutch East Indies, which was unreliable due to their exposure to Japanese attack. If the United States was "unfriendly," additional imports from Venezuela, Iraq, Bahrain, and Romania were necessary. The tanker situation, however, was surprisingly reassuring. Against the 407 and 436 required if the United States was "friendly" or "unfriendly," respectively, Britain would enjoy a surplus of ninety-seven and sixty-eight by 1940.[90] The Oil Board warned, however, that although Britain could find enough oil to meet its overall requirements, supplies of aviation fuel were at risk. The most important source in the Far East was the Burmah refinery at Rangoon, which produced 216,000 tons—125 percent of Britain's requirements east of Suez. During a Far Eastern war, the Rangoon refinery and neighboring facilities would present "attractive" targets for Japanese forces. The loss of this refinery and the East Indies would eliminate Britain's sources of aviation fuel in the Far East, and the Oil Board advised that accumulating local stockpiles be of the "highest importance."[91]

A few months later, the Petroleum Department revised its original January 1937 study of Britain's requirements during a Far Eastern war assuming a "closed" Mediterranean and the loss of the Rangoon refinery—that is to say, if Britain was fighting in two theaters.[92] The empire's total requirements were now

[87] Crookshank, "[...] Closed Mediterranean," December 11, 1936, O.B. 183 (Section B), attached to: Ormsby-Gore, "Effect on Supplies [...]," February 22, 1937, O.B. 189, CAB 50/6.

[88] Board of Trade, "Allocation [...]," November 1937, attached to: Norfolk, note, November 23, 1937, O.B. (P.R.) 94, CAB 50/15.

[89] De La Warr, "Oil Tankers," May 9, 1938, O.B. 251, CAB 50/7.

[90] Crookshank, "Sources [...]," May 26, 1937, O.B. 195, CAB 50/6.

[91] Crookshank, "Sources [...]," February 8, 1938, O.B. 233, CAB 50/6.

[92] Petroleum Department, "Sources [...]," January 1937, O.B. (P.R.) 52, CAB 50/15.

23.08 million tons if the United States was "friendly" and 23.16 million if it was "unfriendly." If the Mediterranean was "closed," rationing would be required in Britain, which would reduce demand by 2.5 million tons. There was sufficient crude oil east of Suez thanks to Iran, and even more if Britain could export Iraqi oil—the problem was getting enough to consumers west of Suez. This would be no problem if the United States was "friendly," but if it was not, at least 452,000 tons would have to travel from Abadan around the Cape to Britain, while another 1.6 million tons from Latin America would be required east of Suez. Since Abadan still did not produce aviation fuel, its supply would be precarious if the Rangoon refinery was not operational, with Trinidad providing 37 percent of Britain's needs east of Suez (65,000 tons of 177,000 tons) and 40 percent west of Suez (85,000 tons of 217,000 tons).[93]

Although Trinidad was a negligible supplier globally (1.85 million tons in 1936), it still accounted for 40 percent of imperial production, with 92 percent (1.67 million tons) exported. Its role as a supplier of aviation fuel gave it immense strategic value. Trinidad's refineries would supply one-quarter of the annual requirements of the Royal Air Force (RAF) for high-octane aviation fuel in a European war, and all of Britain's requirements west of Suez during a Far Eastern war. Unlike Rangoon, Trinidad was secure against enemy attack. It was also one of the few sources of 100-octane gasoline within the British Empire. Finally, it was the closest source of oil to Britain, and production there was immune from "political interference."[94]

One could not say the same about the United States. By the mid-1930s, Britain had made great progress in reducing its U.S. oil imports—from 33.9 percent of the total in 1922 to 10.6 percent by 1935. There was little progress, however, in reducing imports from the Western Hemisphere—they remained at 61.9 percent as of 1935. Venezuela had replaced Mexico, and the only new supplier from the Eastern Hemisphere (38.1 percent) by 1935 was Iraq.[95] The closure of the Mediterranean would not cripple Britain if it could import the bulk of its requirements from the Western Hemisphere. Britain needed U.S. oil more than ever, but the passage of the first Neutrality Act in August 1935 reopened the question of U.S. reliability.

Isolationist sentiment rose following the revelations of the U.S. Senate's Special Committee on Investigation of the Munitions Industry between 1934 and 1936, which claimed that U.S. financiers and arms manufacturers had provoked U.S. belligerence in 1917. Anger over European defaults on wartime debts also prompted passage of the Johnson Act of 1934, which prohibited U.S. loans to any nation that had defaulted on its official war debts—including Britain, which owed $4.4 billion when it suspended repayment in 1933. President Franklin D. Roosevelt supported neutrality legislation for a variety of reasons. Besides a

[93] Petroleum Department, "Estimated Requirements [...]," November 1937, attached to: Crookshank, "Estimated Requirements [...]," November 18, 1937, O.B. (P.R.) 95, CAB 50/6.

[94] De La Warr, "Importance [...]," November 25, 1937, attached to: Nicholl, "Trinidad," November 30, 1937, O.B. 226, CAB 50/6. Shell's wartime production in Trinidad peaked at 750,000 tons. Stephen Howarth and Joost Jonker, *Powering the Hydrocarbon Revolution, 1939–1973*, vol. 2 of *History of Royal Dutch Shell* (Oxford: Oxford University Press, 2007), 37.

[95] Ormsby-Gore et al., "Eleventh Annual Report," December 31, 1936, O.B. 184, CAB 50/6.

desire to outflank radicals calling for the nationalization of the U.S. arms industry and allow the White House to set the conditions of trade with belligerents, Roosevelt wished to avoid having the United States dragged into another war to defend neutral rights. He requested that any legislation allow him to distinguish between aggressors and victims, but isolationists forced him to sign a bill in August 1935 imposing a blanket six-month arms embargo upon belligerents.[96] Roosevelt did not intend to use the legislation against Britain—but not all Americans felt the same way. The U.S. Military Attaché in London urged Washington to use the Neutrality Act and Latin America's oil "to exert a major influence" over affairs in Europe.[97]

In January 1936, Britain's Industrial Intelligence Centre, responsible for gathering economic intelligence for the CID, assessed the draft bill the White House had submitted to Congress to replace the 1935 legislation. It gave the President wide latitude to limit exports of items "used for war purposes" such as oil to "normal" levels—a measure directed against Italy, but also one that could harm Britain.[98] Isolationists opposed the White House's proposal because it gave the President the power to apply export restrictions selectively (for example, limiting exports to prewar levels)—they preferred, instead, a blanket embargo on all trade with belligerents. Roosevelt's efforts to amend the 1935 Neutrality Act failed, and he had to settle for a new one in March 1936 that included both the arms embargo and a ban on loans or credits to belligerents.[99]

Neither side gave up. Roosevelt tried to use the neutrality legislation to assist Abyssinia following the Italian invasion of October 1935 by imposing an arms embargo, which disproportionately affected Italy and, the President hoped, would encourage the League of Nations to impose sanctions. Isolationists and peace groups, meanwhile, still wanted to extend the arms embargo to include strategic commodities including oil. They achieved a limited level of success with the 1937 Neutrality Act, which excluded raw materials from the contraband list but included export restrictions. As the Foreign Office explained, the 1937 legislation distinguished between "implements of war," whose sale was prohibited, and goods which were "essential for the conduct of a war" but not weapons (including oil), whose sale was restricted. Section II prohibited the export of "essential" items to belligerents on U.S. ships. Such goods could, however, be exported on belligerent ships after title was transferred. This so-called "cash and carry" provision would expire after two years and allowed the United States to profit from international trade in wartime without incurring the risks that contributed to U.S. belligerence in 1917.[100]

While preferable to a blanket embargo, "cash and carry" was still a "nuisance" because it required Britain to organize the shipment of goods and would deplete

[96] Divine, *Illusion*, 81–121. For an overview, see: George Herring, *From Colony to Superpower: U.S. Foreign Relations since 1776* (New York: Oxford University Press, 2008), 502–19.

[97] Lee, "Oil [...]," December 11, 1935, NARA, RG 59, 841.6363/413.

[98] Morton, "Some Notes [...]," January 16, 1936, ICF/558, attached to: Clarke, "U.S.A. Neutrality Bill," January 22, 1936, O.B. 155, CAB 50/5.

[99] Divine, *Illusion*, 122–61.

[100] Divine, *Illusion*, 162–99.

its foreign exchange reserves.[101] Nevertheless, British access to U.S. oil exports was safe—at least until the "cash and carry" provision expired in May 1939. Nor was there any reason to doubt that U.S. supplies would be plentiful. When the PPR asked whether the United States could export more than it had done in 1929 (twenty million tons), the Petroleum Department reported that there was nothing to fear.[102] Whitehall was confident that, following the adoption of pro-rationing (limiting production to market demand), the U.S. oil industry was capable of rapidly boosting output.[103] Finally, by mid-1937, the Petroleum Department considered it unlikely that the United States would be "unfriendly" during any future war, and a "friendly" United States reduced the number of neutral tankers required if the Mediterranean was "closed" to less than sixty-six—a manageable deficit considering that 150 neutral tankers (not including those under U.S. registry) would be available by 1940.[104] The CID concurred but decided to revisit the matter in 1939 following the expiration of the "cash and carry" provision.[105]

The 1937 Neutrality Act was a mixed bag for Britain. It eliminated—at least temporarily—fears that the United States might embargo oil exports to Britain. Oil would be available—but only under the terms of "cash and carry." By preventing the chartering of U.S. tankers and forcing Britain to pay cash in advance, U.S. legislation imposed on Whitehall additional expenditure of two items in short supply: tanker tonnage and foreign exchange. Britain was caught in a vice. On the one hand, addressing the shortage of foreign exchange by increasing imports of Middle Eastern oil would increase the tanker deficit. On the other, saving tonnage by drawing more of Britain's imports from the Western Hemisphere would accelerate the drain of its foreign exchange reserves, since much of this oil required payment in dollars. While supplies from the United States appeared secure after 1937, they were also insufficient—Britain could not dispense with Latin America, either, but its position soon came under assault in Mexico.

NATIONALISM IN MEXICO, 1938

Italy's hostility did not reduce the amount of oil available to Britain so much as complicate its logistics. Any closure of the Mediterranean would force Britain to fall back on the Western Hemisphere since it could not transport enough oil from the Middle East around the Cape. The situation in Venezuela appeared stable and Bolivia's expropriation of Jersey's subsidiary in 1937 attracted attention only in the

[101] Foreign Office, "[...] United States Neutrality," May 10, 1937, A 4581/448/45, POWE 33/765.

[102] Petroleum Department, "Capacity [...]," November 1937, O.B. (P.R.) 91, CAB 50/15.

[103] "Brief Survey [...]," no date or author, attached to: Norfolk, note, June 26, 1936, O.B. (P.R.) 14, CAB 50/14.

[104] Petroleum Department, "U.S.A. Neutrality Legislation," July 1937, O.B. 208 (O.B. (P.R.) 69), attached to: Nicholl and Norfolk, "Note [...]," July 21, 1937, O.B. 208, O.B. (P.R.) 70, CAB 50/6. See also: Babij, "Japan," 100–1.

[105] Nicholl, "Sources [...]," March 15, 1938, O.B. (P.R.) 118, CAB 50/16.

United States.[106] Mexico's nationalization of its oil industry on March 18, 1938 transformed the situation. This was, in the words of the Petroleum Department, "[an] event of considerable importance," with potentially "serious repercussions" for Britain's energy supplies.[107]

Mexico's rise and fall was a global oil producer was dizzying. During World War I, Mexico's annual production increased from five to nine million tons, and it became Britain's second-largest supplier.[108] Production trebled between 1918 and 1921 and exports peaked the following year at 180.87 million bbl but shriveled to only 10.53 million bbl by 1928, before recovering slightly to 18.25 million bbl in 1937.[109] The exact reasons why are unclear—some argue that a combination of political uncertainty and inadequate technology caused the decline after 1921, while others contend that the companies overproduced during and after World War I to compensate for the global shortage and ceased investment once new supplies from the United States, Venezuela, and the Middle East caused a glut.[110] Whatever the reasons, by 1937, production was only one-quarter of its 1921 peak.[111] Mexico was still the world's seventh-largest supplier in 1937, at 2.5 percent of global production.[112] While it is unfair to say Mexico had "declined to relative insignificance," its best days seemed behind it.[113] In Britain's case, Mexico still exported 1.1 million tons there in 1935, but only 631,000 tons in 1937 (10.1 percent of British oil imports to 5.5 percent), although it remained the fourth-largest supplier. By 1937, Whitehall had come to expect that Mexico would supply only 1.8 million tons of oil to Britain in wartime, and the Oil Board concluded that the loss of these supplies was manageable.[114]

Ironically, Britain only recovered its commercial predominance in Mexico after its production had plummeted. By the eve of nationalization, U.S. oil companies were only minor players in Mexico, accounting for 20 percent of production. By contrast, following Mexican Eagle's discovery of the Poza Rica oilfield in 1930,

[106] The Bolivian dispute was resolved in 1942. George Philip, *Oil and Politics in Latin America* (Cambridge: Cambridge University Press, 1976), 193–8; Wilkins, *Maturing*, 225 and 230.

[107] Petroleum Department, "Expropriation [...]," April 8, 1938, O.B. 247, Annex to: De La Warr, "Expropriation [...]," May 9, 1938, O.B. 252, CAB 50/7. The following discussion draws from: Clayton Koppes, "The Good Neighbor Policy and the Nationalization of Mexican Oil: A Reinterpretation," *Journal of American History* 69: 1 (1982): 62–81; Noel Maurer, "The Empire Struck Back: Sanctions and Compensation in the Mexican Oil Expropriation of 1938," *Journal of Economic History* 71: 3 (2011): 590–615; Lorenzo Meyer, "The Expropriation and Great Britain," in: *The Mexican Petroleum Industry in the Twentieth Century*, ed. Jonathan Brown and Allan Knight (Austin: University of Texas Press, 1992), 154–72. The diplomatic wrangling is covered in Lorenzo Meyer, *Mexico and the United States in the Oil Controversy, 1917–1942* (Austin: University of Texas Press, 1977), 149–228.

[108] De La Warr, "Expropriation [...]," May 9, 1938, O.B. 252, CAB 50/7.

[109] Haber, *Politics*, 199.

[110] María Del Mar Rubio, "The Role of Mexico in the First World War Oil Shortage: 1918–1922, An International Perspective," *Journal of Iberian and Latin American Economic History* 24: 1 (2006): 69–95.

[111] DeGolyer, *Historical Data*, 5; Wilkins, *Maturing*, 225.

[112] Llewellin et al., "Thirteenth Annual Report," January 24, 1939, O.B. 294, CAB 50/7.

[113] Jonker and Zanden, *History of Shell*, 261.

[114] De La Warr, "Expropriation [...]," May 9, 1938, O.B. 252, CAB 50/7. See also: Jonker and Zanden, *History of Shell*, 453.

British firms controlled two-thirds of Mexican production.[115] Mexican Eagle also enjoyed a secure political position, having signed an agreement in November 1937 guaranteeing Mexico 35 percent of the production from the Poza Rica oilfield. With 500 million bbl of reserves, 41 percent of Poza Rica lay within public lands.[116] Mexican Eagle also agreed to produce oil there using conservation measures that would lengthen the lifespan of the field—something the Mexicans had been demanding since 1925.[117] Most significantly, the company recognized the government's ownership of subsoil resources under Article XXVII of the 1917 Constitution.[118] Mexican Eagle had signed the agreement against the opposition of the U.S. oil companies.[119] It squandered any goodwill the following year, however, when it sided with the U.S. companies against the Mexican oil workers' union and government.[120]

There are many myths shrouding Mexican nationalization. Whitehall believed it was a pretext to seize Poza Rica.[121] In fact, there is no evidence that President Lázaro Cárdenas orchestrated nationalization;[122] rather, the impetus came from organized labor.[123] Cárdenas only chose nationalization after most of the companies refused to abide by the government labor board's 1937 ruling, upheld by the Supreme Court, granting wage and benefit increases to the Petroleum Workers Union. The companies reacted by ceasing to pay their workers, who retaliated by halting production. The imminent collapse of the oil industry forced Cárdenas' hand, and he nationalized the industry on March 18.[124] Ironically, the government proved to be a tough employer. It refused to implement fully the 1937 ruling, and although workers' nominal wages increased, inflation had wiped out any gains by 1944.[125]

Another myth is that political instability following the 1910 Revolution discouraged investment in Mexico. In fact, taxes on the oil companies declined in real terms after 1922, keeping Mexican oil cheaper than U.S. oil. The cause of the collapse in Mexican oil production during the 1920s was geological rather than political. Firms gradually became disappointed with their rate of return on investment; the percentage of new wells that was productive—which fell from 64 percent to 28 percent between 1921 and 1929; and the output of these new wells—1,900 bpd in 1927, compared with 24,800 bpd in 1920. Investment by U.S. firms crested at

[115] De La Warr, "Expropriation [...]," May 9, 1938, O.B. 252, CAB 50/7. See also: Wilkins, *Maturing*, 225–6.

[116] Maurer, "Empire Struck Back," 595; Myrna Santiago, *Ecology of Oil: Environment, Labor, and the Mexican Revolution, 1900–1938* (Cambridge: Cambridge University Press, 2006), 287–8.

[117] Santiago, *Ecology*, 256–90 and 354.

[118] Article XXVII replaced a series of four laws passed between 1884 and 1909 designed to attract private investment by granting to individuals or firms ownership of subsoil resources on private lands. Haber, *Politics*, 193–4, and 207; Hall, *Oil*, 18; Santiago, *Ecology*, 62–4.

[119] Jonker and van Zanden, *History of Shell*, 455.

[120] Meyer, "Expropriation," 168.

[121] "Mexican Oil Dispute," Appendix V (A) to: Ministerial Oil Committee, "Oil Policy," no date, M.O.C. (44) 5, POWE 33/1399.

[122] Wilkins, *Maturing*, 226.

[123] Santiago, *Ecology*, 291–341.

[124] Only Gulf Oil, with 4 percent of Mexican production, was spared. Wilkins, *Maturing*, 226 and 228.

[125] Maurer, "Empire Struck Back," 611–12; Santiago, *Ecology*, 349–53.

$569 million in 1924 (in Mexican Eagle's case, in 1931), while the number of wells drilled peaked in 1926. The discovery of Poza Rica resulted in a modest increase in oil production after 1932, but investment continued to plummet, even as the companies drilled a higher percentage of successful new wells (70 percent by 1930).[126] Mexico would not return to its 1921–2 levels of production until the 1970s, thanks to new technologies that enabled drilling at greater depths or offshore.[127]

According to Whitehall, nationalization would do irreparable damage to the Mexican oil industry. The British believed the Mexicans lacked the skills to boost production or even maintain it. What petrified the British was the precedent set for other producers. Britain could not afford to do without the Latin American oil to which it enjoyed secure access—twenty-eight million tons in 1936 just between Venezuela, Colombia, and Peru.[128] Most of the oil companies sympathized with the British position and boycotted Mexican exports. Besides Shell, they had little at stake in Mexico but wanted to preserve their more valuable investments in Venezuela.[129] Their Mexican revenues in 1937—$7 million, not including depreciation—were less than the cost of the new agreement mandated by the labor board. This was too much for Jersey, the leader of the U.S. companies, which did not even want to be in Mexico—it had only purchased Indiana's Mexican subsidiary in 1932 to secure the lucrative Venezuelan one.[130] The U.S. firms demanded $200 million, even though their own figures suggested the value of their assets was less than $70 million.[131] Shell, which had the most to lose, demanded "restitution" for lost future profits rather than just "compensation" for current assets.[132] Shell's hardline position was designed to protect its investments in Venezuela, which provided 40 percent of its worldwide production by January 1940—twelve out of thirty-one million tons. If Venezuela followed Mexico's example, the consequences would be catastrophic for both Shell and Britain.[133]

Whitehall agreed and opposed the resumption of diplomatic relations with Mexico, which were severed in April 1938, until there was a settlement with the companies.[134] Britain's approach clashed with the United States' "Good Neighbor" policy.[135] Although it shared Britain's desire that the companies return to Mexico,

126 Haber, "Law," 1–3 and 9–16, and Politics, 190–2 and 198–229.

127 Brown, "Domestic Politics," 401.

128 Petroleum Department, "Expropriation [...]," April 8, 1938, O.B. 247, Annex to: De La Warr, "Expropriation [...]," May 9, 1938, O.B. 252, CAB 50/7. See also: Yergin, *Prize*, 276–7.

129 Mexico's share of U.S. oil investment since 1929 had dropped from 18.5 percent to 6.4 percent by 1936. McBeth, "Industry," 436.

130 Maurer, "Empire Struck Back," 601–3 and 596–9. In 1931, Indiana's Mexican properties had a daily production of 16,000 bpd, while those in Venezuela yielded 88,000 bpd. Larson, *New Horizons*, 49.

131 Koppes, "Nationalization," 68 and 72; Wilkins, *Maturing*, 229.

132 Shell's official history maintains that the company would have succeeded but for U.S. pusillanimity. Howarth and Jonker, *History of Shell*, 39–40.

133 Shell, "Memorandum," January 16, 1940, CAB 63/117; Howarth and Jonker, *History of Shell*, 34–5. In fact, Venezuela opposed nationalization since it was dependent on oil revenues (one-quarter of tax revenues by 1928). Maurer, "Empire Struck Back," 602–3; McBeth, "Industry," 429; Rabe, *OPEC*, 60–1.

134 Gallop, "Mexican Oil Dispute," October 7, 1940, A4486/57/26, T 160/1263.

135 Howarth and Jonker, *History of Shell*, 39–40; Wilkins, *Maturing*, 228–9.

the Roosevelt Administration believed this was possible through negotiation.[136] Unlike Washington, Whitehall did not accept the legality of nationalization without due cause.[137] Britain would try to coerce Mexico into undoing nationalization through a combination of diplomatic and economic measures, such as blocking financing, forbidding British merchants from trading expropriated oil, and discouraging firms from supplying the Mexican oil industry.[138] The objective, according to the Petroleum Department, was to offer "a deterrent" to the rest of Latin America.[139]

World War II did not change Whitehall's perspective. In 1941, the Petroleum Department still argued that acquiescence would undermine the position of British oil companies around the world.[140] In fact, the British only succeeded in boxing themselves into a corner. Their uncompromising stance was supposed to impress Venezuela and Iran, but neither of those countries actually desired a settlement that would allow Mexico to resume exports.[141] The U.S. companies broke ranks in 1942 and signed the Cooke–Zevada Agreement, under which they received $24 million in compensation plus $5 million in interest over four years.[142] Eden urged his colleagues to convince Shell to accept the same terms as the Americans.[143] There was still some opposition across Whitehall. Since Mexican oil was "not of vital importance," the Admiralty nonetheless wished to make an example of Mexico.[144] Eden, by contrast, wanted to adopt a conciliatory posture to secure Washington's assistance in defending British interests elsewhere around the world.[145] Britain and Mexico finally restored diplomatic relations on October 22, 1941, but the dispute over nationalization lingered. Whether or not the embargo was working was irrelevant, according to the Petroleum Department, since British policy was a matter of "principle."[146] The impasse did not end until 1947. Shell would collect $81.25 million retroactive to nationalization at 3 percent interest.[147] The real value of the settlement in 1938 dollars was $43.6 million, which was 3.6 times greater than Mexican Eagle's market value in 1938 and 2.6 times greater than

[136] Koppes, "Nationalization," 68–81; Painter, *Oil*, 22–31.

[137] Petroleum Department, "Expropriation [...]," April 8, 1938, O.B. 247, Annex to: De La Warr, "Expropriation [...]," May 9, 1938, O.B. 252, CAB 50/7. The key diplomatic exchanges between London and Mexico City are reprinted in: *Correspondence* [...], Cmd. 5758 (London: HMSO, 1938).

[138] Llewellin et al., "Thirteenth Annual Report," January 24, 1939, O.B. 294, CAB 50/7.

[139] Petroleum Department, "Expropriation [...]," April 8, 1938, O.B. 247, Annex to: De La Warr, "Expropriation [...]," May 9, 1938, O.B. 252, CAB 50/7. The CID approved the recommendations of the Oil Board and the Petroleum Department. Nicholl, "Expropriation [...]," O.B. 260, May 26, 1938, CAB 50/7.

[140] Petroleum Department,' "Mexican Oil Dispute," September 7, 1941, ADM 116/4815.

[141] Meyer, "Expropriation," 161; Maurer, "Empire Struck Back," 607–8.

[142] Wilkins, *Maturing*, 229–30. Although the final figure was less than it had demanded, Jersey still received as much (in 1938 dollars) as it had paid to acquire Indiana's Mexican subsidiary in 1932 ($19 million). Maurer, "Empire Struck Back," 607 and 609.

[143] "Proposed Agreement [...]," no date or author (on or after August 28, 1941), ADM 116/4815.

[144] Cunningham to Pound, September 8, 1941, ADM 116/4815.

[145] Eden, "Resumption [...]," September 6, 1941, W.P. (41) 215, CAB 66/18/38. For the Cabinet's approval, see: "Conclusions [...]," September 8, 1941, W.M. (41), 91st Conclusions, CAB 65/19/27.

[146] Petroleum Department, "Question [...]," February 28, 1942, T 160/1263.

[147] Meyer, "Expropriation," 163–8.

the company's book value.[148] Not surprisingly, Shell considered the agreement "very satisfactory."[149]

Whitehall could afford to be inflexible toward Mexico because of Venezuela. Besides being the world's largest exporter, Venezuela was also Britain's most important supplier, providing 39 percent of its imports in 1937—4.42 million tons of 11.25 million tons. In wartime, it would supply the British Empire with 9.63 million tons annually assuming that the Mediterranean was "closed," or 6.3 million tons if was "open," but these were manageable figures considering that production in 1937 was 27.7 million tons.[150] Ironically, Venezuela would fall short during World War II, but for an unexpected reason. Thanks to Venezuela, Caribbean output increased from 624,100 bpd in 1938 to 824,100 bbl in 1944. Overall, however, Allied production beyond the United States and Soviet Union was actually lower in 1944 than in 1941 (1.74 million bbl versus 1.8 million bbl).[151] The problem was not one of supply but rather quality, as Venezuela's heavy oil was unsuitable for refining into high-octane gasoline.[152]

PREPARING FOR A "DUAL CONTINGENCY," 1938–1939

Mexico's nationalization of its oil industry in 1938, coming six years after the Shah canceled the D'Arcy concession, suggested Britain could not assume that ownership of foreign oilfields would guarantee secure supplies. The Italian threat to passage through the Mediterranean since 1936 also presented a logistical handicap that was insurmountable without increasing imports from the Western Hemisphere. Under such circumstances, it was imperative that Britain develop domestic sources of petroleum. Unfortunately, British crude oil production was minuscule until the discovery of the North Sea oilfields in the 1960s. Despite Anglo-Persian buying out the six remaining firms in 1919, the Scottish shale industry remained unprofitable. The last straw came in 1925, when Anglo-Persian tried to entice the Admiralty with a supply contract for the industry's total production of fuel oil. The company was banking on the government rescuing an unprofitable industry for security reasons but miscalculated. The Admiralty saw no benefit in preserving a trivial supply of fuel that cost more than double the price of imported U.S. oil, while the rest of Whitehall was fearful of creating a precedent for the coal industry to claim public assistance.[153] Production soon collapsed into insignificance.[154]

148 Maurer, "Empire Struck Back," 607–10. 149 Howarth and Jonker, *History of Shell*, 40.

150 "Supply [...]," no date or author (circa 1938), ADM 1/10072.

151 John Frey and Chandler Ide, *History of the Petroleum Administration for War* (Washington, DC: GPO, 1946), 438.

152 Howarth and Jonker, *History of Shell*, 36. Before the war, 60 percent of the output of the refineries at Curaçao and Aruba had been fuel oil. "Brief Survey [...]," no date, attached to: Norfolk, note, June 26, 1936, O.B. (P.R.) 14, CAB 50/14.

153 Worthington-Evans, "Report [...]," March 27, 1925, C.P. 184 (25), CAB 24/172; Cunliffe-Lister, "Scottish Oils Limited," June 3, 1925, C.P. 275/25.

154 Brown, "Fuel Supplies," 249–50; Payton-Smith, *Oil*, 19.

The only alternative left was petroleum synthesized from coal. Whitehall established a Fuel Research Board in 1917 to study synthesizing petroleum. Of particular interest was one process for liquefying coal discovered in the early 1920s—hydrogenation. In 1927, Imperial Chemical Industries acquired the patent and marketing rights for hydrogenation in the British Empire.[155] Thereafter, the government levied tariffs on imported gasoline. In 1938, however, the government decided against developing an indigenous synthetic fuel industry.[156]

Although no one challenged oil's superiority to coal after World War I, oil's strategic liabilities provoked much debate, thanks in part to the lobbying of the British coal industry.[157] Many analysts lamented that the transition to oil had robbed Britain of the leverage it once enjoyed over international trade through its control of coaling stations. As late as 1923, one analyst cautioned against completely converting to oil, since "a temporary loss of command of the sea might very possibly mean a permanent loss" if coal was no longer a substitute.[158] Another complained that oil had "reduced our independence," and that competition for it created new "political entanglements" and military obligations for an overstretched British Empire.[159] Still others supported financial support for research into synthesizing petroleum.[160] The Admiralty responded that naval technology had advanced too far to revert to coal.[161] The former Engineer-in-Chief of the Royal Navy warned that returning to coal would "render the fleet useless," but as late as 1939 the Admiralty was still swatting down proposals to partially revert to coal.[162]

There was still the possibility of synthetic alternatives. Whitehall first considered whether to lend official support to the synthetic fuel industry in November 1929, when it established a CID subcommittee to report on the viability of various synthetic processes. After reviewing the subcommittee's report, the Oil Board concluded hydrogenation was the only worthwhile option.[163] It was unlikely that a British synthetic fuel industry could develop organically since synthetic fuel cost at least double the price of imported oil. The government would have to impose tariffs on imported oil and bankroll the construction of synthetic plants. Supporters stressed the domestic benefits, such as stimulating employment in the coal and steel industries. The reduction in unemployment benefits would also offset the reduction in revenues from duties on imported oil. Opponents countered that the cost of synthetic fuel would be prohibitive, which meant that government support

155 Anthony Stranges, "Friedrich Bergius and the Rise of the German Synthetic Fuel Industry," *Isis* 75: 4 (1984), 663–5.

156 More, *Black Gold*, 62–5. Payton-Smith, *Oil*, 19–24; Anthony Stranges, "From Birmingham to Billingham: Synthetic Fuels in Great Britain, 1910–1945," *Technology and Culture* 26 (1985): 726–57.

157 Brown, "Fuel Supplies," 245ff.

158 Wilkinson, "Influence of Oil [. . .]," *RUSI* 68: 469 (1923): 109–14.

159 Ross, "Imported Oil or Native Coal?" *RUSI* 72: 485 (1927): 137–43.

160 Brame, "Power Fuel [. . .]," *RUSI* 76: 501 (1931): 64–77.

161 Ammon, "Coal or Oil [. . .]," *RUSI* 76: 503 (1931): 634–7.

162 Skelton, "Coal versus Oil [. . .]," *RUSI* 79: 514 (1934): 241–55; Brown, "Fuel Supplies," 253–7.

163 Annexure B to: Thomson et al., "Fifth Annual Report," July 1, 1930, O.B. 51, CAB 50/3; Thomson, "Production [. . .]," July 22, 1930, C.P. 267 (30), CAB 24/214.

was likely to become permanent. A protected synthetic fuel industry would also harm British oil companies and related industries such as shipbuilding.[164]

German analysts envied Britain's luxury of choice—Britain could afford to import oil since, compared to Germany, the "problem of foreign exchange" was "unknown" there.[165] Developing a synthetic fuel industry was not simply an economic matter—it also affected the strategic position of the British Empire. The former Managing Director of Shell Transport and Trading, Andrew Agnew, pointed out in 1935 that producing enough gasoline from coal to cover Britain's annual requirements would require an expenditure of £160 million, plus the loss of an £40 million in tax revenues. Domestic production of synthetic fuel also did nothing to satisfy the needs of the rest of the British Empire. Synthetics were a terrible investment when, for a "mere fraction of the cost," the government could instead "provide defensive armaments in a the way of cruisers and aircraft ample to protect not only our own supplies of oil, but those required for the peoples of the whole Empire."[166]

Starting in 1928, Whitehall began placing duties on imported gasoline, which rose to 9*d* by 1938, and in 1934, the government passed the Hydrocarbons Oil Production Act, which granted synthetic producers preferential treatment vis-à-vis imported petroleum cumulatively worth 36*d* over nine years. In October 1935, Imperial Chemical Industries constructed a plant in Billingham to liquefy coal into 100,000 tons of petroleum per year.[167] The question whether the government should adopt a more-active policy was only resolved in 1937, when a CID subcommittee studied the economic and strategic viability of synthetic fuel. The so-called Falmouth Committee doubted that synthetic production could rise quickly during an emergency, whereas importing additional oil was straightforward. Imported supplies were also less vulnerable thanks to Britain's naval strength, and domestic stockpiles could be concealed. The same did not apply to hydrogenation plants, which were "extremely vulnerable to air attack." The most damning factor was the cost. One synthetic fuel plant capable of producing 150,000 tons of fuel a year would cost £8 million to construct. Although it would employ 6,000 workers directly or indirectly, it did so at a cost of £5 per man each week, which was greater than the average weekly wage.[168]

Whitehall concurred with the committee's findings. It was willing to retain duties on imported petroleum to fund research in synthetic fuel—but nothing more.[169] Unlike in Germany, in Britain the debate over whether to sponsor a synthetic fuel industry focused on commercial considerations. Strategic factors were not absent but they were of secondary importance, for several reasons. Unlike other oil-poor

[164] The arguments for and against may be reviewed in T 160/536.

[165] M.B., "Öl aus Kohle in England," *Vierjahresplan*, 1937/VII.

[166] Agnew, "Empire Oil Supplies [...]," *RUSI* 80: 518 (1935): 278–92.

[167] Payton-Smith, *Oil*, 21–2; Stranges, "Bergius," 664–5. German analysts were unimpressed by the quality of work at Billingham. M.B., "Öl aus Kohle in England," *Vierjahresplan*, 1937/VII.

[168] CID, "Production [...]," November 29, 1937, 272-A, attached to: Inskip, "Report [...]," February 3, 1938, C.P. 19 (38), CAB 24/274. See also: Brown, "Fuel Supplies," 252–3.

[169] Crookshank, "Consideration [...]," May 10, 1938, enclosed with: Stanley, "Production [...]," May 13, 1938, C.P. 114 (38), CAB 24/276.

industrial nations, Britain enjoyed relatively easy access to foreign oil thanks to its maritime power. The fact that British companies owned many of these oilfields allowed Britain to pay for much of its oil imports in sterling or earn foreign exchange. Britain, unlike Germany, was not an aggressive power bent on a program of conquest that would threaten its access to overseas imports. Finally, autarky was never a viable option for Britain. Even if synthetic fuel had proven commercially viable, oil was one of only many items Britain had to import. No matter what, Britain's prospects during either European or Far Eastern war would still depend upon preserving its sea lines of communication.

That task became harder after 1939. Whitehall was no longer planning for an either/or situation. Britain could have coped with the closure of the Mediterranean in a European war, but the task of waging a global war in Europe and the Pacific simultaneously revealed its inability to guarantee the energy needs of the empire without U.S. support. During its 1937 planning for a war against Japan, PPR had not considered the possibility that the Mediterranean might be "closed."[170] The closure of the Mediterranean during a war against Japan "could only arise from the intervention of a hostile European Power"—a "dual contingency" involving simultaneous conflicts against Germany and Japan. In February 1938, the CID abandoned assessing Britain's oil position during a European war and a Far Eastern war separately.[171] The Oil Board could not start planning for a "dual contingency" until the other CID subcommittees drafted their requirements, and the necessary authorization only arrived in March 1938.[172]

By the beginning of 1939, Britain's predicament was serious. Japan had overrun much of northeast China in 1937–8, and it and Germany were poised to go on the offensive in Southeast Asia and Europe. Italy had already broken with Britain and France over Abyssinia in 1935. It joined Germany in sending aid to the Nationalists in the Spanish Civil War in 1936, abandoned its defense of Austrian sovereignty, and signed the Anti-Comintern Pact with Germany and Japan in 1937. In April 1939, the Nationalists triumphed in the Spanish Civil War. During the conflict, the Nationalists, backed by German and Italian air and naval forces, blockaded Spain's Mediterranean coast.[173] If Spain formed an alliance with Germany and Italy, it would jeopardize British control of Gibraltar as well as access ingress and egress from the Mediterranean. Finally, U.S. material assistance to Britain and France became doubtful in July 1939, when the Roosevelt Administration failed to repeal the embargo on arms exports and place all trade with belligerents on a "cash and carry" basis. Following the Sudeten crisis, Roosevelt had adopted a more-aggressive policy. Besides asking for funds for rearmament, he wanted to eliminate the 1937 Neutrality Act's arms embargo, which he and his supporters argued only benefited aggressor nations. Unfortunately, his administration mishandled

[170] Crookshank, "Sources [...]," May 26, 1937, O.B. 195, CAB 50/6.

[171] The following discussion draws from: Payton-Smith, *Oil*, 45–74.

[172] Crookshank, "Sources [...]," February 8, 1938, O.B. 233, CAB 50/6; Nicholl, "Sources [...]," March 15, 1938, O.B. (P.R.) 118, CAB 50/16.

[173] Michael Alpert, "The Spanish Civil War and the Mediterranean," *Mediterranean Historical Review* 13: 1–2 (1998): 150–67.

Congress, which allowed "cash and carry" to expire in May but retained the arms and loan embargos. Roosevelt only rectified this blunder in November, after the outbreak of World War II created additional political support for revision. While he managed to eliminate the arms embargo and restore "cash and carry," the prohibition on loans or shipping goods to belligerents on U.S. ships remained.[174]

British oil-supply planning papers reflected the deteriorating strategic outlook. The armed services revised requirements in April and July 1938. The Admiralty estimated the Royal Navy's requirements at 8.68 million tons with the Mediterranean "open" or "closed." The Admiralty also requested that its fuel oil reserves increase from 7 to 8.5 million tons, and that this reserve should never fall below 50 percent capacity.[175] Accordingly, the Petroleum Department estimated civilian and military requirements throughout the empire in a "dual contingency" at 28.56 million tons, against 28.34 million tons in a European war and 29.17 million tons in a Far Eastern war. Both the Admiralty and the army had higher oil requirements during a "dual contingency" than a one-ocean war, while the RAF required 10,000 tons less in a "dual contingency" than during a European war, since it would pursue a defensive strategy in the Far East. Consumption in Britain was identical during either a "dual contingency" or a European war (8.95 million tons), and higher during a Far Eastern war (11.42 million tons) due to the absence of rationing. Consumption throughout the rest of the empire was identical during either a "dual contingency" or a Far Eastern war (9.36 million tons), and almost one million tons less than during a European war (10.34 million tons).[176] As for supplies, higher production in Venezuela would offset the loss of Mexico, while the possibility of an "unfriendly" United States was exceedingly "unlikely." Provided these assumptions held, Britain would manage even with a "closed" Mediterranean.[177]

The PPR presented its findings to the CID in December 1938. After rationing, civilian requirements in Britain had dropped to 18.32 million tons. Import requirements for the empire came to twenty-five million tons. Iran, the United States, and Venezuela would be the most important suppliers, as shown in Table 3.4.

Table 3.4. Major oil suppliers in "dual contingency"

Source	Mediterranean "open"	Mediterranean "closed"
Iran	8,900,000 tons	7,200,000 tons
United States	9,300,000 tons	10,600,000 tons
Venezuela	4,600,000 tons	4,300,000 tons

[174] David Reynolds, *From Munich to Pearl Harbor: Roosevelt's America and the Origins of the Second World War* (Chicago: Ivan R. Dee, 2001), 42–68.

[175] "Navy's Oil Requirements [...]," July 9, 1938, O.B. (P.R.) 144, CAB 50/16. The other services' requirements—O.B. (P.R.) 128 (dated April 13, 1938) and O.B. (P.R.) 129 (April 29, 1938)—are also included in this file.

[176] In the Oil Board's final annual report, the figures were slightly revised: 28,625,000 (European war), 30,298,400 tons (Far Eastern war), and 29,109,360 tons ("dual contingency"). Llewellin et al., "Thirteenth Annual Report," January 24, 1939, O.B. 294, CAB 50/7.

[177] Petroleum Department, "Estimated Requirements [...]," no date (circa July–December 1938), O.B. (P.R.) 146, CAB 50/16.

The supply of aviation fuel was tight—more than half of the empire's requirements (340,000 tons) would come from the refineries of the Dutch West Indies. The tanker situation had improved due to the decision to draw additional imports from the Western Hemisphere. The number of tankers required with an "open" Mediterranean was now 364 or 332 if it was "closed." The PPR expected to have a comfortable surplus (114 tankers) irrespective of the state of the Mediterranean.[178]

The productive capacity of the Middle East was waxing even as its strategic value was waning. Northbound oil shipments through the Suez Canal had exceeded southbound ones starting in 1914. Initially, much of the northbound traffic came from the East Indies, but by 1922, 88 percent of the tanker tonnage originated in the Persian Gulf (75 percent from Persia). Four years later, oil accounted for the largest share of northbound shipments (19.35 percent), and by 1930, oil tankers comprised 16.5 percent of the canal's throughput. The greatest single factor behind these developments was Iranian oil, half of which traveled through the canal.[179]

The rise in Iranian production during the 1920s and 1930s was so great that by 1937 the Persian Gulf was the busiest oil-transit artery in the Eastern Hemisphere, handling 15 percent of all overseas exports (twelve million tons).[180] To process the increased output, the capacity of the Abadan refinery increased to 13–14 million.[181] Capacity was so high that the refinery could not operate at 100 percent if the Mediterranean was "closed," as there was no way to allocate the oil. It was theoretically possible for Britain to eliminate most of its requirements from the Western Hemisphere by running Abadan at 100 percent, but this would require that Britain import 4.62 million tons of Iranian oil on an additional twenty-five tankers if the Mediterranean was "open" and 3.31 million tons on 105 tankers if it was "closed." In case of the former, the tanker surplus would be eighty-nine, but during the latter, it would only be nine. What this meant in practice was that, despite two decades of effort, the United States would remain Britain's largest supplier in wartime. The U.S. and Venezuelan allocation with an "open" Mediterranean exceeded Iran's allocation by five million tons (13.9 million tons versus 8.9 million) and was twice as large if the Mediterranean was "closed" (14.9 million tons versus 7.2 million tons).[182]

While the Western Hemisphere offered a solution to the perennial shortage of tankers and the logistical headaches created by Italian hostility, it created an entirely new challenge for which Whitehall was completely unprepared—that of payment. The impossibility of finding a balance between transporting oil it could not afford and paying for oil it could not move not only proved the death knell for Britain's oil strategy but also accelerated the collapse of the empire as an autonomous actor in great power politics.

[178] Crookshank, "Sources [...]," December 23, 1938, O.B. 290 (Revise), attached to: Llewellin, "Sources [...]," February 22, 1939, O.B. 304, CAB 50/8.

[179] Farnie, *Suez*, 566–72 and 614.

[180] "Ölwirtschaft [...]," *Vierjahresplan*, 1938/X.

[181] Bamberg, *History*, 69–75.

[182] Crookshank, "Sources [...]," December 23, 1938, O.B. 290 (Revise), attached to: Llewellin, "Sources [...]," February 22, 1939, O.B. 304, CAB 50/8.

4

The Price of Failure
1939–1942

The British strategy after 1919 was essentially negative—to preserve the valuable, sprawling empire it had previously acquired at such great cost. Whitehall's challenge was to figure out how to defend these gains against multiple enemies in different theaters in an era of constrained resources. Air and naval power were indispensable to this task—but planes, ships, and expeditionary forces required infrastructure and, most importantly, oil.

On the eve of World War II, Whitehall could take some solace in developing a variety of new sources of oil under British commercial control and securing a position of virtual parity with the United States around the world in ownership of proven reserves (see Illustration 9). Although the Americans remained ascendant in the Western Hemisphere, the British position in the Middle East was strong (see Illustrations 10 and 11). While Bahrain and Saudi Arabia had slipped from their fingers, British firms still had a monopoly over production in Iran, a half-share in Kuwait, and more than a quarter-share in Iraq (see Map 5). Whitehall had also shepherded the growth of one major British oil company (Anglo-Iranian) while deepening ties with another (Shell) to challenge U.S. companies in global markets. British strategy nevertheless fell short of the objective Whitehall had established in 1918–19: freeing Britain from foreign-controlled sources of oil. In fact, Britain's position in 1939 was in many respects inferior to that of 1914.[1]

This failure had less to do with geology than geography. Britain's ideal source of supply from an imperial and financial perspective was the Persian Gulf, which was 1,500 nautical miles farther away than the Gulf of Mexico. Ideally, Britain would pipe the Iranian and Iraqi oil to Haifa to save on tanker tonnage and Suez Canal tolls, but this assumed both the security of the Iraq Petroleum Company (IPC) pipelines running from Iraq to Palestine or Lebanon and safe passage through the Mediterranean. If the latter was closed, however, the cost of shipping Iranian or Iraqi oil around the Cape would be greater than from any other major oil-export terminal in the world.

The British also based their oil strategy on the premise that there would not be any threat to their supply lines. This was a reasonable proposition in 1919 but not after 1935. As a result, the elaborate infrastructure Britain had created to satisfy its oil requirements—powerful oil companies supplied from a British-dominated Middle East—proved less useful than its friendship with the United States.

[1] Payton-Smith, *Oil*, 24–5.

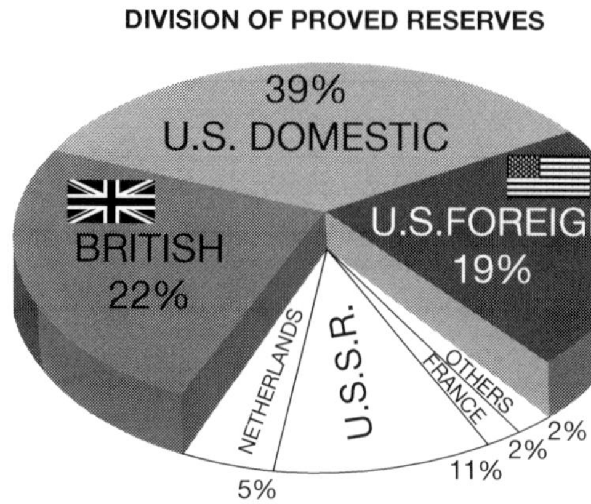

Illustration 9. "Division of Proved Reserves," February 22, 1945
Source: LOC, Harold L. Ickes Papers, Box 221.

A HOUSE BUILT ON SAND, 1939–1941

Whitehall had spent two decades ensuring that British firms were poised to reap the benefits of the Middle East's oil windfall. While its efforts had succeeded in preserving Anglo-Iranian's independence and safeguarding its position in Iran and Iraq, the company contributed little to the Allied war effort west of Suez following the closure of the Mediterranean in 1940, precisely because its only oilfields were in the Middle East.[2] Escalating tanker losses in 1940–1 forced Britain to readopt the "short-haul" policy in March 1941 and shift all tankers traveling between the Persian Gulf and Britain via the Cape to the transatlantic route. This decision slashed Iran's oil exports to Britain from 1.53 million tons in 1940 to 324,000 tons by August 1941, when Britain and the Soviet Union occupied Iran.[3]

Shell was a far more valuable partner. On the eve of the war, it was cooperating with Whitehall to construct another refinery to produce aviation fuel, and one executive promised the group would circumvent any U.S. legislation restricting exports to Britain.[4] By 1940, Shell produced more than thirty-one million tons annually, which was larger than the British Empire's projected consumption during the first year of either a European or a Far Eastern war. Unlike Anglo-Iranian, Shell was not restricted to the Eastern Hemisphere, possessing secure oilfields in the United States and Latin America with tanker requirements significantly lower than oilfields in the Middle East. Shell also owned thirty-six refineries with an annual output of 35.75 million tons, controlled the world's largest tanker fleet, and was the largest supplier of high-octane aviation fuel. The group repatriated 97 percent of its

[2] The Admiralty closed the Mediterranean to Allied shipping in August 1939. Two weeks after the invasion of Poland, it lifted the ban, but the collapse of French resistance in 1940 forced it again to reroute traffic around the Cape. The Allies did not reopen the Mediterranean until after July 1943. Bamberg, *History*, 232.

[3] Bamberg, *History*, 216–20; Payton-Smith, *Oil*, 177–80.

[4] Kent, *Moguls*, 153.

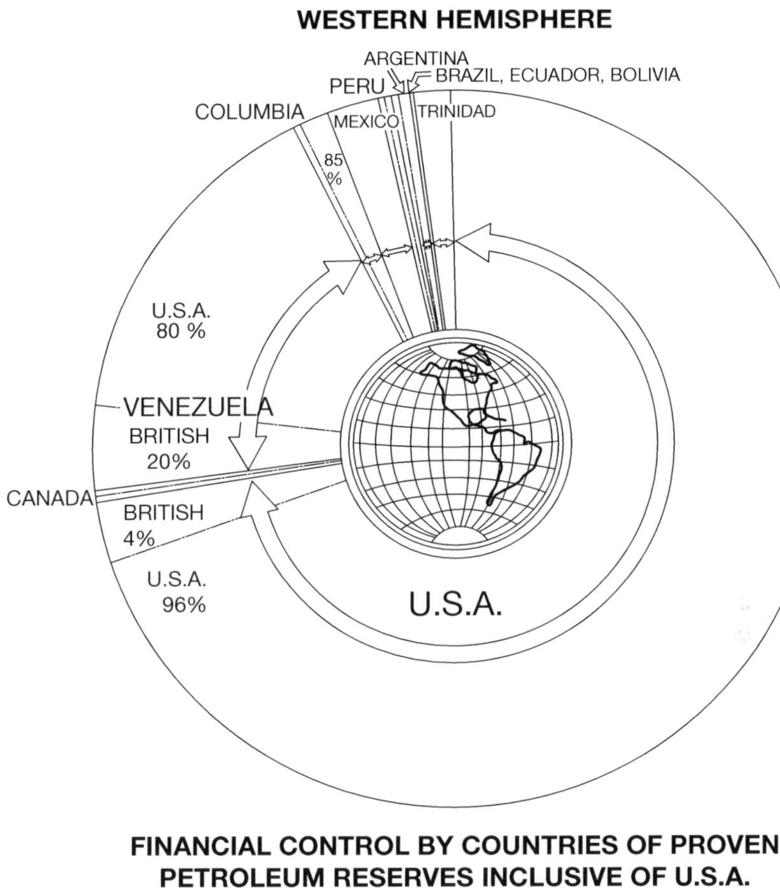

Illustration 10. "Western Hemisphere: Financial Control by Countries of Proven Petroleum Reserves Inclusive of U.S.A.," no date (circa 1938)
Source: NARA, RG 107, Entry 141, Box 251.

earnings to London—£260 million per annum by 1940—and assured Whitehall that it was maximizing purchases within the sterling area, members of which abided by British exchange controls, thus easing the pressure on British foreign exchange reserves. The remaining 3 percent went to the Royal Dutch shareholders as dividends.[5] Shell, however, was overselling its utility—any firm that operated beyond the sterling area would have at least some foreign exchange costs. Before World War II, roughly 60 percent of the British purchases of oil equipment and materials, as well as royalties on U.S. patents and wages for U.S. employees, were denominated in dollars. In Venezuela, Shell needed $58 million in the twelve

[5] Shell, "Memorandum," January 16, 1940, and "Shell Memorandum," January 18, 1940, both in: CAB 63/117.

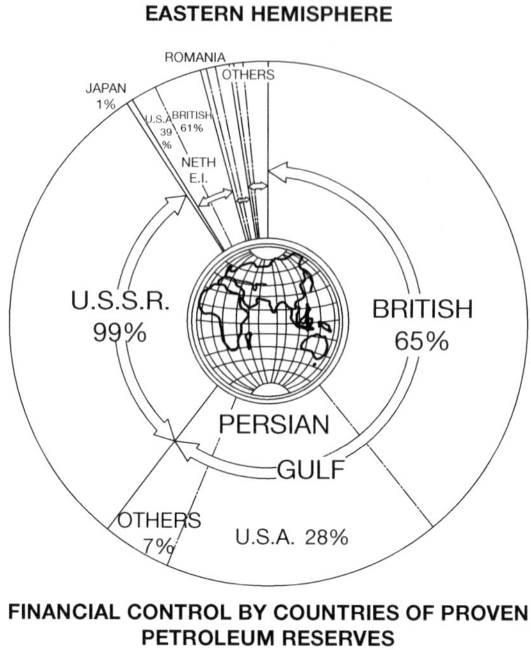

Illustration 11. "Eastern Hemisphere: Financial Control by Countries of Proven Petroleum Reserves," no date (circa 1938)
Source: NARA, RG 107, Entry 141, Box 251.

months after June 1940—half of Britain's dollar expenditures on purchases from U.S. firms.[6]

Whitehall's preoccupation with Shell's "foreign" ownership was misplaced. Despite some resistance to giving British employees executive power during an emergency, the Dutch management and employees were committed to the Allied cause. Britain's enemies never doubted where the group's loyalties lay. Royal Dutch began purging its pro-German staff before Germany invaded Holland in May 1940, and few of the Dutch personnel left in Europe collaborated.[7]

This should not imply that the oil companies were government stooges. Whitehall granted the companies substantial autonomy to guarantee adequate supplies for civil and military consumption in exchange for relatively higher costs, but it received little in return that the companies did not consider in their interest to give.[8] During the 1930s, the government failed to convince the companies to undertake precautions such as constructing protected storage.[9] Even Anglo-Persian/Anglo-Iranian did not have any particular affection for its majority shareholder. The company billed British consumers the same inflated prices and freight rates that it charged

[6] Payton-Smith, *Oil*, 149–50.

[7] Howarth and Jonker, *History of Shell*, 10–11, 31–4, and 78–86.

[8] T.A.B. Corley, "Oil Corporations and Public Policy: A US–UK Comparison, 1900–1975," *Business and Economic History* 21 (1992): 142ff.

[9] Coordination did improve following the Sudeten crisis. Kent, *Moguls*, 152.

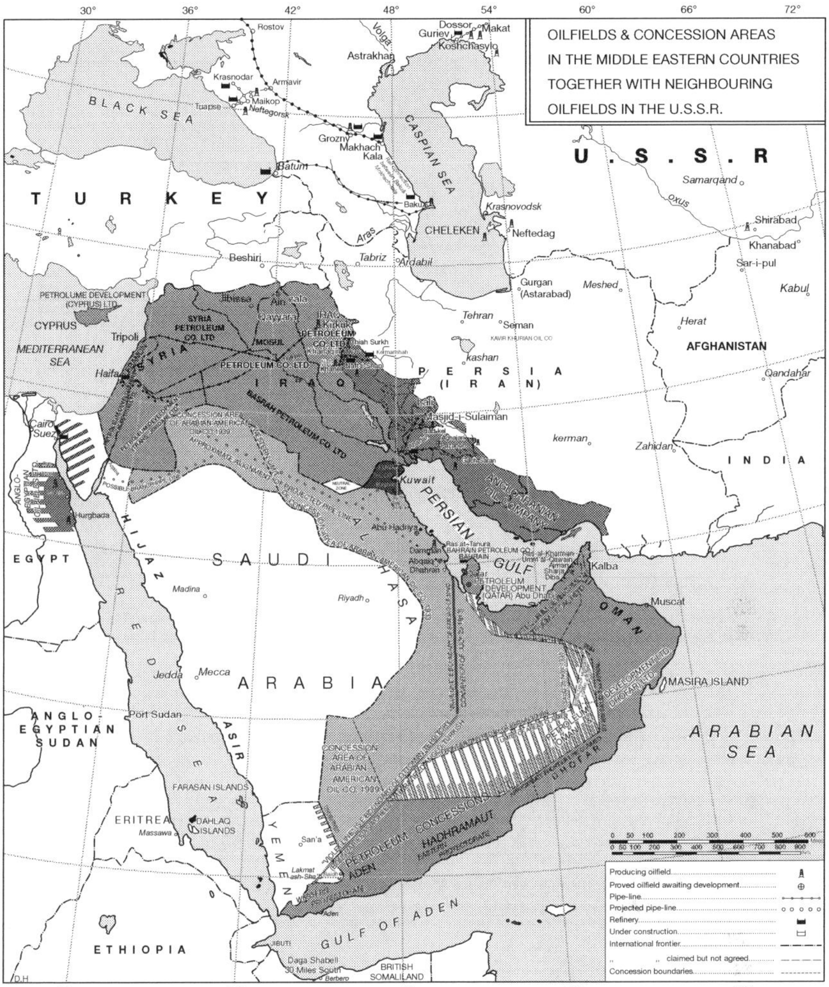

Map 5. "Oilfields & Concession Areas in the Middle Eastern Countries Together with Neighbouring Oilfields in the U.S.S.R.," March 1945

Source: British National Archives, FO 371/45274.

all others until Whitehall found out during World War II.[10] Far from challenging Shell and Jersey, Anglo-Persian joined them in 1928 by signing the Achnacarry Agreement, as well as partnering with Shell in India and Africa. In 1931, it even merged its domestic marketing operations (British Petroleum) with Shell's primary marketing subsidiary (Shell-Mex).[11]

[10] Blair, *Control*, 113–14.

[11] Bamberg, *History*, 183–8. By 1938, three firms accounted for 85 percent of all downstream operations in Britain—Shell-Mex: 40 percent; BP: 15 percent; Jersey's subsidiary: 30 percent. Payton-Smith, *Oil*, 43–4.

Even Anglo-Persian's relations with its closest customer, the Admiralty, were not always smooth. The Admiralty often crowed about its warm relationship with the company, in contrast with its "businesslike" one with Shell.[12] This was a fantasy, for Anglo-Persian never went beyond fulfilling its contractual obligations, and even then, it could be recalcitrant. For instance, the company resented having to pay a rebate after oil prices fell in the mid-1920s. Conversely, the Admiralty was convinced Anglo-Persian was using duplicitous bookkeeping to depress its profits. One official fumed in 1925 "that the Company are exerting every effort and seeking very possible device to pare down the rebate which is properly due to the Admiralty." It was only in 1928 that the two sides agreed to abandon the rebate in exchange for a lower price for fuel oil.[13] Thereafter, Anglo-Persian and the Admiralty butted heads over the route of the IPC pipeline and the company's desire to cooperate with the Americans in Kuwait.

Overall, Anglo-Persian/Anglo-Iranian and Shell each offered tangible benefits but also considerable disadvantages to Whitehall. The former was flush with oil from territories under British influence, but the logistical requirements were excessive. The latter compensated by offering oil that was relatively cheaper to transport but had a larger foreign exchange cost. In other words, the companies' contribution mirrored precisely the relative strengths and weaknesses of the Middle East and the Western Hemisphere as British oil suppliers. Juggling Britain's tight supplies of tankers and foreign exchange would have been a difficult task in peacetime—and it proved impossible under the exigencies of war without U.S. assistance.

Britain ended up needing far more than just tankers from the United States. At the beginning of 1939, the Oil Board estimated that in wartime the empire would have to import twenty-five million tons of oil a year, 90 percent of which came from three sources: Iran, the United States, and Venezuela.[14] The number of consumers had increased exponentially since the last war—not just because of the emergence of new weapons platforms, but also as a result of the continuing conversion of coal-fired ships to oil.[15] Assuming Britain retained access to its foreign supplies, there was still the matter of paying for and transporting them. Britain entered the war with over 25 percent of the world's tanker tonnage (3.01 million tons out of 10.72 million tons), with much of the remainder held by friendly neutrals—primarily the United States, at 2.76 million tons, and Norway, at 1.98 million tons.[16] It took the opportunity to prioritize imports through the Mediterranean while it was still open. Losses were initially low, but this did not last.[17] Between September 1939 and January 1945, Anglo-Iranian lost forty-four tankers to enemy action (46 percent

[12] Note to Amery, "Mr. Churchill's Replies [...]," November 14, 1923, ADM 1/8658/55.

[13] Brown, "Fuel Supplies," 185–7; Jones, *Industry*, 227–8 and 238–9.

[14] Crookshank, "Sources [...]," December 23, 1938, O.B. 290 (Revise), attached to: Llewellin, "Sources [...]," February 22, 1939, O.B. 304, CAB 50/8.

[15] During World War I the bulk of the merchantmen sailing to Britain were coal-fired. By 1937, half of the world's merchant fleet was oil burning. Payton-Smith, *Oil*, 46–7.

[16] Llewellin et al., "Thirteenth Annual Report," January 24, 1939, O.B. 294, CAB 50/7.

[17] German Embassy, Washington, to Foreign Office, "Amerikanische Stimmen [...]," February 22, 1940, BA-B, R 901/116640.

of its prewar fleet).[18] Shell Transport and Trading lost sixty-six tankers (40 percent of its prewar fleet), while Shell lost forty-seven ships, which left it with a smaller fleet at the end of the war than at the beginning.[19]

The problem of payment was truly pernicious. Britain's financial position depended upon consuming oil denominated in sterling—primarily from the Middle East. The only alternative was dollar oil controlled by U.S. firms. Strangely, the question of how Britain should pay for dollar oil was not resolved before World War II.[20] As early as 1938, half of Britain's oil imports (by value) required payment in hard currency—£27,380,000 of £46,040,000.[21] Every dollar Britain spent on oil was in turn unavailable for other necessary imports, and Britain could not fight a war without imports. Whitehall therefore had to conserve foreign exchange. Despite the downturn in global trade, Britain increased its hard-currency reserves from £121 million in 1931 to £825 million six years later thanks to capital flight from Europe.[22] Starting in spring 1938, however, reserves declined by 17 percent in just six months thanks to capital flight and rearmament. Under U.S. pressure, Whitehall sacrificed hundreds of millions to preserve sterling's value, but by August 1939, the pound had depreciated by $1, while Britain's foreign exchange reserves declined by half. This crisis destroyed Britain's ability to finance a protracted conflict against the Axis without U.S. financial assistance.[23] Much of the gold and foreign exchange that might have enlarged Britain's war-making capacity had migrated, instead, to the United States to build factories that sold war materiel to Britain.[24]

The writing was on the wall before the invasion of Poland, as rearmament began undermining Britain's economic stability.[25] In July 1939, the Treasury had concluded that the basis of Britain's strategy against Germany—slow economic strangulation—was flawed. Britain "could only afford a *short* war," while "the Chiefs of Staff stated she could only win a *long* one."[26] At war's outbreak, the Treasury warned that Britain's supply of foreign exchange and securities would cover expenditure for a force of fifty-five divisions and 2,500 aircraft for only two years.[27] The failure to reconcile the burden of rearmament with Britain's financial position before

[18] Bamberg, *History*, 216. [19] Howarth and Jonker, *History of Shell*, 57.

[20] Payton-Smith, *Oil*, 147–8. [21] Eady to Padmore, "Oil," June 13, 1944, T 161/1195.

[22] Murray, *Balance*, 52–5, and "Collapse," 403–4.

[23] Paul Kennedy, "Strategy versus Finance in Twentieth-Century Great Britain," *International History Review* 3: 1 (1981), 54–7; R.A.C. Parker, "The Pound Sterling, the American Treasury and British Preparation for War, 1938–1939," *English Historical Review* 98: 387 (April 1983): 261–79; Murray, *Balance*, 72–3.

[24] Christopher Price, *Britain, America and Rearmament in the 1930s: The Cost of Failure* (Basingstoke: Palgrave, 2001), 184 and 186–8.

[25] G.C. Peden, "A Matter of Timing: The Economic Background to British Foreign Policy, 1937–1939," *History* 69: 225 (1984): 15–28.

[26] Correlli Barnett, *Strategy and Society: The Spenser Wilkinson Memorial Lecture, 1974* (Manchester: Manchester University Press, 1975), 7. Emphasis in the original. See also: R.A.C. Parker, "Economics, Rearmament and Foreign Policy: The United Kingdom before 1939—A Preliminary Study," *Journal of Contemporary History* 10: 4 (October 1975): 637–47. For discussion of the Anglo-French strategy of economic strangulation, see: Murray, *Balance*, 311–14.

[27] Parker, "Pound Sterling," 262–4.

September 1939 was "perhaps [the] most egregious error" in British planning before World War II.[28]

Throughout the interwar period, British officials took for granted that, even if the United States did not supply Britain directly, it would take care of Canada and not disrupt exports from other Western Hemispheric suppliers. As war drew near, the British assumed in their planning for a "dual contingency" that the United States would be "friendly." Although Washington was incapable of providing financial assistance due to the Johnson Act, Foreign Secretary Lord Halifax was confident in the summer of 1939 that "the attitude of the United States would be sufficiently favorable to us to enable us to win the war."[29] Halifax's prediction proved prescient, but everything depended upon Roosevelt occupying the White House and using his considerable political skills to outmaneuver isolationists.[30] The price Whitehall paid, moreover, was steep: the forfeiture of Britain's strategic independence to the United States, which began in earnest in 1941 with the United States using Lend-Lease aid repayment to smash Imperial Preference (the policy since 1932 of stimulating trade within the empire through reciprocity or free trade agreements while discriminating against foreigners) and the sterling area.[31]

Whitehall should have been better prepared. Despite London's position as the center of global finance, the British had devoted little thought to the financial demands of waging a protracted great power conflict before 1914.[32] This lack of preparation was nearly catastrophic. The Allied war effort during World War I depended on imported U.S. finished goods and raw materials, including oil. The cost of meeting its own requirements and serving as the Allies' banker strained Britain's finances to the breaking point, such that Britain became dependent upon U.S. private, then public financing.[33] Rather than heed their experience and husband their financial reserves, the British squandered their foreign exchange cushion between 1938 and 1940, and they reached the end of their financial tether even faster than during World War I. Once again, U.S. economic and industrial power rescued Britain after its financial reserves had evaporated. The sterling area's dollar expenditure on oil products, services, and equipment between 1941 and 1945 amounted to \$505.4 million. This was dwarfed by the amount supplied under Lend-Lease: \$320 million in 1941–2 and \$1.5 billion just to Britain by June 1945.[34]

In some respects, "the lessons of the First World War had been learned and absorbed." The planning and coordination process underwent a welcome degree of

[28] Before the passage of the Lend-Lease Act in 1941, British hard-currency reserves stood at \$12 million, thanks only to Canadian charity. Kennedy, "Net Assessment," in: Millet, Calculations, 54–6; "Finance," 60.

[29] Parker, "Economics," 644–5.

[30] Roosevelt intimated in the spring of 1939 that he planned to render financial assistance to the Allies. Parker, "Pound Sterling," 276–8.

[31] Warren F. Kimball, *The Juggler: Franklin Roosevelt as Wartime Statesman* (Princeton: Princeton University Press, 1991), 43–61; Kathleen Burk, "American Foreign Policy and Lend Lease," in *The Rise and Fall of Lend Lease*, ed. Anne Lane and Howard Temperley (New York: St. Martin's, 1995), 43–68.

[32] Kennedy, "Finance," 49–50.

[33] Kathleen Burk, "Great Britain in the United States, 1917–1918: The Turning Point," *International History Review* 1: 2 (1979): 228–45.

[34] Payton-Smith, *Oil*, 195–8 and 467–72.

"professionalization," and the Oil Board transitioned effectively from the benign geopolitical circumstances of the 1920s "to the more complex threat environment of the 1930s."[35] It identified the three major "variables" affecting Britain's oil supply: the status of the Mediterranean, the attitude of the United States, and the demands of fighting wars simultaneously in Europe and the Pacific.[36] When the war finally came, the British transitioned smoothly to a new administrative structure. In January 1939, the Committee of Imperial Defence (CID) decided that, in wartime, the Oil Board's executive functions would migrate to the Petroleum Department.[37] In November, however, Prime Minister Chamberlain established the Oil Control Board, staffed by civilian and military experts.[38] The board never functioned as a policymaking agency. Rather, it delegated tasks to expert subcommittees and adjudicated between competing parties.[39]

Wars, however, are not won by administrative reorganizations. The true measure of Britain's prewar strategy was how effectively it fulfilled Allied requirements in a global conflict against multiple adversaries. Within this context, the reality of Britain's failure becomes evident. In January 1940, the Petroleum Department estimated that half of Allied consumption between December 1, 1939 and November 30, 1940 (21.53 million tons) would come from just the United States and Venezuela.[40] Of the oil Britain had imported by mid-1940, 60 percent was priced in sterling and the remainder in dollars. Although Whitehall tried to prioritize imports of sterling oil at the war's outbreak, the tanker shortage prevented this. As one Admiralty official remarked in December 1940, "It is the tanker tonnage situation that decides what we *have* to draw from other [non-sterling] sources."[41] Consequently, the sterling area's oil-related expenditures of dollars and Dutch guilders during the first six months of the war (not including freight) reached a rate of $200 million per annum.[42]

Following the surrender of France, the British foreign exchange position became truly disastrous. In February 1940, the Treasury estimated that Britain's dollar reserves could last two years, but by August it concluded that expenditures over the coming year would reach $32 billion—against only £490 million in reserves and U.S. securities.[43] To make matters worse, the British could not curtail purchases of dollar oil—now costing an estimated $170.5 million for the twelve months after September 1940—due to the tanker shortage. All the British could do was pray for Roosevelt's reelection and that their hard-currency reserves did not evaporate before Washington came to the rescue. Whitehall's hands were tied even if it tried to substitute sterling oil for dollar oil. The Bank of England warned in November 1939 that the major oil companies "would not co-operate willingly," and the possible

[35] Babij, "Oil Supply," 101; More, *Black Gold*, 65–8. [36] Kent, *Moguls*, 148–9.

[37] Llewellin et al., "Thirteenth Annual Report," January 24, 1939, O.B. 294, CAB 50/7.

[38] Chamberlain, "Oil Control," November 9, 1938, (G) (39) 78, CAB 67/2/29. The Petroleum Department (renamed the Petroleum Division in 1942) moved to the Board of Trade in 1940 and then the Ministry of Fuel and Power two years later. Payton-Smith, *Oil*, 114–18; Bamberg, *History*, 206–10.

[39] Payton-Smith, *Oil*, 80–1, 120, and 251–2.

[40] Petroleum Department, "Allied Requirements of Petroleum," January 17, 1940, BP 37134.

[41] Coxwell to Rowan, December 14, 1940, T 161/1425. Emphasis in the original.

[42] Payton-Smith, *Oil*, 147–53. [43] Barnett, *Collapse*, 12–14.

savings in 1940/1—$50 million at most—would not make much of a dent in its current account deficit. Even the modest step of substituting purchases of U.S. equipment with British goods was impossible, as British industry was incapable of meeting the additional demand. Not surprisingly, Britain's share of the industrial requirements of its oil companies dropped during the war, from 40 percent to 10 percent.[44]

Britain's outlook would have been brighter had France not succumbed and Italy not entered the war in June 1940. Perhaps the Allies would have been better off launching a preemptive attack against Italy before May 1940, but the Allies' preference for Italian neutrality was not mere pusillanimity—a belligerent Italy left Britain and France without a secure line of communication to the Middle East. When the Royal Navy finally took action in 1940, its successes backfired. Starting in 1941, modest German forces, with Italian assistance, pushed the British back into Egypt and chased the Royal Navy to its harbors. While it is true that Germany provided significant amounts of economic assistance to Italy, especially petroleum—3.57 million tons between 1940 and 1943, or roughly 67 percent of Italian military consumption during that period—the Mediterranean also soaked up British resources required in other theaters, especially the Far East.[45]

By 1939, the British could not even guarantee the security of one of their most important suppliers in the Western Hemisphere. Half of Britain imports originated in the Caribbean—some of which could be purchased in sterling—as well as significant refining capacity, but British defenses in the region were woefully inadequate, and the Admiralty had to suggest that the U.S. Navy undertake the defense of Aruba.[46] The security of the West Indies took on an additional degree of urgency in May 1940—the Chiefs of Staff now agreed with the British Ambassador in Washington that the United States should "share in the occupation" of Aruba and Curaçao.[47] Luckily, Hitler was determined not to provoke the United States until he had achieved his objectives in Europe and forbade the German Navy from attacking the refinery at Aruba in May 1940.[48]

Venezuela and the United States allowed Britain to write off the Middle East once it no longer had to choose between saving tanker tonnage or dollars.[49] Although the United States was still the world's largest oil producer on the eve of World War II, the supply/demand balance was not as favorable as before World War I. Reserves had almost trebled between 1918 and 1940, but production had quadrupled during that period.[50] As a result, the share of U.S. production

[44] Payton-Smith, *Oil*, 147–53.

[45] Gerhard Schreiber et al., *Mediterranean, South-East Europe, and North Africa, 1939–1941*, vol. 3 of *Germany and the Second World War*, ed. Militärgeschichtliches Forschungsamt (New York: Oxford University Press, 1995), 25–39; Simpson, "Superhighway," 61ff.

[46] "Oil Supplies [...]," no date, I.D.C. 9, attached to: Binney, "Defence [...]," October 6, 1939, ADM 1/9982.

[47] Pound et al., "Protection [...]," no date, W.P. (40) 148, CAB 66/7/28.

[48] Holger Herwig, "Prelude to *Weltblitzkrieg*: Germany's Naval Policy toward the United States of America, 1939–41," *Journal of Modern History* 43: 4 (1971): 653.

[49] Bridgeman to Winnifrith, November 13, 1941, T 161/1425.

[50] DeGolyer, *Historical Data*, 17.

exported had shrunk from 20 percent to 10–12 percent since 1913. For the United States to supply 80 percent of the Allies' wartime oil requirements, while covering a 20 percent rise in domestic consumption, production would have to increase by more than 30 percent from the level of 1938. On a "purely economic-technical" level, German analysts concluded in 1939, such an increase was eminently feasible.[51] The United States entered the war with roughly one million bpd of spare capacity and managed to increase production by 30 percent between 1940 and 1945, which again allowed it to satisfy roughly 80 percent of the Allies' wartime oil consumption.[52]

Between May 1940 and April 1941, the Oil Control Board estimated that the United States could supply 23 percent of the empire's requirements—6.3 million of 27.76 million tons. Britain had already claimed all available non-U.S. sources of high-octane gasoline, and the United States would be the only major oil producer left if Britain lost access to one of its existing suppliers, as happened in 1940 when Iran ceased exports west of Suez and in 1942 when the Japanese overran the East Indies.[53] British imports of U.S. oil rose from 1.86 million tons in 1940 to 13 million tons in 1944, and those from Venezuela peaked at 6.71 million tons in 1944. Imports from those two countries comprised two-thirds of the total in 1940 and 93 percent of the total in 1944.[54] Even these figures are "slightly misleading," for Venezuelan production at war's end was only 200,000 tons higher than when it began. Although Venezuela was supposed to be Britain's largest wartime supplier, it proved far less important than the United States because of the U-boat threat and the difficulty of refining its oil into high-octane gasoline.[55] Venezuela's contribution still eclipsed that of the region Whitehall expected would make Britain energy independent.

FAILURE IN THE MIDDLE EAST, 1939–1942

Seldom has a strategy promised so much yet yielded so little as Britain's efforts in the Middle East following World War I. In the case of Iraq, even after the completion of the IPC pipeline in 1934, Iraq's percentage share of British imports remained in the single digits before World War II, since it exported only crude oil, which suited France but not Britain.[56] The main reason why Iraq's output was nowhere near its potential was the machinations of the world oil cartel. Iraq had renegotiated the IPC's concession in 1931.[57] The new agreement rescued Baghdad from the brink of insolvency by providing it with a steady source of "dead rent" (advances against future royalties).[58] At a time of oversupply, however, these payments killed any

[51] "Der amerikanische Kontinent," *Vierjahresplan*, 1939/XVI. [52] Yergin, *Prize*, 377–9.

[53] Oil Control Board, "American Oil Policy [...]," August 15, 1940, O.C.B. (40) 71, attached to: Kirsch to Seymour, August 15, 1941, FO 371/25212.

[54] "Petroleum Statistics [...]," no date, BP 60205.

[55] Howarth and Jonker, *History of Shell*, 36. [56] Kent, *Moguls*, 155–7.

[57] Penrose, *Iraq*, 69–71. [58] Sluglett, *Iraq*, 74–5 and 135–40.

incentive to increase Iraqi output. Instead, the IPC used its monopoly power to curtail Iraqi output to sustain global prices.[59] They also stymied efforts to expand local pipeline and refinery capacity.[60] Anglo-Iranian's share of Iraqi production therefore rose only slightly, from 1.11 million tons in 1935 to 1.24 million tons in 1938. Worse was to come—between 1939 and 1941 Anglo-Iranian's share dropped from 1.18 million tons to 570,000 tons, and British troops sabotaged all but six wells in northern Iraq the following year.[61] Shell's share of IPC production also declined by two-thirds by 1941 (to 320,000 tons).[62]

Whitehall had, in fact, written off Iraq long before. In 1939, the IPC decided to double the capacity of its pipeline. War began two months later, and the Allies had to decide whether to allow the project to proceed. The French were in favor—Iraq was their only significant overseas supplier, and exports from Tripoli to France required a journey of only 1,700 miles compared to 5,400 miles from the Gulf of Mexico. Although pipeline expansion was expensive (£6 million of goods from the United States), Paris argued that higher purchases of sterling oil from Iraq would offset the initial dollar costs—every one million tons imported from Iraq saved Allies $12 million in foreign exchange. Whitehall disagreed. Britain was already annoyed by France's insistence on importing Iraqi crude oil, which inflated France's requirements for tanker tonnage and hampered the sharing of refined products. As for the proposed expansion of the IPC pipeline, France's expected savings in tonnage were negligible—3.75 tankers per one million tons—and Britain could neither spare the foreign exchange nor risk angering Venezuela to accommodate Iraq. The British therefore blocked the IPC's application for hard currency. The day after Italy entered the war, Britain instructed the IPC to shut down the pipeline to Tripoli and reduce the throughput to Haifa to 800,000 tons. Iraq's exports collapsed by 80 percent and Baghdad's oil revenues by 33 percent. The only reason why Britain did not dispense with Iraq altogether was the shortage of tankers to increase liftings from Iran.[63]

Iran was even more disappointing. It alone could have theoretically satisfied the Royal Navy's annual wartime consumption.[64] By the end of 1940, however, Admiralty stocks in Britain had dropped by 800,000 tons since the war began, but with Venezuela, Trinidad, and Haifa exporting at maximum capacity, there were insufficient tankers (forty) to lift the required quantities from Abadan.[65] Overall, Iranian exports dropped from 9.12 million tons in 1938 to 5.42 million tons in 1941

[59] Walter Adams et al., "Retarding the Development of Iraq's Oil Resources: An Episode in Oleaginous Diplomacy, 1927–1939," *Journal of Economic Issues* 28: 1 (1993): 69–93; Blair, *Control*, 80–5; Nowell, *Cartel*, 270–5.

[60] Daniel Silverfarb, *Britain's Informal Empire in the Middle East: A Case Study of Iraq, 1929–1941* (New York: Oxford University Press, 1986), 97–8. The Anglo-Iranian/Shell refinery in Haifa could process less than half of Iraq's production in 1939. Llewellin, et al., "Thirteenth Annual Report," January 24, 1939, O.B. 294, CAB 50/7.

[61] Bamberg, *History*, 168 and 219–23.

[62] Howarth and Jonker, *History of Shell*, 37.

[63] Payton-Smith, *Oil*, 122–3; Silverfarb, *Informal Empire*, 94–105 and 118–19.

[64] Payton-Smith, *Oil*, 16.

[65] "Purchases [...]," no date or author, enclosed with: Coxwell to Rowan, December 6, 1940, and Admiralty, "Purchases," December 17, 1940, 170 (E.R.C.), both in: T 161/1425.

(40 percent). Although they recovered by 1943, exports went largely to Africa, South Asia, the Far East, and the Soviet Union, and Iran exported virtually no oil west of Suez between the Anglo-Soviet invasion of August 1941 and the opening of the Mediterranean in 1943.[66] Anglo-Iranian claimed that the reason it did not export more to Britain in the period 1940–3 was a shortage of tankers.[67] This is misleading—there was no absolute shortage of tankers, only a shortage to send exports to Britain via the Cape.

The Iranians were naturally unhappy with this arrangement—they were still unhappy with the renegotiation of the Anglo-Persian concession in 1933 and clashed with Britain over Bahrain.[68] Tehran demanded that Anglo-Iranian provide "make-up" payments to cover the difference between royalties on projected versus actual production, and Whitehall supported a settlement to avoid trouble.[69] Anglo-Iranian and Tehran signed an agreement in August 1940 whereby the company agreed to pay £1.5 million for lost royalties in 1939 and whatever was required to raise Iran's royalties in 1940 and 1941 to £4 million per annum. Whitehall initially covered half the cost, but when Anglo-Iranian renewed the agreement in 1943, the government withdrew its support. The payments for 1942 and 1943 totaled £572,067 and £382,083, respectively, and they finally ceased in 1944/5 when royalties exceeded £4 million.[70] These outlays were in addition to the £3 million Iran could convert into gold every year under the terms of a June 1940 agreement. Although Britain could not afford the pressure on sterling, it had no option.[71]

The oil of the Middle East proved less valuable to Britain than its refineries. The Anglo-Iranian refinery at Abadan was one of the few sources of 100-octane gasoline within the empire, and it was relatively secure compared to the other major one east of Suez (Rangoon). In 1940, there was only one source of 100-octane gasoline within Britain (Stanlow).[72] The others were Abadan (after 1940), Trinidad, the Dutch West Indies (Aruba and Curaçao), the Dutch East Indies, and the United States.[73] The Royal Air Force had consumed 77-octane gasoline and switched to 87-octane in 1935. By 1938, the newest aircraft required 100-octane gasoline, and Whitehall established a committee to determine how to increase the supply of the required additives (iso-octane).[74] It made sense to develop Abadan as a supplier east of Suez due to the vulnerability of the East Indies, but the government did not make any substantial investment to expand Abadan's capacity until October 1939, and the Air Ministry did not certify Abadan's production until June 1940. Exports of 100-octane gasoline from Abadan commenced in July but production for the rest

[66] Bamberg, *History*, 220 and 230ff.

[67] Anglo-Iranian, "Notes [...]," December 22, 1943, BP 63800; Anglo-Iranian, "Memorandum [...]," February 4, 1944, BP 71146.

[68] "Das Bahrein-Öl," *Militär-Wochenblatt* (November 8, 1940).

[69] German analysts concluded that Britain's weak strategic position had compelled Whitehall to appease Tehran. "Die Anglo-Iranian [...]," *Vierjahresplan*, 1940/XXII.

[70] Bamberg, *History*, 218–19 and 230–57.

[71] Payton-Smith, *Oil*, 151.

[72] Howarth and Jonker, *History of Shell*, 12.

[73] Gavin Bailey, "The Narrow Margin of Criticality: The Question of the Supply of 100-Octane Fuel in the Battle of Britain," *English Historical Review* 123: 501 (2008): 402–9; Kent, *Moguls*, 151–2.

[74] Bamberg, *History*, 182–3; Payton-Smith, *Oil*, 53–7.

of 1940 was only 23,000 tons. Major expansion of output did not start until after 1941; production then rose twelvefold, from 67,000 tons to 858,000 tons by 1944.[75]

By July 1941, the British military were warning that Britain could not afford to lose Abadan and Iranian oil, because this would leave the empire east of Suez dependent on the East Indies for oil—an unappetizing prospect considering Japan's hostile disposition.[76] In the winter of 1941/2, Allied intelligence concluded that Germany was poised to exploit the Allies' vulnerability.[77] Japan's successes during the first half of 1942 suggested that the Axis vice was closing on the Middle East. In April 1942, Churchill warned Roosevelt that if the Allies lost Abadan, the British position in the Indian Ocean would become untenable.[78] Churchill was not exaggerating—a few days earlier, U.S. Army Chief of Staff General George Marshall, then on a visit to London, asked Washington to evaluate the consequences of Abadan's loss. His chief of operations, Major General Dwight Eisenhower, concluded that losing Middle Eastern oil "would very quickly result in the collapse of our entire defenses in the Middle East and India."[79]

The British did not develop contingency plans until the summer of 1942, when the Afrikakorps was in striking distance of the Suez Canal and Army Group A was about to enter the Caucasus. In May, the Petroleum Department warned that replacing lost production in Iraq and Iran with U.S. imports would increase Britain's tanker requirements by 2.5 million tons, which "could not be met." Even a 15 percent reduction in Britain's civilian consumption (one million tons per year) would save only 300,000 tons of tonnage, and a 20 percent reduction in overseas consumption (1.25 million tons) only 500,000 tons. Projected increases in military consumption would also wipe out any savings extracted from the civilian economy. Furthermore, the loss of Abadan's 100-octane gasoline would "be little short of catastrophic," since global supplies were already tight.[80] One official worried that the war "cannot be waged unless we can rely on supplies from Abadan," and urged his colleagues to reconsider the existing "Scorched Earth policy" if Britain had to withdraw.[81] Considering the consequences for the Allied war effort, "there should be a maximum concentration on the military effort needed to safeguard the Persian oilfields."[82]

[75] Bamberg, *History*, 241–6; Payton-Smith, *Oil*, 270–2.

[76] Joint Planning Staff, "Situation [...]," July 23, 1941, J.P. (41) 580, attached to: Howkins, "Arab Federation," November 13, 1941, J.P. (41) 961 (8), CAB 95/1.

[77] Donovan, "Oil and German Strategy," March 5, 1942, Franklin Delano Roosevelt Library (FDRL), Papers of Harry L. Hopkins, Box 154.

[78] Document C-69 in: Warren Kimball, ed., *Churchill and Roosevelt: The Complete Correspondence* (Princeton: Princeton University Press), i: 452–4.

[79] Peter Hoffmann, "The Gulf Region in German Strategic Projections, 1940-1942," *Militärgeschichtliche Mitteilungen* 44: 2 (1988): 67–8.

[80] Ministry of Fuel & Power, "Oil Supplies [...]," July 1, 1942, Ex. O.C.B. (42) 16, POWE 33/1349.

[81] "Vital Importance [...]," no date or author, POWE 33/1349. The British went ahead with plans to destroy the Iraqi and Iranian oilfields, but due to the opposition of U.S. oil companies, the Saudi oilfields were off limits. Daniel Silverfarb, "Britain, the United States, and the Security of the Sa'udi Arabian Oilfields in 1942," *Historical Journal* 26: 3 (1983): 719–26.

[82] Ministry of Fuel & Power, "Oil Supplies [...]," July 1, 1942, Ex. O.C.B. (42) 16, POWE 33/1349.

In July, the Oil Control Board calculated that, in the event that Germany overran Iran and Iraq, Britain would lose 13.63 tons of the 14.59 million tons (93 percent) it expected to receive from the Middle East excluding Bahrain and Egypt over the next year. Even if production in Bahrain doubled, Britain would still require an additional 10.67 million tons from the Western Hemisphere, plus 200 tankers to transport it. If Bahrain ceased production, the amount of oil required from the Western Hemisphere would increase by another 2.75 million tons, plus seventy tankers. The United States and Venezuela could replace most of the oil but could not supply additional tankers. Accordingly, Britain would have to impose a 20 percent cut in consumption even though previous reductions "represented the utmost that could be achieved without adverse effect to our war effort."[83] The fact that Britain was short of tankers rather than oil is a vital distinction. To claim that the closure of the Mediterranean was a "nuisance" and that the Middle East was just a "convenient source of supply" overlooks the lack of tonnage to haul additional supplies from the Western Hemisphere.[84] The Oil Control Board had no doubts as to the danger confronting Britain in the summer of 1942—it assessed the loss of Iran as "calamitous."[85]

The board could have applied the same judgment to Whitehall's entire oil strategy since World War I. The only way for Britain to get the oil it needed without depending on foreigners or ruining its financial position—relying on the Middle East—was unavailable because of the closure of the Mediterranean and the shortage of tankers to ship oil around the Cape. Britain could—and did after 1940—import oil from the Western Hemisphere to save tanker tonnage, but its inability to pay the financial costs exposed the shallow foundations of its imperial pretensions following World War I. Britain paid dearly for putting its faith in oil supplies it could neither protect nor access safely, but it was not the only great power to make that mistake.

[83] Lloyd, "Oil Supplies [...]," July 21, 1942, C.O.S. (42) 352, attached to: Bridges, "Oil Supplies [...]," August 4, 1942, W.P. (42) 338, CAB 66/27/18.

[84] Alan Levine, "Was World War II a Near-Run Thing?" *Journal of Strategic Studies* 8: 1 (1985): 58–9.

[85] Lloyd, "Oil Supplies [...]," July 21, 1942, C.O.S. (42) 352, attached to: Bridges, "Oil Supplies [...]," August 4, 1942, W.P. (42) 338, CAB 66/27/18.

PART II

GERMANY

5

Making Do with Less
1914–1935

Based on their experience during World War I and assessment of future trends in warfare, German leaders recognized that they had to redress their petroleum inferiority vis-à-vis their great power adversaries.[1] Since defeat in 1918 had deprived Germany of its oil assets in Eastern Europe and the Middle East, the solution to its energy needs had to come from home. For this reason, there was widespread backing for the synthetic fuel industry long before the National Socialists came to power. This support cut across political lines, as synthetic fuel was a potential solution not only for Germany's military vulnerability but also for its economic weakness. Even though their ambitions were modest in the 1920s, military planners understood that synthetic fuel was the only reliable source of petroleum for Germany's war machine. German firms, meanwhile—especially the industrial behemoth IG Farben—saw in synthetic fuel a way to transform Germany's entire economic and financial outlook, as well as to build fruitful partnerships within other transnational firms.

After 1933, the National Socialists adopted this foundation even as they expanded its scope. The Third Reich eventually devised a multifaceted strategy to secure the petroleum it needed to realize Hitler's geopolitical "program"—securing the mastery of Europe following a series of localized campaigns before transitioning to an inter-generational struggle against the United States for world domination.[2]

In peacetime, the Third Reich's petroleum strategy included incentives to increase domestic production of crude oil, accumulating reserves, and acquiring foreign concessions. The most important decision was to harness the power of Germany's chemical industry to synthesize important commodities.[3] Chemistry, coal, and necessity would make Germany "independent" of most imported raw materials.[4] Synthetic fuel offered a solution to various economic challenges, including shortages

[1] Unless otherwise indicated, all German-language sources cited in chapters five through eight are from NARA, Record Group 242: Foreign Records Seized (RG 242). I have used the following citation format: Microfilm Publication No./Reel No. (Item No.).

[2] Andreas Hillgruber, *Hitlers Strategie: Politik und Kriegsführung, 1940–1941* (Frankfurt am Main: Bernard & Graefe Verlag für Wehrwesen, 1965). See also: Milan Hauner, "Did Hitler Want a World Dominion?" *Journal of Contemporary History* 13: 1 (1978): 15–32.

[3] Besides petroleum, German chemists also synthesized ammonia, methanol, rubber, acetylene, ethylene, benzol, and toluene. Nuernberg Military Tribunal, *Trials of the War Criminals* (Washington, DC: GPO, 1953), viii: 1253–4 (hereafter cited as *NMT*).

[4] Hedler, "Kampf um Rohstoffe," *Militär-Wochenblatt* (October 27, 1939).

of foreign exchange and even exhaustion of inexpensive oil sources or future demand growth in China.[5] It would also obviate Germany's dependence on international trade, which Hitler complained had "plunged" the country into "the mire" of the Great Depression.[6] Synthetic fuel could not make Europe energy independent due to its high cost and low production ceiling, but that did not detract from its political utility.[7] From the perspective of German strategy, synthetic fuel was always a means to the wider end of making German a world power.

OIL AND THE OUTCOME OF WORLD WAR I

During both world wars, Germany fought at significant numerical, economic, financial, and materiel disadvantages. In both cases, the Allies also proved incapable of quickly exploiting German deficiencies. During World War I, the main reason for this had to do with the fact that, as a continental power, Germany was not as vulnerable to economic warfare as the British had expected. Before the Great War, a blockade would have deprived Germany of only 8 percent of its national income, and the Germans could compensate by seizing the resources of neighbors.[8] Regarding petroleum, the Germans were fortunate that their prewar consumption was so small: there were only 61,000 cars in Germany in 1913, compared to 91,000 in France and 209,000 in Britain.[9] That year, Germany imported 1.3 million tons of petroleum valued at 183 million RM, 93 percent of which came from the sources shown in Table 5.1.

Table 5.1. German oil imports, 1913[10]

Source	Quantity (tons)
United States	693,000
Austria-Hungary	232,000
Russia	158,000
Romania	117,000

[5] Seidenzahl, "Erschöpfung [...]?" *Die deutsche Volkskraft* (December 9, 1937); Ruprecht, "Ölsuche und Ölgewinnung [...]," *Militärwissenschaftliche Rundschau* 3: 3 (1938); Ruprecht, "Erschöpfung der Erdölvorräte," *Deutsche Wehr* (September 27, 1940).

[6] *Table Talk*: No. 45 (October 18, 1941).

[7] In 1937, synthetic fuels accounted for only 3 percent of global consumption of light petroleum products—2.83 million of 89.15 million tons. Kiewitt, "Öl und Ölersatzstoff [...]," *Deutsche Wehr* (October 27, 1939).

[8] Paul Kennedy, "Mahan versus Mackinder: Two Interpretations of British Sea Power," *Militärgeschichtliche Mitteilungen* 16 (1974): 50–2.

[9] Rainer Karlsch and Raymond Stokes, *Faktor Öl: Die Mineralölwirtschaft in Deutschland, 1859–1974* (München: C.H. Beck, 2003), 93; Jonker and van Zanden, *History of Shell*, 173.

[10] Armaments Office, "Deutschlands Rohstofflage," April and May 1920, T-84/142 (EAP 66-c-12-62/6); Foreign Office, "Die Weltlage [...]," June 20, 1920, PAAA, R 97787; Friedensburg, *Erdöl*, 70; Karlsch, *Faktor Öl*, 93.

Germany started World War I with a reserve of 343,000 tons—one-quarter of annual peacetime demand, or 10 percent of its total wartime consumption. Gasoline stocks amounted to 76,000 tons against an estimated annual wartime demand of 105,000 tons.[11] Once it had exhausted these stockpiles, Germany had to "anchor" its fuel supply on benzol, a byproduct of the coking process that was either used singly or blended with gasoline or alcohol.[12] Benzol production increased from 130,000 tons in 1913 to 230,000 tons in 1917, which equaled 80 percent of Germany's gasoline imports in 1913. The balance between the consumption and production of kerosene, lubricants, and diesel was less favorable. Against the approximately 800,000 tons of kerosene, 300,000 tons of lubricants, and 130,000 tons of bunker fuel imported in 1913, Germany's domestic producers that year could muster only 50,000 tons each of kerosene and lubricants, plus 90,000 tons of bunker fuel.[13] Germany could also draw from two nearby suppliers: Romania and Austria-Hungary (Galicia). In 1909, the latter was briefly the world's third-largest producer, at two million tons a year.[14] Romanian production reached 1.89 million tons in 1913, of which 1.04 million tons was exported.[15] That year, nearly 40 percent of Romanian petroleum exports went to Britain and France, and less than 20 percent to Germany and Austria-Hungary.[16] Since combined German–Austrian consumption peaked at 2.32 million tons in 1918, the Central Powers appeared to be in a strong position.[17]

Things did not work out as planned. The Russians occupied Galicia in September 1914. When they retreated in May 1915, the Russians burned 350,000 tons of stocks. Inexplicably, the Russians left 480,000 tons of stocks and the refineries untouched. The Austrians estimated that the occupation had caused one million tons of damage and lost production—a significant quantity, since German–Austrian consumption was 1.13 million tons in 1915—but it could have been worse.[18] A larger problem was that Galician production peaked in 1909 at two million tons per annum and declined rapidly thereafter—by 1913, it was only 1.11 million tons. The Austrians had been producing oil since the 1850s but due to the proliferation of new technologies, small firms, and an unusual legal regime authorizing private ownership of mineral resources, they wasted large amounts. Since firms often shared ownership of a single oilfield, they competed with each other to extract as

[11] In 1913, two-thirds of Germany's imports came from the United States and Russia, while Galicia and Romania supplied 27 percent. Sußdorf, "Das Feldkraftfahrwesen," in: *Der große Krieg, 1914–1918*, ed. M. Schwarte (Leipzig: Johann Ambrosius Barth, 1921), viii: 340 and 346; Friedensburg, *Erdöl*, 67–80; Leonhardt, "Der kriegsentscheidende Treibstoff," *Militär-Wochenblatt* (December 22, 1939).

[12] The utility of alcohol was limited by the fact that the feedstock (potatoes) also had to feed the population. Sußdorf, "Feldkraftfahrwesen," 353–4; Holger Herwig, *First World War: Germany and Austria-Hungary, 1914–1918* (London: Bloomsbury, 2014), 253–4.

[13] Armaments Office, "Deutschlands Rohstofflage," April 1920, T-84/142 (EAP 66-c-12-62/6).

[14] Alison Frank, *Oil Empire: Visions of Prosperity in Austrian Galicia* (Cambridge: Harvard University Press, 2005).

[15] Friedensburg, *Erdöl*, 26 and 21.

[16] Friedensburg, *Erdöl*, 52 and 51.

[17] Friedensburg, *Erdöl*, 14.

[18] Sußdorf, "Feldkraftfahrwesen," 362–3 and 384; Friedensburg, *Erdöl*, 126–7; Frank, *Oil Empire*, 185–9; Herwig, *World War*, 240.

much oil as possible, thereby ruining the health of the oilfields.[19] Between 1914 and 1918, Galicia averaged only 850,000 tons a year.[20] The Austrians also hastened depletion by encouraging the consumption of oil, which was cheaper than coal.[21]

Not surprisingly, the Germans and Austrians feuded constantly over oil. Berlin opposed giving Vienna any meaningful role in either the Romanian or the Ottoman oil industry.[22] Vienna reciprocated by demanding ever higher financial and material (coal) compensation during the quarterly negotiations concerning Austrian petroleum exports to Germany.[23] Germany's price for Austrian fuel oil increased from 6 RM before the war to more than 58 RM by August 1918.[24] The German Navy's costs doubled between January 1917 and July 1918 alone.[25]

Romania also proved to be an unreliable supplier. Germany and Austria-Hungary were desperate for Romanian oil after the loss of Galicia, but the subsidiaries of the U.S. and British companies in Romania refused do business with Germany, and Bucharest, under Allied pressure, forbade the export of either crude oil or gasoline soon after the war's outbreak. This led to a drop in exports from 1.04 million tons in 1913 to 429,000 tons in 1915, most of which went to Germany and Austria-Hungary—155,000 tons and 227,000 tons, respectively.[26] The bulk of these supplies came from German firms. Between July 1914 and August 1916, the largest German oil company in Romania, Steaua Romana (controlled by Deutsche Bank since 1903), sent 161,000 tons of refined petroleum to Germany and Austria-Hungary, and another 319,000 tons during the rest of the war to the German occupation authorities.[27] This was more than 5 percent of Germany and Austria-Hungary's total oil consumption between 1914 and 1918.[28]

Romania entered the war on the side of the Allies in August 1916, but the Central Powers counterattacked and quickly overran most of the country. In December, and just ahead of advancing German troops, Allied saboteurs destroyed most of Romania's drilling rigs and tanks, as well as blocking 80 percent of the wells, although they left many of its refineries intact.[29] Romanian production dropped from 1.24 million tons in 1916 to 517,000 tons in 1917 but recovered to

[19] Frank, *Oil Empire*, 53 and 140–72.

[20] Friedensburg, *Erdöl*, 20–5.

[21] Frank, *Oil Empire*, 173–204.

[22] Kühlmann to Foreign Office, March 6, 1917, Wedel to Bethmann, May 16, 1917, and Austro-Hungarian Foreign Ministry, "Notiz," May 7, 1917, all in: DBA, O-1063; Rodern to Helfferich, "Friedensvertrag [...]," March 2, 1918, and Kühlmann to Hertling, March 26, 1918, both in: BA-B, R 3101/1422.

[23] See BA-B, R 3101/1403–1404.

[24] Economics Office to Commissar for Coal-Distribution, August 11 and 29, 1918, BA-B, R 3101/1403.

[25] Navy Office to Economics Office, August 26, 1918, BA-B, R 3101/1403.

[26] Friedensburg, *Erdöl*, 52 and 27.

[27] Deutsche Bank, "Die Bedeutung [...]," May 23, 1921, DBA, P-8364.

[28] Friedensburg, *Erdöl*, 14–15.

[29] Stuchlik, "Die Wirtschaftliche Bedeutung [...]," 1918, BA-B, R 3101/884; "Die anglo-rumänische [...]," 1918, no author, BA-B, R 3101/1405. The Allies estimated that they had destroyed five million bbl of petroleum. Norton-Griffiths, "[...] Roumanian Oilfields," January 22, 1917, G.T. 25, BNA, CAB 24/6; Norton-Griffiths, "Supplementary [...]," January 27, 1917, G.T. 136, CAB 24/7. See also: Yergin, *Prize*, 179–82; Herwig, *World War*, 213 and 216.

1.21 million tons in 1918.[30] Exports in 1918 reached 827,000 tons, of which 609,000 tons went to Germany and the remainder to Austria-Hungary.[31] Total exports to the Central Powers during the twenty-month occupation amounted to 1.14 million tons, 78 percent (890,000 tons) of which went to Germany, and monthly deliveries peaked at 110,000 tons.[32]

The control of foreign oilfields arguably became a war aim earlier in Berlin than in Whitehall. In 1916, the German government expressed interest in working with Deutsche Bank to produce oil in Mesopotamia, to be refined for use by local German and Ottoman forces.[33] Over the long run, Berlin hoped Mesopotamia would become a major supplier of oil to Germany, particularly its navy and merchant marine. Optimists were convinced that Mesopotamia's productivity far surpassed that of Romania, and if Mesopotamian oil was exported from Alexandretta, it was closer to Europe than oil traveling from either Russia or the East Indies.[34] In 1917, Germany dispatched a team of military engineers to northern Mesopotamia, which found and refined oil at several locations.[35] The following decade, British engineers carried on the Germans' work, often drilling in the same locations and making a huge discovery at Baba Gurgur in 1927.[36]

The Germans had more immediate success in Romania. Considering its large supplies of oil and food, Berlin was determined to cement German economic control over Romania soon after occupying it.[37] Two years later, as compensation for the cost of repairing Allied sabotage, the Germans secured a petroleum agreement supplementing the Treaty of Bucharest giving them a controlling stake in Romania's oil industry with only nominal Austrian and Romanian participation.[38] In a precursor to German efforts after 1941, the Reich established a joint public–private firm, the Central European Crude Oil Company, to oversee its oil interests—both those originally belonging to German firms and those expropriated from Allied nationals—in Romania during and after the war. The long-term objective was to monopolize as much of Romania's oil production as possible while restricting

[30] "Bericht [. . .]," no date (circa 1918), BA-B, R 3101/884; Friedensburg, *Erdöl*, 26.

[31] Maurice Pearton, *Oil and Romanian State* (Oxford: Clarendon Press, 1971), 76–95.

[32] Ro V, "Notiz [. . .]," February 14, 1940, T-77/123 (Wi/IF 5.533); Wi VII, "Sitzung [. . .]," January 29, 1940, T-77/400 (Wi/IF 5.3063).

[33] Lehner to State Secretary of the Navy Office, June 28, 1916, Lehner and Meissner to Prussian War Ministry, August 15, 1916, "Besprechung im Kriegsministerium [. . .]," April 17, 1918, Lehner to Stauss, August 14, 1918, file note, August 15, 1918, author unclear, Prussian War Ministry to Deutsche Bank, October 19, 1918, and Deutsche Bank to Prussian War Ministry, October 30, 1918, all in: DBA, O-1067.

[34] Seidl, "Gründe [. . .]," June 1, 1917, enclosed with: letter to Stauss, July 10, 1917, author unclear, and Seidl, "Zur mesopotamischen Erdölfrage," December 6, 1917, enclosed with: Seidl to Günther, December 6, 1917, both in: DBA, O-1067.

[35] "Akten-Notiz [. . .]," November 17, 1917, Prussian War Ministry, "Cajara," December 31, 1917, Wrisberg, memorandum, June 19, 1918, and Military Plenipotentiary at German Embassy, Contantinople, production statistics, May 25, 1918, enclosed with: Prussian War Ministry to Deutsche Bank, July 26, 1918, all in: DBA, O-1067; Continental Oil, "Wer erschloß Mossul-Öl?" May 5, 1941, T-580/907 (Box 8, Ordner 56); "Der Deutsche Anteil [. . .]," no date, attached to: Wi VI d, "irakisches Erdöl," May 27, 1941, T-77/1399 (Wi/IIA 5.1–2).

[36] Keating, *Mirage*, 26, 36, and 175.

[37] Chancellor to Quartermaster-General, October 31, 1916, and Quartermaster-General to Chancellor, November 6, 1916, both in: BA-B, R 3101/884.

[38] See: BA-B, R 3101/1422.

Romanian consumption and refining the oil in Germany, thereby breaking Standard Oil's stranglehold over Germany's petroleum supplies.[39]

What the Germans prized most, however, was Baku. In July 1918, during a Crown Council meeting, General Erich Ludendorff, Germany's de facto military dictator, demanded Baku's oil as a "matter of life and death," without which Germany could neither exploit the Ukrainian harvest nor hold the Western Front in 1919.[40] Finally, in August, even as the German lines on the Western Front buckled under Allied counteroffensives, Germany signed a supplementary agreement to the Treaty of Brest-Litovsk with the Bolshevik regime, receiving one-quarter of the oil produced in the Caucasus.[41] That month, German officials agreed that Germany had to take possession of the Mesopotamian oilfields after the war because Romanian output was insufficient to meet future demand.[42]

It is unclear whether Germany's efforts in 1918 to secure control of the Caucasus were the product of a predetermined strategy or simply a manifestation of wartime exigencies, euphoria over operational successes, and Ludendorff's megalomania.[43] German firms were ambivalent in view of the "insecure" environment.[44] In any event, Germany extracted only paltry gains from its eastern conquests. Far from enriching Germany, the commander of the expedition to Georgia had to beg for oil from Romania plus grain and coal from the Ukraine, lest the local economy implode.[45] Such aggravations did not dampen German expectations. The war had demonstrated that even great powers could only satisfy the demands of a protracted conventional conflict with external raw material supplies. The former head of the General Staff's Railway Section, General Wilhelm Groener, stressed to German officers in September 1918: "The conduct of war is not simply a question of manpower... it is also a question of fuel, and if we don't exploit Baku and its oilfields for own purposes, we can't manage it."[46] Furthermore, the British blockade, by exposing the liabilities of Germany's dependence on overseas trade, provided the Germans with a powerful incentive to secure a "landbridge" linking Europe to Central Asia, while also stabilizing the Ottoman lines in northern Mesopotamia and threatening British interests in Persia and India.[47] Even assuming such schemes

[39] Hamlin, *Romania*, 283–93 and 316ff.

[40] Winfried Baumgart, *Deutsche Ostpolitik: Von Brest-Litowsk bis zum Ende des Ersten Weltkrieges* (Wien: Oldenbourg, 1966), 197, n. 84, and "Das 'Kaspi-Unternehmen'—Größenwahn Ludendorffs oder Routineplanung des deutschen Generalstabs: Erster Teil," Jahrbücher für Geschichte Osteuropas 18: 1 (1970): 53, n. 37.

[41] Reprinted in: U.S. Senate, *Bolshevist Movement in Russia*, 66th Congress, 2nd Session, Document No. 172 (Washington, DC: GPO, 1920), 49–55.

[42] Fritz Fischer, *Germany's Aims in the First World War* (New York: W.W. Norton, 1967), 583–6.

[43] Compare Fischer, *Germany's Aims*, 550–7, with Baumgart, *Ostpolitik*, 174–207, and "Kaspi-Unternehmen," 52–73.

[44] Central Purchasing Company to War Ministry, "Kaukasus," May 1, 1918, BA-B R 3101/1317.

[45] Baumgart, *Ostpolitik*, 194–5, and "Kaspi-Unternehmen," 56, n. 50.

[46] Baumgart, *Ostpolitik*, 197, n. 84.

[47] Robert Koehl, "A Prelude to Hitler's Greater Germany," *American Historical Review* 59: 1 (1953): 43–65; Baumgart, *Ostpolitik*, 192–3, "Kaspi-Unternehmen," 62ff; Werner Zürrer, *Kaukasien,*

were feasible, they depended on German battlefield success, which was no longer in the cards after the Western Front collapsed in August 1918.[48]

Two weeks after the signing of the Armistice, Lord Curzon boasted that "the Allies floated to victory on a wave of oil."[49] Allied supplies dwarfed those of the Central Powers and allowed them to exploit the opportunities afforded by the internal combustion engine.[50] By 1918, Allied consumption was 2.5 times that of the Central Powers (two million versus nine million tons).[51] Motor vehicles played important roles at several key points in the war. They were particularly important to the French war effort. In 1914, the French used 600 Parisian taxis and plus thousands of trucks and assorted vehicles to ferry to troops to the Battle of the Marne. Two years later, at Verdun, French trucks moved 90,000 troops and 50,000 tons of supplies weekly, since the only rail line was inadequate. Finally, from March to May 1918, the French shifted moved 1.4 million troops to help stem the Germans' last-ditch offensives. Conversely, the German advance in 1918 sputtered in part due to a lack of supplies, since the Germans lacked trucks and could not construct new railway lines fast enough.[52]

The Germans and Austrians also appreciated the value of trucks. Germany deployed 4,000 during the execution of the Schlieffen Plan in 1914, but 60 percent of them broke down during the campaign.[53] Shortages of gasoline and rubber limited their use thereafter. The Germans built 40,000 trucks during the war but had to reduce their load capacity by 20 percent to save rubber. By war's end, the Germans were using iron and Austrians leather in lieu of rubber tires.[54] The shortage of motorized transport also strained German railways and diverted horses from agriculture.[55] At sea, the coal-fired German High Seas Fleet was limited to a cruising radius of only eleven days and could not operate safely beyond the North Sea.[56] By 1918, the disparity in tanks, motorized transport, and aircraft was overwhelming.[57] In the air, the Germans were almost even with the Allies—3,670 airplanes to 4,500. The gap on land was enormous—40 German tanks and 23,000 trucks against 800 Allied tanks and 100,000 trucks.[58] By war's end, German estimates put the Allied advantage in the air at 3:1 and that for tanks at 67:1.[59] Even had Germany survived

1918–1921: Der Kampf der Großmächte um die Landbrücke zwischen Schwarzem und Kaspischem Meer (Düsseldorf: Droste, 1978), 161–2.

48 Baumgart, "Kaspi-Unternehmen," 115–18.

49 Quoted in: Anton Mohr, *Oil War* (New York: Harcourt, Brace and Co., 1926), 155–6. See also: Yergin, *Prize*, 183.

50 Yergin, *Prize*, 170ff.

51 Friedensburg, *Erdöl*, 14–16; Jonker and Zanden, *History of Shell*, 163–73.

52 Herwig, *World War*, 188; Laux, "Trucks," 65ff; Yergin, *Prize*, 168–70.

53 Laux, "Trucks," 64.

54 Herwig, *World War*, 101, 237, 253.

55 Jonker and Zanden, *History of Shell*, 173–8.

56 W.G. Jensen, "The Importance of Energy in the First and Second World Wars," *Historical Journal* 11: 3 (1968): 544.

57 Jonker and Zanden, *History of Shell*, 175–7.

58 Herwig, *World War*, 389 and 405; Laux, "Trucks," 67–8.

59 Leonhardt, "Treibstoff."

the defeats of the autumn of 1918, the future was bleak. Under the so-called Plan 1919 drafted by Colonel J.F.C. Fuller, the Allies were preparing a massive onslaught backed by tens of thousands of tanks and aircraft.[60]

A former mining engineer and political opponent of National Socialism, Ferdinand Friedensburg, challenged Curzon's claims in 1939. The Allies did indeed have a large quantitative advantage, but this was not by itself decisive.[61] German military consumption was increasing sharply before the war—in 1911–12 the army estimated that it would require 20,000 tons per year in wartime, but this figure had tripled by 1913–14. Nonetheless, Germany was still able to cover 80 percent of its requirements during the war.[62] As the conflict progressed, motorization, higher than expected operational consumption, and civilian requirements produced shortages by the summer of 1915, which forced the High Command to impose army rationing in September.[63] Nevertheless, Germany still possessed large stocks at war's end. The U-boat fleet had enough diesel fuel to continue operations for several months, and army and air force supplies were "adequate."[64] According to one postwar assessment, benzol production and imports from Romania and Galicia had "sufficed to meet the most urgent requirements."[65] The German position only became intolerable after the collapse of Bulgaria severed imports from Romania, which also forestalled any recommencement of hostilities following the signing of the Armistice.[66]

Friedensburg disputed "any direct connection between the collapse of 1918 and the supply of petroleum."[67] Rather, he showed that the consequences of the Central Powers' shortage of oil were more insidious. Inadequate supplies of lubricants and synthetic alternatives contributed to the decrease in industrial productivity.[68] The shortages also harmed morale.[69] Ludendorff recalled that the effects of the kerosene shortage were catastrophic.[70] The problem was partially one of the Germans' own making. As Romanian oil production resumed in 1917, the Reich decided to concentrate on refining crude into diesel fuel rather than kerosene to fuel the U-boat campaign, hoping that it would bring about victory before the German economy ground to a halt. By year's end, civilian officials were warning of the danger

[60] Jensen, "Energy," 542–5.

[61] Friedensburg, *Erdöl*, 121–8. See also: Tryon, "Raw Materials [...]," June 16–29, 1930, NARA, Record Group 107: Records of the Secretary of War (RG 107), Entry 191, Box 45; Karlsch, *Faktor Öl*, 96–112.

[62] Karlsch, *Faktor Öl*, 93 and 111.

[63] Suβdorf, "Feldkraftfahrwesen," 353–4 and 363.

[64] Friedensburg failed to mention that Germany had to use benzol mixed with gasoline or alcohol, which resulted in lower performance for German aircraft. Erich Ludendorff, *Meine Kriegserinnerungen* (Berlin: Ernst Siegfried Mittler und Sohn, 1919), 273. Benzol also froze in cold weather unless mixed with toluol and gasoline. Bronk, "Deutschlands Erdöl-Selbstversorgung, [...]," *Militär-Wochenblatt* (October 25, 1935); Suβdorf, "Feldkraftfahrwesen," 354.

[65] Armaments Office, "Deutschlands Rohstofflage," May 1920, T-84/142 (EAP 66-c-12-62/5).

[66] Friedensburg, *Erdöl im Weltkrieg*, 124. See also: Ludendorff, *Kriegserinnerungen*, 579; Suβdorf, "Feldkraftfahrwesen," 394.

[67] Friedensburg, *Erdöl*, 122. This was also the opinion of German planners on the eve of World War II. "Stellungnahme [...]," no date, author unclear, T-77/526 (Wi/I.156).

[68] Herwig, *World War*, 237.

[69] Friedensburg, *Erdöl*, 123–4.

[70] Ludendorff, *Kriegserinnerungen*, 274.

that kerosene shortages posed for German economic productivity, especially agricultural output. Throughout the winter and spring of 1918, the military held firm, but the pressure became unbearable.[71] With the monthly deficit running at 7,000 tons, Chancellor Georg Hertling warned during a conference on August 30, 1918 that a crisis was looming. The Finance Ministry demanded the navy release supplies from its stocks. The navy's representative complained that these supplies were vital for future operations, but Hertling overruled him—the imminent collapse of the Germany economy posed a greater threat than the inability to continue the U-boat campaign.[72] If Germany could last until the spring of 1919, however, railway officials were confident that they could import 30,000 tons of petroleum per month from the Caucasus.[73]

The military situation did not merit such confidence. In July 1918, Ludendorff warned his field commanders that Germany would by "no later than the spring of 1919 enter a very dangerous situation" unless the military eliminated all "unnecessary consumption." Austria-Hungary could not increase deliveries, withdrawals from inventories in Romania were outpacing production, and exports from the Caucasus were logistically impossible.[74] Years later, veterans recalled their "bitter" experience of coping with fuel shortages.[75] The effects were cumulative rather than sudden—"the less perceptible minute accumulation of little items piling one on top of the other until at some unknown point the mass of accumulated actions may be large enough to be critical."[76] "Even if the oil crisis in Germany in no way reached the most severe magnitude," Friedensburg concluded that "it was doubtless one of the primary causes for the gradual and then ever faster breakdown in strength." Of even greater significance were the lessons of the conflict: "The connection between the conduct of war and the economy, most notably the importance of raw materials questions and in particular the importance of petroleum supplies" was now indisputable.[77] German officers were avid readers of Friedensburg's work, but they needed no prompting from him about oil's importance after World War I.[78]

PICKING UP THE PIECES, 1919–1932

Before 1914, Germany seemed poised to break the Anglo-American hold over the European petroleum market. German oil companies—backed by their banking

[71] Hamlin, *Romania*, 220–30.

[72] "Besprechung [...] 30 August 1918[...]," BA-B, R 3101/1422.

[73] Bereckheim to Hertling, August 23, 1918, PAAA, R 22361.

[74] "Zusammenstellung [...]," attached to: Ludendorff to Army Groups et al., July 10, 1918, reprinted in: Erich Ludendorff, *Urkunden der Obersten Heeresleitung über ihre Tätigkeit, 1916/18* (Berlin: Mittler & Sohn, 1921), 172–3.

[75] Scholz Roesner, "Die Treibstoff-Frage [...]," *Militär-Wochenblatt* (May 25, 1934).

[76] J.C. Wylie, *Military Strategy: A General Theory of Power Control* (New Brunswick: Rutgers University Press, 1967), 119.

[77] Friedensburg, *Erdöl*, 128.

[78] Gieche, "Mineralölversorgung [...]," February 1, 1940, attached to: Heemskerck to Becht, February 2, 1940, T-77/231 (Wi/IF 5.1180). See also the review in: *Militär-Wochenblatt* of July 14, 1939.

patrons, Deutsche Bank and the Disconto-Gesellschaft—posed a serious threat to Standard Oil's predominance in European markets.[79] Germany enjoyed a strong position in Romania: by 1907, German capital had controlled 62 percent of the Romanian oil industry.[80] Steaua, majority owned by Deutsche Bank, was the second-largest oil company in terms of production by 1913 and the largest in both capitalization and refining capacity.[81] Another German firm, Concordia, a subsidiary of Disconto, occupied fifth place in 1913. German banks had also been building partnerships with Russian suppliers. These efforts culminated in 1906 with the formation of the European Petroleum Union (EPU), whereby Deutsche Bank, in partnership with the two largest Russian companies, owned by the Nobels and French Rothschilds, would market Caucasian oil in Europe.[82] When efforts to drive out Standard Oil failed, the EPU formed a cartel with it, receiving 20 percent of the European kerosene trade.[83]

The future was also promising beyond Europe. Deutsche Bank had a strong presence in the Ottoman Empire since 1888 through the construction of the Anatolian and Baghdad railways, and in 1904, the bank received an oil concession to the provinces of Mosul and Baghdad. Deutsche Bank did not exercise its option after receiving an inconclusive geological report. The bank eventually parlayed its lapsed rights into a one-quarter share of the Turkish Petroleum Company (TPC) in 1912. The following year, Foreign Secretary Grey broached the idea of an Anglo-German oil condominium in Mesopotamia.[84] The Germans leaped at the opportunity. In the spring of 1914, as part of a wider settlement of Anglo-German differences in the Ottoman Deutsche Bank ceded directorship of the TPC to the British in exchange for equal supplies of Mesopotamian fuel oil for British and German navies, as well as access to the British capital market to finance the TPC's operations.[85] The British also agreed to use Deutsche Bank's disputed concession as the basis for the TPC's claim to a new concession.[86] The TPC was on the verge of acquiring a concession for Mosul and Baghdad when World War I began.[87] Once the Ottoman

[79] The firms merged in 1925 while their patrons followed suit in 1929. Lothar Gall et al., *Deutsche Bank* (London: Weidenfeld & Nicolson, 1995), 64–7, 146–50, 171–2, and 181. The following discussion draws from: Hamlin, *Romania*, 54–69.

[80] "Rumäniens Öl [...]," *Militär-Wochenblatt* (December 27, 1940); W.F., "Neuordnung [...]," *Deutsche Wehr* (November 28, 1941).

[81] Memorandum for Fuchs, February 21, 1934, author unclear, and "Steaua Romana," July 10, 1940, author unclear, both in: DBA, P-7068.

[82] Fritz Seidenzahl, *100 Jahre Deutsche Bank, 1870–1970* (Frankfurt: Deutsche Bank, 1970), 205–24.

[83] The agreement broke down in 1912 after Deutsche Bank joined an abortive effort to establish an official kerosene monopoly in Germany. Chandler, *Scale*, 435–41; Fursenko, *Battle*, 107–25; Nowell, *Cartel*, 59–65.

[84] Lichnowsky to Bethmann, July 16, 1913, DBA, O-1062.

[85] Lichnowsky to Bethmann, March 17, 1914, DBA, O-1062; Gwinner and Helfferich to Foreign Office, May 20, 1914, DBA, P-8364/1; "Notiz," June 22, 1914, no author, DBA, O-1062.

[86] Lichnowsky to Bethmann, April 22, 1914, DBA, P-8364/1.

[87] Overview, January 1916, enclosed with: Günther to Tyralla, February 18, 1918, DBA, O-1067; Deutsche Bank, "[...] Verhandlungen [...]," July 9, 1932, DBA, P-8364; Seidenzahl, Deutsche Bank, 224–7.

Empire entered the war on the side of the Central Powers, both Berlin and Deutsche Bank moved to terminate the agreement with the TPC and sign an exclusive concession with the Sublime Porte.[88]

The Versailles Treaty left Germany with nothing. During the Armistice, monthly supplies dropped to 7,500 tons—one-quarter of prewar consumption. In May 1919, with the country in the midst of civil war and over the objections of the War Ministry, the Economics Ministry informed the army that its allocation (including for rightwing paramilitary units) would drop by half to ensure adequate supplies for the agricultural and railway sectors.[89] The situation did not improve much afterwards—as late as 1922, the Economics Ministry requested that the Interior Ministry instruct police to use inferior benzol because the country could not afford to import U.S. gasoline.[90] The loss of its colonies was not especially harmful, but Germany's new borders reduced its already meager production—it lost 50,000 tons of the 120,000 tons produced in 1913 when Alsace reverted to France.[91] More harmful still were the abrogation of the punitive treaties with Romania and Russia, and the expropriation of Deutsche Bank's assets across Europe and the former Ottoman Empire. Foreign companies also took control of the German market. As shown in Table 5.2, as late as 1938, half of the petroleum trade in Germany remained in the hands of the subsidiaries of the major oil companies.

In 1920, the army completed several assessments of Germany's fuel situation. During the war, Germany's requirements had averaged 28,000 tons per month, half of which was army consumption. In early 1920, the supply of benzol was 85,000 tons per annum, but 35,000 tons went to the Allies as reparations, leaving only 10,000 tons available for common consumption after industrial requirements were subtracted. Imports of gasoline stood at 20,000 tons per month. Only a fraction of total civilian and military demand could be satisfied from the existing fuel supply of 20,000 tons available in January. The army estimated the monthly requirements of a 200,000-man force at 1,200 tons per month, and the requirements

Table 5.2. Foreign shares of the German petroleum market, 1938[92]

Subsidiary (parent company)	Share
Deutsche-Amerikanische Petroleum Gesellschaft (Jersey)	26.1%
Rhenania-Ossag Mineralölwerke (Shell)	22%
OLEX (Anglo-Persian)	9.6%

[88] Memorandum for Wangenheim, May 4, 1915, author unclear, and "Schutz [...]," no date or author, enclosed with: Gwinner and Wassermann to Foreign Office, July 3, 1918, both in: DBA, O-1063; "Entwurf [...]," no date, DBA, O-1067.

[89] The government only gave the army equal access to fuel supplies in January 1920. See BA-B, R 3101/5114, esp. War Minister to High Command, May 31, 1919.

[90] Economics Minister to Interior Minister, June 17, 1922, and Interior Minister to Economics Ministry, August 15, 1922, BA-B, R 3101/5115.

[91] Löhr, "Deutsche Treibstoffversorgung," *Militär-Wochenblatt* (June 18, 1933).

[92] Chandler, *Scale*, 519–21.

for a 100,000-man force at 700 tons. Deliveries, however, totaled only 500 tons in January and 600 tons in February.[93]

The army lamented that, before the war, there had been little attention paid to the problem of Germany's dependence on imports. With regard to oil, nature had been a "stepmother" to Germany. Prewar production was only 120,000 tons, nearly half of which was lost to France under the Versailles Treaty. Imports, primarily from the United States, were twelve times larger. On the plus side, postwar imports of kerosene had dropped because of electrification, while Germany's territorial losses had reduced overall consumption.[94] The situation was nevertheless grim. The supply of benzol was not the problem, although its availability was tied to the production of hard coal. The situation with diesel and lubricants was even worse. Only half of the 180,000 tons of diesel Germany required in 1913 was produced internally through low-temperature carbonization of coal. Synthetic fuels had proven inadequate during the previous war, but that could change, as R&D was being pursued "tirelessly." Assuming it was possible to reach a solid to liquid fuel conversion rate of 7 percent, 23.5 million tons of coal would "suffice to cover our entire demand for oil."[95]

The connection between oil and great power status was unmistakable. According to one widely circulated assessment in 1920, oil-poor states would "lose the ability to compete economically" and "could no longer conduct war." The most useful step Germany could take was to develop synthetic substitutes.[96] Luckily, German scientists developed the Bergius (1913—otherwise known as hydrogenation) and Fischer–Tropsch (1923) processes. The latter yielded diesel and low-quality gasoline, as well as lubricants, but required higher-quality coal and supplied products with less military demand.[97] The former was the handiwork of the German Nobel Laureate, Friedrich Bergius.[98] Through hydrogenation, pulverized coal—usually a blend of lignite and bituminous—was liquefied using hydrogen under high pressure, although the process could also use coal tar or crude oil as a feedstock. The resulting liquid hydrocarbons could then be refined into high-octane gasoline and aviation fuel by adding tetraethyl lead or iso-octane.[99] If German scientists could scale up the process and make it cost-competitive, they would transform Germany's economic and military position.

[93] Armaments Office, "Uebersicht [...]," February 1920, T-84/142 (EAP 66-c-12-62/6).

[94] The cost of imports, however, had increased due to the loss of Germany's oil tankers. Foreign Office, "Die Weltlage [...]," June 20, 1920, PAAA, R 97787.

[95] Armaments Office, "Deutschlands Rohstofflage," May 1920, T-84/142 (EAP 66-c-12-62/5).

[96] Schlawe, "Das Erdöl [...]," July 8, 1920, PAAA, R 94461.

[97] Koppenberg, "Mineralölgewinnung aus Kohle," *Vierjahresplan*, 1937/V; British Intelligence Objectives Subcommittee, Final Report No. 1697, "Interrogation of Dr. Bütefisch," Fischer-Tropsch Archive; Arnold Krammer, "Fueling the Third Reich," *Technology and Culture* 19: 3 (1978): 394–9; Raymond Stokes, "The Oil Industry in Nazi Germany, 1936–1945," *Business History Review* 59: 2 (1985): 264–72.

[98] Stranges, "Bergius," 642–67.

[99] Anthony Stranges, "Germany's Synthetic Fuel Industry, 1927–1945," in: *German Chemical Industry in the Twentieth Century*, ed. John Lesch (Dordrecht: Kluwer, 2000), 192–6.

Progress during the war had been disappointing both for technical reasons and because Germany put its resources into rehabilitating Romania's oilfields.[100] The Germans had already confounded expectations once before by developing the Haber–Bosch process for the synthetic fixation of nitrogen in 1913. This allowed Germany to prolong World War I even after it lost access to imported Chilean nitrates, and U.S. military analysts warned that "[the] next war may find the Germans able to make their own oil as the last one found them able to make their own nitrates."[101] Germany could eliminate both its need for oil imports—thus saving 500 million RM annually during the 1920s—and its unemployment problem.[102] Moreover, hydrogenation yielded better results with low-quality lignite (brown coal) or bituminous coal than expensive anthracite. Between 1924 and 1933, Germany provided 20 percent of the world's coal production but 75 percent of its lignite production.[103]

In 1921, Bergius established a partnership with Shell through the International Bergin Company.[104] He made some progress but had to sell control of the company and its German patent rights to Badische Anilin- und Soda-Fabrik (BASF) in 1925 for 2.85 million RM.[105] BASF's engineers improved upon Bergius' work by finding sulfur-resistant catalysts and increasing the yield of lighter fuels that could be refined into high-octane gasoline. In December 1925, BASF merged with Bayer, Hoechst, Agfa, and two other companies to form IG Farben, and the following year the new conglomerate moved forward with plans to build a pilot plant at its Leuna facility, originally constructed in 1917 to mass-produce ammonia.[106]

Around this time, the Weimar Republic took the first steps toward lending public support to Germany's synthetic fuel industry. In March 1925, the Labor Ministry hosted a conference to determine whether to assist Bergius in building a factory in Lower Saxony.[107] The Labor Ministry was determined to reinforce the coal industry, and synthetic fuel was appealing both as a new source of demand for coal and as a new supply of petroleum. The ministry claimed that hydrogenation was capable of converting 100kg of ground coal into 50kg of liquid hydrocarbons.[108] The Defense Ministry agreed that Bergius "deserved the strongest support," and

[100] Stranges, "Fuel Industry," 150.

[101] Tryon, "Raw Materials [...]," June 16–29, 1930, NARA, RG 107, Entry 191, Box 45.

[102] Stranges, "Fuel Industry," 175–6.

[103] Thomas Parke Hughes, "Technological Momentum in History: Hydrogenation in Germany, 1898–1933," *Past and Present* 44: 1 (1969): 116.

[104] Stranges, "Bergius," 660–3. Over the next two years, Shell invested 9.8 million RM in Bergius' Rheinau facility. Stranges, "Fuel Industry," 189.

[105] Stranges, "Bergius," 665–6. Shell sold 60 percent of the international patent rights to BASF in 1925/6 and the remainder in 1931. Karlsch, *Faktor Öl*, 135–6; Jonker and van Zanden, *History of Shell*, 340.

[106] Stranges, "Fuel Industry," 148, 172, and 173ff.

[107] Weigert to Finance Minister et al., "Förderung [...]," March 31, 1925, T-77/82 (Wi/IF 5.372).

[108] The director of the Silesian Coal Research Institute modified this calculation, estimating that the conversation rate of up to 40 percent. Hofmann, "Gutachtliche Aeußerung [...]," no date, T-77/82 (Wi/IF 5.372). See also: "Das Berginverfahren [...]," *Kulturbeiträge* 94 (November 26, 1925), T-77/82 (Wi/IF 5.372).

the Labor Ministry endorsed a six-year, 2.5 million RM loan. Other officials were reluctant, but there was agreement that public support was justified if private capital was unavailable.[109] Although the Defense Ministry could not provide Bergius with any direct assistance, it did take his case to the state government of Baden, where Bergius already had a plant in operation. The Defense Ministry assured Baden that hydrogenation could achieve a coal-to-liquid fuel conversion rate of 60 percent. This was of "great significance for the oil-poor Reich," as the Defense Ministry already hoped that hydrogenation could "make the Reich independent of oil imports."[110] Baden's Interior Minister thereafter decided to inspect the pilot plant Bergius had constructed in Rheinau in 1915.[111] The Defense Ministry also approached the Bavarian coal industry to determine whether its low-grade coal was suitable for hydrogenation.[112] These negotiations fell apart, but the following year the Defense Ministry was once again pressing the Finance Ministry to provide financial support or tax incentives for the construction of largescale hydrogenation facilities, and especially one using hard coal as a feedstock, since IG Farben was already using soft coal at Leuna.[113]

The driving force within the Defense Ministry came from the Ordinance Office, which oversaw the military's economic requirements until 1934 and had determined by 1925 that hydrogenation (as opposed to other methods such as the Fischer–Tropsch process) was the most effective means of meeting Germany's petroleum requirements. The process could yield as much as 520 kg of refined petroleum for every one ton of coal—more than five times as much liquid fuel as older methods such as carbonization without requiring high-quality coal feedstock. The Ordinance Office estimated that seven million tons of synthesized coal—a trivial amount considering that Germany was producing 140 million tons of lignite annually—could satisfy Germany's entire petroleum consumption. The army was also confident that synthesis could be commercially remunerative, and estimated a profit of 30 RM per ton of coal synthesized into petroleum.[114]

By the early 1920s, in fact, Bergius was already converting one ton of coal at a cost of 71 RM into 650 kg of liquid fuel worth 141 RM.[115] Such figures make no allowance, however, for the fact that coal was also necessary for supplementary tasks, including power generation and hydrogen production. The actual cost was unclear. In 1937, according to public statements, synthetic fuel producers required 22 tons of brown coal or 4.5 tons of hard coal to produce one ton of gasoline.[116] Following the war, German engineers told Allied interrogators that 1.6 tons of hard coal yielded one ton of petroleum through hydrogenation, but if one included

[109] Labor Minister, "Förderung [...]," April 11, 1925, T-77/82 (Wi/IF 5.372).
[110] Wurzbacher to Baden Legation, July 7, 1925, T-77/82 (Wi/IF 5.372).
[111] Baden Legation to Defense Ministry, August 15, 1925, T-77/82 (Wi/IF 5.372).
[112] Wurzbacher to Bergius, August 20, 1925, T-77/82 (Wi/IF 5.372).
[113] See the correspondence of 1926/7 in: T-77/315 (Wi/IF 5.1844).
[114] "Das Bergius-Verfahren," no date or author (circa 1925; probably Ordinance Office), T-77/82 (Wi/IF 5.372).
[115] Stranges, "Bergius," 663, and "Fuel Industry," 174–5.
[116] Koppenberg, "Mineralölgewinnung aus Kohle," *Vierjahresplan*, 1937/V.

the coal burned for power generation and hydrogen production, seven tons of coal was actually required to produce a single ton of petroleum. The cost worked out to 260–310 RM per ton of liquid fuel including fixed capital expenditures. The Fischer–Tropsch process was even more expensive—eight tons of coal yielded one ton of petroleum at a cost of up to 360 RM.[117] The cost of feedstocks varied from 10 to 21.5 RM per ton of coal, 40 RM per ton of tar, and 123 RM per ton of petroleum residues. Costs also varied depending on the facility—aviation fuel produced at the joint Shell–Jersey–IG Farben facility at Pölitz cost 360–90 RM per ton, compared to 190 RM at Leuna, due to the higher cost of the coal feedstock.[118]

Price was one thing that did not especially concern synthetic fuel's most avid supporter in the government—the military. Among the lessons German strategists drew from the Great War was that Germany could not afford to fight another protracted war. The only way to avoid one was to undertake sufficient preparation in advance to ensure that the country was fully mobilized the moment hostilities began.[119] In 1924, the army established a Supply Staff—renamed the Economic Staff in 1929—within the Ordinance Office to handle economic mobilization. Before 1933, the army had also collaborated with clandestine organizations bankrolled by private industry, as well as various think tanks and informal contacts.[120] Besides creating an administrative structure to oversee wartime raw materials policy, the military worked closely with firms such as IG Farben.[121] Unlike before World War I, the military was not oblivious to economic affairs. As early as 1928, Major Georg Thomas was advising his superiors to provide financial support for production of "indispensable" raw materials including gasoline and lubricants.[122] One of the interwar army's elite staff officers, Thomas had joined the Ordinance Office in 1927, and quickly became one of the military's leading advocates for economic planning and mobilization.[123]

No amount of planning, however, could compensate for Germany's abysmal raw materials position. In wartime, Germany could only rely upon assets located between the Rhine and Oder rivers—"rump" Germany. One 1925 assessment of the country's requirements for a sixty-three-division army concluded that the

[117] B.I.O.S. Final Report No. 1697, "Interrogation of Dr. Bütefisch," Fischer-Tropsch Archive. See also: Stokes, "Oil Industry," 269–70; Karlsch, *Faktor Öl*, 137. Stranges, "Bergius," 666 and "Fuel Industry," 208, puts forward a figure of 190 RM per ton of fuel (or $0.24 per gallon of gasoline), but he does not incorporate ancillary costs.

[118] Labor costs are difficult to calculate due to the varying treatment of free, forced, and slave workers (one-third of the synthetic fuel industry's labor force by 1944). Stranges, "Fuel Industry," 197, 204–6, and 208–12; Yergin, *Prize*, 343–6.

[119] Dorothea Fensch and Olaf Groehler, "Imperialistische Ökonomie und militärische Strategie: Eine Denkschrift Wilhelm Groeners," *Zeitschrift für Geschichtswissenschaft* 19: 3 (1971): 1167–77.

[120] In 1934, the Economic Staff moved to the War Ministry. Georg Thomas, *Geschichte der deutschen Wehr- und Rüstungswirtschaft, 1918–1943/1945* (Boppard am Rhein: Boldt, 1966), 54–61; Titus Kockel, "Geologie und deutsche Ölpolitik, 1928 bis 1938" (Ph.D. Dissertation, Technische Universität, Berlin, 2003), 226–8.

[121] Ordinance Office, "Rohstoffe," 1928, T-77/640 (OKW/2050).

[122] "[...] Wirtschaftliche Aufstellungsvorarbeiten," November 22, 1928, reprinted in: Thomas, Geschichte, 488ff. See also: Kockel, "Geologie," 229–31.

[123] Thomas, *Geschichte*, 1–5 and 51–3.

armed forces would, upon mobilization, require 91,123 tons of benzol, 2,121 tons of gasoline, and 11,078 tons of lubricants. Annual military and civilian demand for motor fuel (benzol and gasoline) after rationing was 1.3 million tons—one million tons for the military, the remainder for the civilian economy. In the case of a twenty-one-division army, the former figure would be 918,711 tons, plus 643,055 tons of diesel and fuel oil for the navy and civilian economy. Existing motor fuel production—180,000 tons in 1924—was grossly inadequate and vulnerable, since benzol and natural gasoline production within "rump" Germany was only 8,500 tons. The situation was slightly better in terms of diesel and fuel oil—471,000 tons (largely from coal)—but again only a fraction (85,000) was located within "rump" Germany. Imports were out of the question due to the shortage of tanker capacity. Replacing benzol with gasoline was impossible, since annual production from crude oil was 4,000 tons—half of the 8,751 tons required. While it was possible "to stretch" production of lubricants to 100,000 tons per year against 116,000 tons of military demand (57,000 tons with a smaller army), all 200,000 tons of civilian requirements would be unfilled. Hydrogenation was a promising development but required additional time. Until it was ready, Germany could expect to produce only 200,000 tons of fuel per year for the coming decade.[124]

Based on to these figures, building a modern, self-sufficient army was impossible. Accordingly, military planners during the 1920s determined that Germany could only fight a continental rival (France) with supplies from the United States and Britain, which had to be "financially enticed" to break any blockade.[125] Although the military agreed that Germany had to preserve its connections to the global economy, it also insisted on stockpiling or developing domestic alternatives whenever possible.[126] Synthetic fuel production tripled between 1928 and 1931—from 40,000 tons to 120,000 tons.[127] Much of the new output was also geographically secure—by 1929, 70 percent of production took place within Germany's interior.[128]

While hydrogenation never lived up to its billing in terms of cost competitiveness, the German military was convinced of its strategic value. According to one analyst, writing in 1928, Germans had not appreciated the precariousness of their fuel situation before the war and paid a heavy price for this. In the future, European powers would have only two options: they could either acquire foreign concessions or find "other, suitable replacement fuels in order to avoid dependence on foreigners." The growth in postwar consumption was steep. In 1927, there were 1,074,000 cars and motorcycles in Germany, a 30 percent increase from 1926. Oil consumption in 1926 had been 1.1 million tons, and it would rise by 300,000 tons in 1927. Germany's supply position had not improved. Only one-third of its gasoline supply

[124] "Denkschrift [...]," attached to: Nahe, note, December 15, 1925, NARA, T-73/4 (RMfRuK/69); "Graphische Übersicht [...]," no date or author, enclosed with: Ludwig, note, January 12, 1927, T-77/331 (Wi/IF 5.2034, 2222).

[125] Ordinance Office, "Deutschlands Lage [...]," no date (circa 1925), T-77/98 (Wi/IF 5.422).

[126] Geyer, *Aufrüstung*, 184–8.

[127] Löhr, "Deutsche Treibstoffversorgung," *Militär-Wochenblatt* (June 18, 1933).

[128] Ordinance Office, "Preisaufgaben [...]," March 12, 1929, T-77/296 (Wi/IF 5.1601).

(250,000 tons) and half of its lubricants (170,000 tons) was produced internally, and Germany was still far behind other industrialized nations in terms of motorization—in 1928, there were 173 people per vehicle in Germany, against 43 in Britain and France. As imports increased faster than production, the financial burden of those imports—as much as 300 million RM per year by 1930—would also increase. Germany's only hope was its chemists and engineers, who had to improve existing methods of synthesizing fuel, develop new processes, and introduce more efficient engines. Hydrogenation was clearly the most promising synthetic option, with a coal-to-liquid fuel conversion rate of almost 50 percent. The new hydrogenation facility in Leuna promised great things—but only if oil prices held steady.[129] By contrast, the military was indifferent to the survival of the independent German oil companies because the domestic crude oil industry had so little room for growth, even with state support.[130]

During the 1920s, demand growth existed largely in the civilian sector. Although Germany did not have an air force, civilian airline consumption was 30,000 tons. Consumption of fuel oil was 330,000 tons, having doubled over the preceding five years, with 60 percent going to the merchant marine and 25 percent to the navy.[131] The German automotive market on the eve of the Great Depression was tiny—equal to that of the United States in 1911—but production more than doubled between 1925 and 1928.[132] By 1927, Germany's automobile fleet was growing at a rate of 25 percent per annum, and the petroleum import bill reached 238 million RM.[133] There was some growth in military needs—2.12 million tons, according to one military study of 1932. After subtracting production and reserves, Germany would have to import 1.28 million tons annually. This figure was "vanishingly small" relative to global production and presented no difficulties if Germany could import from overseas. If not, Germany's only overland options were the Soviet Union and Romania. Both countries could theoretically satisfy Germany's needs, although German planners advised that the "possibility of deliveries requires closer attention."[134] According to the navy, the notion of significant deliveries of oil and grain along the Danube was extremely unlikely.[135] These assessments proved prophetic by 1939, but they were of little concern during the Weimar era.

The global economic downturn halted motorization and reversed global growth in petroleum production and consumption.[136] Between 1930 and 1932, German

[129] "Preisaufgabe [...]," March 1928, T-77/418 (Wi/IF 5.3329). See also: "Grundsätze [...]," no date or author, enclosed with: Ludwig to Troop Office, April 24, 1929, T-77/364 (Wi/IF 5.2507).

[130] Memorandum of conversation, October 29, 1925, and Ordinance Office to Defense Ministry, February 5, 1929, both in: T-77/314 (Wi/IF 5.1844).

[131] "Das deutsche Ölmarkt," no date or author (circa 1931), T-77/183 (Wi/IF 5.762).

[132] Hughes, "Momentum," 115.

[133] Zornig, "German Motor Fuel Situation [...]," January 22, 1929, in: University Publications of America, *U.S. Military Intelligence Reports, Germany 1919–1941* (Frederick: University Publications of America, 1983), Reel 8.

[134] Ordinance Office, "Planstudie," July 23, 1932, T-77/205 (Wi/IF 5.1029).

[135] Navy Office to Ordinance Office, December 2, 1932, T-77/205 (Wi/IF 5.1029).

[136] "Zur Lage [...]," no date or author (circa 1932), BA-B, R 2/16815.

consumption dropped from 3.45 million tons to 2.66 million tons (23 percent), and demand for gasoline and benzol from 2 million tons to 1.5 million tons (25 percent). Imports also fell from 2.81 million tons to 2.02 million tons (28 percent). Although domestic production increased its share of consumption from 19 percent to 24 percent, the volume held steady at 640,000 tons.[137] The Weimar Republic, like governments around the world, reacted by hiking tariffs—on imported gasoline, they went from 8.5¢ to 14.3¢ per gallon. The following year, it tasked the Economics Ministry with overseeing imports and raw materials allocations.[138] The National Socialists would inherit and expand this system after January 1933. They would also find an eager collaborator in the firm that had staked its future on synthetic gasoline—IG Farben.

CREATING A GLOBAL HYDROCARBON CARTEL

The Great Depression, which followed a series of major oil discoveries in the United States, Venezuela, and Iraq and the recovery of Soviet production, accelerated the decline in oil prices—by as much as 50 percent between 1928 and 1931 in the United States.[139] No German firm was at greater risk from these developments than IG Farben.[140] This firm's role in fueling the German war effort was a central component of the sixth Nuremberg Trial in 1947–8. According to the prosecutors, synthetic fuel was at the heart of the conglomerate's partnership with the National Socialists starting in 1931. While in the dock, Carl Krauch, who oversaw Germany's synthetic fuel program between 1938 and 1945, claimed that his initial dealings with the National Socialists had nothing to do with rearmament but rather were about commercial matters.[141] By the late 1920s, IG Farben was in serious trouble—the cost of hydrogenation was increasing even as oil prices tumbled.[142] When BASF first decided to invest in hydrogenation in 1925, the price for imported gasoline was seventeen Pfennigs per liter, against an estimated twenty Pfennigs for synthetic gasoline. By 1931, the figures had diverged considerably: natural gasoline cost only five Pfennigs per liter, while synthetic gasoline cost 40–50 Pfennigs to produce, including amortization, but only sold for 30–35.[143] In 1929 the Leuna

[137] Memorandum, October 21, 1933, in: *Akten der Reichskanzlei: Regierung Hitler, 1933–1938*, ed. Karl-Heinz Minuth and Friedrich Hartmannsgruber (Boppard am Rhein: Boldt, 1983–2008), i: 918–20; Economics Ministry, "Begründung," no date, attached to: Schacht to Lammers, October 27, 1934, T-120/5677. See also: Löhr, "Deutsche Treibstoffversorgung," *Militär-Wochenblatt* (June 18, 1933); "Deutscher Kraftstoff," *Militär-Wochenblatt* (September 25, 1934).

[138] Stranges, "Fuel Industry," 158–60.

[139] DeGolyer, *Historical Data*, 3 and 75–1.

[140] The following discussion draws from: Hughes, "Momentum," 106–32; Peter Hayes, *Industry and Ideology: IG Farben in the Nazi Era* (Cambridge: Cambridge University Press, 1987), 36–42.

[141] *NMT*, vii: 607–14. See also the transcripts of Krauch's interrogations of September 1945 and April 1947 in: NARA, Record Group 238: National Archives Collection of World War II War Crimes Records, National Archives Microfilm Publication T-301, Reel 52, Documents NI-6767 and NI-6768.

[142] Wolfgang Birkenfeld, "Leuna 1933," *Tradition: Zeitschrift für Firmengeschichte und Unternehmerbiographie* 8 (1963): 98–100.

[143] Hughes, "Momentum," 122–3.

facility, with a theoretical output of 300,000 tons per annum at a cost of 220 RM per ton, was only producing 80,000 tons per year at 450 RM per ton.[144]

By 1931, despite investments of 400 million RM, it was clear that hydrogenation was a commercial failure. An internal IG Farben study that year concluded that synthetic gasoline had a price floor of twenty-three Pfennigs per liter, and that IG Farben should cease supporting hydrogenation with the proceeds from more profitable sectors of the conglomerate.[145] Only public assistance could save hydrogenation, and, starting in 1930, the beleaguered Brüning government quadrupled the import duty and reduced the tax on Leuna gasoline from 3.80 RM per 100kg to 1 RM.[146] Raising taxes in the midst of an economic depression provoked public anger, so the conglomerate waged a public-relations campaign.[147] In October 1931, IG Farben invited representatives of the German press and major political parties, including the National Socialists, to Leuna to showcase hydrogenation.[148] The following year, IG Farben executives approached Hitler directly and found their warm reception gratifying. Hitler emphasized that "an economy without oil is not to be thought of in a Germany which wishes to remain independent" and urged the conglomerate to continue its work.[149]

East German and even some Western scholars have emphasized the culpability of IG Farben in the planning and direction of the German war economy.[150] Others have suggested that hydrogenation's unprofitability forced the conglomerate into the role of supplicant—an unloved one, too, since IG Farben's 1933 agreement with the Third Reich did not cover its prior expenditures—as much as 430 million RM—or its net losses by 1936 when hydrogenation finally became profitable (perhaps 150 million RM).[151] Since IG Farben had to recoup its losses by producing synthetic rubber, it depended upon Berlin providing a monopoly and guaranteeing a favorable price. In that sense, IG Farben's support for autarky was a matter of expediency, rather than any affinity for National Socialist ideology.[152]

IG Farben initially pursued hydrogenation for commercial reasons. BASF had excess hydrogen-production capacity through the Haber–Bosch process. Hydrogenation would allow the company to utilize existing physical assets that would

[144] Stranges, "Fuel Industry," 183–5.

[145] Statement of Friedrich Jaehne, May 2, 1947, NARA, RG 238, T-301/52 (NI-6765).

[146] Karlsch, *Faktor Öl*, 135–8. [147] Hughes, "Momentum," 123–4.

[148] Gattineau, "Bericht [...]," October 12, 1931, NARA, RG 238, T-301/123 (NI-15257). IG Farben had to cultivate the National Socialists after falling foul of party radicals. Hayes, *Industry*, 64ff.

[149] Greiling, "25 Years Leunawerke," no date, NARA, RG 238, T-301/116 (NI-14304); *NMT*, vii: 535–54; Hughes, "Momentum," 124ff. The date of meeting is unclear—either June or November 1932. Hayes, *Industry*, 66–7.

[150] Dieter Petzina, "IG-Farben und nationalsozialistische Autarkiepolitik," *Tradition* 13 (October 1968): 250–4; Dietrich Eichholtz, "Zum Anteil des IG-Farben-Konzerns an der Vorbereitung des zweiten Weltkrieges," *Jahrbuch für Wirtschaftsgeschichte* 2 (1969): 83–105.

[151] Birkenfeld, "Leuna," 110; Karlsch, *Faktor Öl*, 135–8 and 158–9.

[152] Hayes, Industry and "Industrie und Ideologie: Die IG Farben in der Zeit des Nationalsozialismus," Zeitschrift für Unternehmensgeschichte, 32: 2 (1987): 124–36. See also: Henry Ashby Turner, *German Big Business and the Rise of Hitler* (New York: Oxford University Press, 1985), 246–9; Gottfried Plumpe, *Die I.G. Farbenindustrie AG: Wirtschaft, Technik und Politik* (Berlin: Duncker & Humblot, 1990), 281–96; Wilkins, *Investment*, 406–14. For an alternative assessment, see: Adam Tooze, *The Wages of Destruction: The Making and Breaking of the Nazi Economy* (New York: Viking, 2006), 227–30.

otherwise remain idle due to lower demand for ammonia-based fertilizers. Although oil prices were already falling by the time IG Farben decided to take the plunge into synthetic fuel, it was confident of making a profit if it controlled the cost of production and raw materials.[153] If crude oil became scarce, hydrogenation would offer the company entrée into the petroleum industry. In the event that oil supplies remained plentiful, the company could still make money if it created a protected market in Germany by promoting hydrogenation as a work-creation measure that would also improve the country's balance of payments. By eliminating the need to import Chilean nitrates, the Haber–Bosch process was already saving Germany 200 million RM in foreign exchange annually since the end of World War I, and the savings through hydrogenation would be even greater. In addition, gasoline consumption was sure to rise as German society motorized.[154] The fact that hydrogenation also served a strategic purpose was an ancillary consideration. The primary aim in the 1920s was, instead, to corner an emerging market while enhancing the profitability of the firm's existing operations. That did not prevent IG Farben from manufacturing a political rationale for public support—in the case of hydrogenation, that of energy independence.[155] Ultimately, commercial self-interest guided IG Farben's actions, as evidenced by the fact that it preserved and expanded partnerships with companies from the very powers against which the Third Reich was rearming.[156]

The most profitable such relationship was with Jersey, beginning around 1925–6. In 1927, the two firms concluded a twenty-five-year agreement regarding Jersey's use of hydrogenation in the United States.[157] The partnership with Jersey was part of a wider campaign waged by the German chemical industry after World War I to recover market share.[158] The German conglomerate's technical prowess made it an invaluable partner to Jersey. Although it was never a commercial success in terms of coal liquefaction, hydrogenation made an invaluable contribution to refining.[159] Whereas in 1926 Jersey required two bbl of crude oil to produce one barrel of gasoline, by 1928, using hydrogenation, one barrel of crude could yield more than a barrel of gasoline. The company also found that synthetic lubricants sometimes outperformed natural equivalents.[160]

The U.S.–German collaboration "fulfilled all expectations," according to an IG Farben assessment.[161] Under a 1929 agreement, IG Farben sold its hydrogenation and oil-refining patent rights beyond Germany to Jersey for 20 percent of the royalties and 546,011 shares in Jersey, making it the largest shareholder in the company behind the Rockefeller family. The two firms also created a jointly owned holding

[153] "Synthetisches Petroleum," no date, enclosed with: Zur Nedden to Schubert, October 15, 1927, PAAA, R 118173.

[154] Hughes, "Momentum," 110–18; Stranges, "Fuel Industry," 170–2 and 175–6.

[155] Boesler, "Leunas Bedeutung [...]," June 1937, BA-B, R 8128/3293.

[156] Nowell, *Cartel*, 1–44 and 280–8.

[157] The following discussion draws from: Larson, *New Horizons*, 152–9; Yergin, *Prize*, 329–31; Nowell, *Cartel*, 223ff; Stranges, "Fuel Industry," 185ff; Wilkins, *Investment*, 238–51 and 328–35.

[158] Mira Wilkins, "German Chemical Firms in the United States from the Late-19th Century to Post-World War II," in: Lesch, Chemical Industry, 302ff.

[159] Plumpe, *I.G. Farbenindustrie*, 295.

[160] Stranges, "Fuel Industry," 186–7 and 191.

[161] IG Farben, "Chronik [...]," December 13, 1944, BA-B, R 8128/817.

company, Standard–IG, which would administer the patents. In 1930, they set up another firm, JASCO (Joint American Study Company), which held the rights to synthetic rubber (Buna). That year, Jersey began licensing hydrogenation in the United States through the Hydro Patents Company, and seventeen U.S. companies bought shares. After 1931, Standard–IG—in cooperation with Shell and Imperial Chemical Industries, owner of the British patent rights to hydrogenation—issued global licenses for these patents through the International Hydro Patents Company. Although Jersey took over leadership of Standard–IG, IG Farben earned substantial revenues through licensing fees, while preserving its protected German market.[162]

These partnerships "cannot be explained as a function of Nazi strategic interests" so much as commercial advantage.[163] Even after the Third Reich began pressuring German multinationals to cease investing earnings by foreign subsidiaries abroad and limit technology sharing, IG Farben maintained its international connections.[164] In fact, as early as 1933, German military analysts warned that IG Farben's production of synthetic gasoline might be "tied to commercial agreements with foreign concerns."[165] Between 1937 and 1940, IG Farben, Jersey, and Shell—who were already partners within Germany through Deutsche Gasolin—subsidized the construction of a synthetic plant at Pölitz with an annual capacity of 600,000 tons.[166] The impetus came not from IG Farben but rather from Jersey and Shell, who used it as a means of investing stranded funds earned from sales in Germany.[167] The outbreak of World War II forced the companies to restructure their partnership, and between September and December 1939 IG Farben transferred 2,000 of its foreign patent rights vested under Standard–IG and JASCO to Jersey to avoid their confiscation.[168] Although Jersey bought IG Farben's shares in Standard–IG and JASCO for nominal fees, the latter would be able to repurchase them and its patent rights after the war ended. The two firms did not terminate their relationship until March 1942, when Jersey signed a consent decree under government pressure. Washington also oversaw the sale of IG Farben's remaining shares in Jersey.[169]

The U.S. Justice Department and Congress blamed the agreements with IG Farben for contributing to the wartime shortage of rubber by delaying Buna production, although opinion among the public and in the press over whether this was deliberate varied based on political persuasion—liberals and antitrust advocates took the harder line. Jersey eventually responded with a sophisticated public-relations

162 Nowell, *Cartel*, 236ff; Stranges, "Fuel Industry," 187–91; Wilkins, *Investment*, 241–2.

163 Nowell, *Cartel*, 241. 164 Wilkins, "Chemical Firms," 313–14.

165 Löhr, "Deutsche Treibstoffversorgung," *Militär-Wochenblatt* (June 18, 1933).

166 The oil companies participated secretly, since outside of Germany they posed as opponents of autarky. Jonker and van Zanden, *History of Shell*, 473–4. Deutsche Gasolin emerged from the ruins of Hugo Stinnes' empire. BASF bought it in 1925 and offered Jersey and Shell each a 25 percent stake. Karlsch, *Faktor Öl*, 138–40.

167 Hans-Jürgen Zetzsche, "Logistik und Operationen: Die Mineralölversorgung der Kriegsmarine im Zweiten Weltkrieg" (Ph.D. diss., Christian-Albrechts-Universität zu Kiel, 1986), 74–6; Plumpe, *I.G. Farbenindustrie*, 289.

168 Larson, *New Horizons*, 405–8.

169 Larson, *New Horizons*, 428–33; Wilkins, *Investment*, 461–70 and 510–16.

offensive that rebutted the most egregious accusations impugning its patriotism.[170] One engineer, part of the first team that visited Germany in 1927 to investigate hydrogenation, pointed out that the patent-sharing agreement provided the Allies with numerous benefits, including a means of producing 100-octane gasoline, as well as synthetic toluene and rubber.[171] IG Farben disputed these claims—neither the production of iso-octane nor that of synthetic toluene had anything to do with hydrogenation. In fact, IG Farben had acquired valuable information such as the method for producing tetraethyl lead.[172] Such knowledge had indeed proven invaluable—but not in the way IG Farben expected when it began its partnership with Jersey.

A NEW DIRECTION, 1933–1935

While IG Farben's efforts during the 1920s and 1930s were not motivated by strategic considerations, they nonetheless facilitated German aggression following the National Socialists' "seizure of power" in January 1933. Self-sufficiency was at the heart of National Socialist economic theory.[173] Within months of taking power, the National Socialists, backed by the military, pressed for the expansion of domestic petroleum output irrespective of cost. The task initially fell to State Secretary Gottfried Feder. An engineer by training, Feder was one of the founders of the National Socialist Party and drafted its 1920 program. He also distrusted large enterprises and was committed to the defense of small businesses.[174] In 1933, he took over petroleum affairs at the Economics Ministry and pledged to utilize all available measures. Refinery construction was an attractive option because of the cost savings and increased employment.[175] This was Feder's preference, but by the autumn of 1933, he had come around to supporting synthetic fuel.[176]

Synthetic fuel was more than three times costlier than imported gasoline—20 RM versus 6 RM per liter—but economic factors alone would not determine strategy. Rather, the new regime would also consider employment and strategic factors, and it would encourage synthetic production and crude oil imports

[170] Larson, *New Horizons*, 433–52; Sampson, *Seven Sisters*, 77–83; Wilkins, "Chemical Firms," 315 and Foreign Investment, 1914–1945, 533–4 and 542–3.

[171] "Secrets [...]," *NMT*, vii: 1297–303.

[172] "Comments [...]," June 6, 1944, attached to: Knieriem to Schmitz et al., "Haslam Article," June 6, 1944, *NMT*, vii: 1303–11.

[173] Organizational Department II, "Wirtschaftliche Grundanschauungen [...]," March 5, 1931, reprinted in: Avraham Barkai, "Wirtschaftliche Grundanschauungen und Ziele der N.S.D.A.P.: Ein unveröffentlichtes Dokument aus dem Jahre 1931," *Jahrbuch des Instituts für Deutsche Geschichte* 7: 2 (1978): 355–85.

[174] Turner, *Business*, 62–4.

[175] As of 1933, 40 percent of Germany's supply of lubricants was imported—112,000 tons out of 250,000 tons. Refining 100 percent would save 13.2 million RM. Foreign Policy Office to the Chancellery, "Massnahmen [...]," January 18, 1934, BA-B, R 43 (II)/486.

[176] Karlsch, *Faktor Öl*, 167. Hitler later claimed that he "broke with Feder" because of his opposition to synthetic fuel. *Table Talk*: No. 68 (November 12, 1941).

"simultaneously."[177] It had no ideological incentive to bail out IG Farben—the National Socialists in fact proclaimed their distaste for state interference in private industry unless necessary.[178] The civilian bureaucracy was, however, risk-averse. The Economics Ministry acknowledged that the foreign exchange burden of imports made "the expansion of domestic production necessary above all on strategic grounds," but argued that this should only happen through "the most necessary interference in the petroleum industry."[179] The ministry was reluctant to assist "high-risk" projects and wished to limit public support to price and marketing guarantees (including one for IG Farben at Leuna), plus tariff reductions to incentivize refinery construction.[180] The military and the Transportation Ministry, in contrast, wanted the emphasis on synthetic fuel, since refineries would be of little use if imports ceased.[181]

"The only worthwhile solution," one military journal declared in 1934, was hydrogenation.[182] Despite the regime's enthusiasm, however, progress was slow—Hitler blamed the Economics Ministry, in particular.[183] Nevertheless, the military, National Socialists, and IG Farben began preparing a "Four-Year Plan" to boost petroleum production in the summer of 1933.[184] The point man for these discussions was an IG Farben executive, Ernst Rudolf Fischer. Fischer was a rising star within IG Farben's gasoline subsidiary who joined the National Socialist Party in April 1933 and moonlighted at its Economic Policy Office.[185] In June, he unveiled to the Ordinance Office and the Aviation Ministry a four-year plan that would raise production from 750,000 tons to 2.1 million tons. By 1937, Fischer estimated that Germany could produce 60 percent of its consumption.[186] A few months later,

[177] "Niederschrift [...]," attached to: the Economics Minister to Finance Minister et al., September 1, 1933, T-120/5677.

[178] Barkai, "Grundanschauungen," 383–4. In September 1933, Feder stated that developing Germany's synthetic fuel capacity was "enormously important" but insisted that the regime not exercise direct control. "Erste Sitzung [...] 20 September 1933," *Akten*, i: 794–5.

[179] "Vermerk [...]," October 20, 1933, *Akten*, i: 918–20; "Vorschläge [...]," no date, attached to: Economics Minister to Finance Minister et al., July 17, 1933, T-120/5677.

[180] "Vermerk [...]," October 20, 1933, *Akten*, i: 918–20.

[181] "Niederschrift [...]," attached to: Economics Minister to Finance Minister et al., September 1, 1933, T-120/5677. Higher crude oil production was a laudable aim, but military analysts worried that little was possible due to the capital requirements. Löhr, "Deutsche Treibstoffversorgung," *Militär-Wochenblatt* (June 18, 1933).

[182] "Deutscher Kraftstoff," *Militär-Wochenblatt* (September 25, 1934).

[183] *Table Talk*: No. 68 (November 12, 1941) and No. 128 (January 27, 1942). The first two years' developments are summarized in Tooze, *Wages*, 115–20.

[184] This is not to be confused with the "first" Four-Year Plan, of 1933, to promote economic recovery. The "second," of 1936, focused on economic self-sufficiency. Alfred-Ingemar Berndt, *Gebt mir vier Jahre Zeit! Dokumente zum ersten Vierjahresplan des Führers* (München: Franz Eher Nachf., 1937), 211–19.

[185] Keppler to Fischer, July 16, 1933, Fischer to Völmicke, June 8, 1933, and Economic Policy Office to the Reich Association of Garage Owners, June 23, 1933, all in: BA-B, R 8128/10352; Avraham Barkai, *Nazi Economics: Ideology, Theory, and Policy* (New Haven: Yale University Press, 1990), 30. For details concerning Fischer's party membership, see: A3340, MFOK, E051; A3340, MFKL, G126. His biography before 1940 is summarized in Titus Kockel, "Eine Quelle zur Vor- und Gründungsgeschichte der Kontinentale Öl AG aus dem Jahr 1940," *Jahrbuch für Wirtschaftsgeschichte* 1 (2003): 176–8.

[186] Economic Policy Office to Ordinance Office, June 21, 1933, and Economic Policy Office to Aviation Ministry, June 27, 1933, both in: BA-B, R 8128/10352.

following an introduction by the industrialist Albert Vögler, Krauch, then leading the IG Farben division responsible for hydrogenation, presented State Secretary Erhard Milch of the Aviation Ministry with "a four-year plan" to increase production by 1.3 million tons.[187] Krauch estimated it would cost 400 million RM and require 46,000 workers. Of the new production, 800,000 tons would come through hydrogenation, and Krauch expected that by 1937 Germany would satisfy two-thirds of its consumption internally.[188] Milch was enthusiastic and forwarded Krauch's plan to the head of the Ordinance Office, General Alfred von Vollard-Bockelberg.[189] The general agreed with Milch, and impressed upon Economics Minister Kurt Schmitt the need to support the construction of hydrogenation facilities and storage facilities.[190] There is also evidence that Hitler supported the plan.[191]

In December 1933, IG Farben and the Third Reich signed the so-called "Gasoline Agreement." The firm agreed to triple Leuna's production by 1935 in return for a guarantee that the government would purchase any unsold stock for ten years at a fixed price (18.5 Pfennigs per liter) that covered the company's past investment plus a five percent profit, while repaying excess profits.[192] As a result, IG Farben tripled Leuna's output from 108,000 tons in 1933 to 332,000 tons by 1936. Ironically, after 1936, the contract became an encumbrance to IG Farben by depriving it of any excess profits. During the first two years of the deal, IG Farben received five million RM from the regime, but once production costs dropped—to 13.5 Pfennigs per liter by 1943—the conglomerate had to recompense Berlin. This amounted to more than ninety-one million RM by 1944—a net loss of eighty-six million RM.[193]

One should not overstate the importance of developments such as the "Gasoline Agreement": in fact, the Third Reich's petroleum strategy before 1939 was not limited to synthetic fuel. Unfortunately, the most popular histories do not convey the full scope of German efforts and ambitions, choosing to focus exclusively on synthetic fuel before 1939.[194] This misconception arose from the fact that many

[187] Vögler to Krauch, August 10, 1933, NARA, RG 238, T-301/44 (NI-5930).

[188] "Die deutsche Treibstoffwirtschaft," June 1933, BASF Archive, IG Farben, M02/1; Krauch to Milch, September 14, 1933, *NMT*, vii: 571–3. I am grateful to Ray Stokes for sharing the former document.

[189] Bockelberg, "Besprechung... 15.9.33," NARA, RG 238, T-301/55 (NI-7123). Thanks to Krauch, IG Farben and the Aviation Ministry negotiated an agreement in June 1936 whereby the former promised to keep Leuna's output of aviation fuel at 200,000 tons per annum until the end of 1950. Kesselring et al., "Vertrag [...]," June 10, 1936, attached to: Heinz to War Minister, September 14, 1936, NARA, RG 238, T-301/64 (NI-7836). For background, see: Hayes, *Industry*, 139–42.

[190] "Vortragsnotizen [...]," November 4, 1933, no author, and Wa Wi, "Ergebnis [...]," November 7, 1933, both in: NARA, RG 238, T-301/64 (NI-7828).

[191] *Table Talk*: No. 68 (November 12, 1941).

[192] "Benzin-Vertrag," December 14, 1933, NARA, RG 238, T-301/9 (NI-881). Total investment in the construction of synthetic facilities before 1933 was 108 million RM. Of IG Farben's expenditure on facilities between 1925 and 1944 (786 million RM), 88 percent took place after 1932—65 percent (586 million RM) between 1941 and 1944. Plumpe, *I.G. Farbenindustrie*, 292–3.

[193] Birkenfeld, "Leuna," 103–10; Hayes, *Industry*, 115–20; Karlsch, *Faktor Öl*, 167–8.

[194] Robert Goralski and Russell Freeburg, *Oil & War: How the Deadly Struggle for Fuel in WWII Meant Victory or Defeat* (New York: Morrow, 1987), 13–26; Yergin, *Prize*, 328–50.

scholars relied on the United States Strategic Bombing Survey (USSBS) and the materials collected for the Nuremberg Trials.[195] The limitations of the latter, collected for proving criminal charges, are self-evident. As for the USSBS, its authors were less interested in understanding how the German economy functioned than with identifying its chokepoints.[196] Things were different on the other side of the Iron Curtain. According to Marxist orthodoxy, the Third Reich was an instrument of German monopoly capitalism, and this was true in particular of IG Farben, the largest German firm before the war and fourth-largest in the world, behind only General Motors, Jersey, and U.S. Steel.[197] In fact, the company's share of German petroleum production between 1933 and 1943—5.9 percent—was meager. Its output peaked in 1942 at 561,000 tons, and after Leuna, IG Farben never constructed another synthetic fuel facility on its own.[198]

Moreover, many of the same interest groups—in particular, foreign and domestic oil companies—that competed for favor during the Weimar Republic remained active after 1933.[199] At the start of the Third Reich, there were still doubts about whether synthetic fuel could live up to expectations.[200] These reservations evaporated once the regime made clear its determination to boost consumption while reducing the foreign exchange burden of imports. In 1928, these reached 438 million RM—they would have been four times larger if Germany had the same number of automobiles as Britain or France.[201] In May 1933, Hitler met with oil industry representatives and Feder to discuss a 100-million RM plan to expand refinery capacity. Hitler and his advisors were concerned about both Germany's import dependence and its reliance on the major oil companies, which controlled the domestic market and 80 percent of existing refinery capacity. Eliminating imports of refined products, as France had done after 1928, would save 200 million RM of foreign exchange—a figure that would rise as consumption increased.[202]

That same day, oil featured prominently in a speech Hitler gave to German industrialists. Much of the talk was devoted to the construction of the Autobahn,

[195] Especially early studies such as Wolfgang Birkenfeld, *Der Synthetische Treibstoff* (Göttingen: Musterschmidt-Verlag, 1964); Dieter Petzina, *Autarkiepolitik im Dritten Reich: Der nationalsozialistische Vierjahresplan* (Stuttgart: Deutsche Verlags-Anstalt, 1968).

[196] United States Strategic Bombing Survey (USSBS), *Effects of Strategic Bombing on the German War Economy* (Washington, DC: U.S. 1945), and *Oil Division: Final Report* (Washington, DC: GPO, 1945).

[197] Turner, *Business*, xvi.

[198] Plumpe, *I.G. Farbenindustrie*, 284 and 287–8.

[199] Aside from the major oil companies, Titus Kockel identifies eight different groups as of 1932. Dietrich Eichholtz and Titus Kockel, *Von Krieg zu Krieg: Zwei Studien zur deutschen Erdölpolitik in der Zwischenkriegzeit* (Leipzig: Leipziger Universitätsverlag, 2008), 114–19.

[200] Lehmann, "Der wehrpolitische Kampf [...]," *Deutsche Wehr* (February 10, 1933); Löhr, "Deutsche Treibstoffversorgung," *Militär-Wochenblatt* (June 18, 1933).

[201] Heberlein, "Deutscher Kraftstoff," *Deutsche Wehr* (February 7, 1934).

[202] Chancellery, "Vermerk [...]," May 31, 1933, NARA, RG 238, T-301/6 (NI-549). See also: Birkenfeld, "Leuna," 102; Karlsch, *Faktor Öl*, 165; Kockel, *Von Krieg*, 120–1. By the end of 1934, French refinery capacity had reached thirty-four million bbl per year, compared to 22.7 million bbl in Britain and 13.8 million bbl in Germany. "Deutscher Kraftstoff," *Militär-Wochenblatt* (September 25, 1934).

which had little strategic value.[203] Rather, Hitler considered it a work-creation measure and economic stimulus.[204] Hitler also hoped that road construction would stimulate both the petroleum and the automotive sectors, the latter of which was in severe difficulties even before 1933. The number of automobiles in service declined between 1930 and 1932 from 81,000 to 41,000. Accordingly, in April 1933, the government eliminated all taxes on the purchase of cars, which led to a threefold increase in registrations by 1938.[205] Nevertheless, it was impossible for Germany to enjoy the same rates of car-ownership as the United States due to foreign exchange considerations.[206] There was not a moment to lose—oil prices were sure to rise, and as Feder warned in 1933, the bill for Germany's oil imports would be 800 million RM if it possessed the same number of automobiles as France.[207]

Three points require emphasis. First, the National Socialists—following the lead of private industry—justified their early support for hydrogenation as a work-creation measure.[208] The proposed contract with IG Farben alone would create employment for 12,000 workers and inject 45 million RM of orders into the economy.[209] Second, the Third Reich did not advertise its objectives to avoid provoking the foreign oil companies and their parent governments.[210] This explains why a "Petroleum Law" proposed by the Economics Ministry in October 1933 went nowhere.[211] The legislation would have established a public–private national oil company to purchase internally produced petroleum at above market prices and would have hindered the expansion of facilities that processed imported oil.[212] After two cabinet meetings, the government shelved the law but authorized the Finance Ministry to underwrite government contracts with private firms to purchase petroleum at fixed prices—the legal basis of the agreement with IG Farben.[213]

Third, the regime adopted the perspective that synthetic fuel was only one leg of a three-pronged approach before the war that included stockpiling imports (especially

[203] The army was unenthusiastic due to the shortage of gasoline and rubber. R.J. Overy, "Transportation and Rearmament in the Third Reich," *Historical Journal* 16: 2 (1973): 390–400.

[204] *Table Talk*: No. 201 (July 18, 1942).

[205] Birkenfeld, "Leuna," 100–1; Tooze, *Wages*, 150–1.

[206] "Besprechung [...] 29 Mai 1933," *Akten*, i: 512–13.

[207] Limiting imports to 1932 levels under such circumstances would require a 13.5-fold increase in production. Roesner, "Die Treibstoff-Frage [...]," *Militär-Wochenblatt* (May 25, 1934).

[208] A point Hitler reiterated during the war: *Table Talk*: No. 68 (November 12, 1941) and No. 128 (January 27, 1942).

[209] "Kabinettssitzung vom 8 Dezember 1933," *Akten*, i: 987. Vögler estimated that a policy of petroleum self-sufficiency could provide "lasting" employment for 150,000 workers. "Erste Sitzung... 20 September 1933," *Akten*, i: 779.

[210] For example, the regime decided not to renew contracts with Soviet exporters expiring in 1934. Moscow was still eager to expand oil exports to Germany, but whereas the War Ministry had supported supply contracts with Soviet firms in 1929, by 1934 military analysts considered them undesirable. Defense Ministry to Foreign Office, "Bezug [...]," May 17, 1929, PAAA, R 118173; Ritter, file note, March 8, 1933, PAAA, R 118176; "Kabinettssitzung vom 8 Dezember 1933," *Akten*, i: 988, n. 11; "Erdöl [...]," *Militär-Wochenblatt* (November 25, 1934).

[211] Kockel, *Von Krieg*, 120–34.

[212] See: PAAA, R 118177, esp. "Entwurf [...]," no date or author, enclosed with: Feder to the Chancellery et al., October 14, 1933; Kiewitz, "Gesetzentwurf [...]," November 2, 1933; Economics Minister to Reich Chancellery, November 17, 1933, PAAA, R 118178.

[213] "Kabinettssitzung vom 1 Dezember 1933," and "Kabinettssitzung vom 8 Dezember 1933," both in: *Akten*, i: 987–9 and 1012–13.

crude versus refined petroleum), and stimulating domestic crude oil production. None of these activities on their own could break Germany's dependence on imported petroleum, which meant the regime could not afford to overlook any of them.[214] As far as crude oil was concerned, Germany's leading petroleum geologist, Alfred Bentz, concluded in 1933 that the most the country could hope for was a minor boost in production to improve its balance of trade.[215] Monthly crude oil production doubled between 1931 and 1935 (19,391 tons to 35,992 tons), but its share relative to supply was unchanged due to higher consumption.[216]

There was quicker progress with synthetic fuel. In the period September–October 1934, the acting Economics Minister, Hjalmar Schacht, forced the nine largest coal companies into a "compulsory association" with IG Farben, known as Braunkohle-Benzin (BRABAG), to expand synthetic fuel production.[217] The managing board agreed that the group would construct three new hydrogenation facilities under license to IG Farben, and later another using the Fischer–Tropsch process.[218] While IG Farben's direct share of petroleum production during the Third Reich was small—5.9 percent (30 percent of synthetic fuel production)—its indirect share through BRABAG (26.9 percent by 1942–3) was much higher.[219] Such actions were unavoidable, as Schacht confessed to Hitler, because Germany's foreign exchange position was dire. The monthly deficit was running at 160 million RM, and Germany might even have to dip into its food and raw materials stockpiles. Petroleum reserves covered only three months of consumption, but new synthetic facilities would not be ready for another year.[220] In December, Schacht proposed new legislation granting the Economics Ministry joint authority with the Finance Ministry to provide guarantees to the textile, petroleum, metallurgical, and chemical industries. The agreement with IG Farben, which provided 350,000 tons of gasoline annually, would serve as the model, and the Economics Ministry was negotiating a further 95,000-ton increase. Schacht also expected that lignite producers would furnish funds through BRABAG for the construction of facilities to add 500,000 tons to Germany's supply.[221]

By the spring of 1934, therefore, the Third Reich had implemented a multifaceted strategy to boost Germany's petroleum supplies. As Feder explained in April to Hans

[214] "Mineralölwirtschaft," no date or author, enclosed with: Herz to Ritter, July 5, 1933, PAAA, R 118177.

[215] Karlsch, *Faktor Öl*, 172.

[216] Pflanz, "Das deutsche Erdöl," *Die deutsche Volkskraft: Beilage zur "Deutschen Wehr,"* 6: 3 (April 16, 1936). See also: "Erdöl- [...]," *Wehrtechnische Monatshefte* 2 (1936), T-77/425 (Wi/IF 5.3444).

[217] Kockel suggests that Schacht undertook this action to head off criticism by the War Ministry regarding the slow pace in increasing petroleum supplies. Kockel, *Von Krieg*, 140–1.

[218] Bütefisch, "Gründung [...]," attached to: "Bericht... 10 Januar 1935," NARA, RG 238, T-301/62 (NI-7669). For the license agreement, see: "Lizenvertrage [...]," June 14/August 22, 1935, NARA, RG 238, T-301/63 (NI-7767). Key documents concerning the foundation of BRABAG are in: NARA, RG 238, T-301/28 (NI-3975). See also: Hayes, *Industry*, 133–5; Karlsch, *Faktor Öl*, 182–4; Plumpe, *I.G. Farbenindustrie*, 281–5.

[219] Plumpe, *I.G. Farbenindustrie*, 281–2 and 287.

[220] "Chefbesprechung [...] 18 Oktober 1934," *Akten*, ii: 105.

[221] "Gesetz [...]" and "Begründung," attached to: Schacht to Lammers, December 1, 1934, PAAA, R 28806. The cabinet gave its approval on December 13, 1934. "Kabinettssitzung [...]," *Akten*, ii: 242 (esp. n. 3).

Lammers, the Chief of the Chancellery (and presumably for Hitler's benefit), IG Farben was expanding production at Leuna and planning to construct two new facilities. Meanwhile, planning for two Fischer–Tropsch facilities was nearly complete. German oil companies were committed to increasing crude oil production and constructing new refineries, and the government would soon provide them with an annual subsidy of five million RM to evaluate German reserves. Finally, Feder was negotiating with the European Storage and Transportation Company (Eurotank), which was controlled by a U.S. businessman, William Rhodes Davis, to barter Mexican crude oil for German finished goods. The Mexican oil would be processed in the cutting-edge refinery Eurotank was building in Hamburg (eventually the third-largest in Germany).[222] A quarter of Eurotank's gasoline production would remain in Germany, while the remainder would be exported to earn foreign exchange. Feder described this project as the "first large-scale exchange of goods along the lines set by the National Socialist economic leadership."[223]

Feder was soon fired and eventually replaced by one of his "close collaborators," Fischer. By 1934, the latter was still an employee of IG Farben, but he was becoming increasingly influential, blurring the line between the public and private sectors. He had already negotiated the "Gasoline Agreement" of 1933 and served as a liaison to the Economics Ministry the following March during negotiations between importers and domestic producers to reorganize the German petroleum market.[224] A few months later, he assessed Germany's petroleum position for his private employers. He expected that transportation consumption would rise by 10 percent per year until 1938 (to 2.3 million tons). In terms of supplies, Fischer dismissed benzol as inefficient and argued it was best to refine German crude oil into lubricants and diesel. The only practical options for meeting future consumption were hydrogenation and the Fischer–Tropsch process. Fortunately, there was no shortage of coal to hold back synthetic production. Reserves stood at twenty-two billion tons of soft coal and ninety billion tons of hard coal. At current rates of production, reserves of soft coal would last for 170 years and hard coal 600 years.[225] Assuming that sixteen tons of soft coal and five tons of hard coal produced one ton of gasoline, IG Farben required only eight million tons of the former and 1.5 million tons of the latter to synthesize the 800,000 tons it planned to produce.[226]

[222] Karlsch, *Faktor Öl*, 168–9.

[223] Feder to Lammers, April 23, 1934, BA-B, R 3101/20320. See also: "Der Davis Plan," no date or author (circa December 1934), Ritter to Ulrich et al., "File Note," December 18, 1934, and Davidsen, note, December 28, 1934, all in: PAAA, R 118179. The Eurotank refinery was complete in 1935 but little Mexican oil reached Germany until 1938. Klaus Volland, *Das Dritte Reich und Mexiko: Studien zur Entwicklung des deutsch-mexikanischen Verhältnisses 1933–1942 unter besonderer Berücksichtigung der Ölpolitik* (Frankfurt am Main: Peter Lang, 1976), 154–6.

[224] Kiewitz to Ritter et al., "Erdöl-Bewirtschaftung," March 28, 1934, PAAA, R 118178.

[225] These figures were actually conservative. Official reserves rose by 1937 to 289 billion tons of hard coal and fifty-seven billion of soft coal, sufficient for 2,000 and 400 years, respectively, at rates of consumption of the time. Reichs-Kredit-Gesellschaft, "Treibstoffwirtschaft [...]," April 1938, T-84/51 (EAP 66-c-2-10/22).

[226] "Der gegenwärtige Stand [...]," September 7, 1934, attached to: Fischer to Ammoniawerk Merseburg, September 13, 1934, BASF Archive, IG Farben, M02/1. I am grateful to Ray Stokes for sharing this document with me. See also: Fischer et al., "Deutschlands Treibstoff- [...]," no date (circa 1935), T-84/194 (EAP 66-c-12-62/130).

The company had yet to deliver on such promises, and until it did, Berlin would not be able to afford to forsake cooperation with traditional suppliers if they could serve the interests of German strategy. Compared with years past, however, by 1934 Germany was in a stronger position vis-à-vis the companies. Across Europe, even before the Great Depression, economic nationalism was rising—virtually every European nation had abandoned economic liberalism and adopted mercantilist policies. For the major oil companies, this could mean operating under capital and trade controls, participating in state-sponsored cartels, expanded exploration efforts, increased stockpiling, supporting synthetic alternatives, or constructing refineries.[227]

The years between 1933 and 1936 were also the apex of Schacht's influence. Schacht is a study in contrasts. A banker with impeccable establishment credentials, cofounder of the liberal German Democratic Party, and protégé of Foreign Minister Gustav Stresemann, he oversaw the stabilization of the German currency following the hyperinflation crisis of 1923/4. Following Stresemann's death in 1929, Schacht renounced his liberal creed, cast his lot with the radical right, and embraced economic nationalism. After becoming de facto Economics Minister in 1934, he sought to expand Germany's civilian and war economy through the "New Plan" of 1934—a complex set of arrangements to conserve foreign exchange by regulating imports and reorienting trade on a bilateral basis.[228] With regard to oil, Schacht wanted the companies to increase deliveries without accepting immediate payment in foreign exchange. Considering the depressed state of the oil market, he was convinced the companies would comply in exchange for financial assistance.[229]

Efforts commenced in 1933. As Deterding explained to a German official in September 1934, he had been willing the previous year to stockpile in Germany, on credit, one million tons of oil—six months of consumption—purchased from the Soviet Union, but this plan had collapsed for reasons unknown even to Deterding.[230] He now complained about a recent Japanese law compelling oil companies to maintain large stockpiles that could be "impounded" by the state.[231] Italy was also considering such measures, and although Deterding sympathized with National Socialism, he urged Germany instead to cooperate with the companies.[232]

This is exactly what the regime was trying to do. In July 1934, the Economics Ministry approached the German subsidiaries of Jersey, Shell, and Anglo-Persian

[227] Wilkins, *Maturing*, 233–7; Yergin, *Prize*, 268–9.

[228] Milward, "Reichsmark Bloc," 355–8; Tooze, *Wages*, 15–16 and 67–98.

[229] RWM, "Bericht…30 September 1934 […]," reprinted in: International Military Tribunal, *Trial of the Major War Criminals* (Nuremberg, 1947–9), xxxvi: 176–8—hereafter cited as: *IMT*; Kockel, *Von Krieg*, 133.

[230] The documentary record is unclear as to why Deterding's plan failed, but there is no doubt he tried. See: "Kabinettssitzung vom 1 Dezember 1933," *Akten*, i: 988; Jonker and van Zanden, *History of Shell*, 485–6.

[231] Wilkins, *Maturing*, 230–1.

[232] "Besprechung […]," September 28, 1934, no author, T-77/228 (Wi/IF 5.1171); Daitz to Rosenberg, "Erdöleinlagerung […]," November 1, 1934, NARA, RG 238, Entry 1, Box 107 (1356-PS). Other German experts were skeptical of efforts to recruit Deterding, as his influence over Shell's management had waned since 1931. W Wi VI, "Aktennotiz […]," June 17, 1938, T-77/683 (Wi/VI. 356). Deterding's flirtation with fascism is detailed in Yergin, *Prize*, 369–70; Jonker and van Zanden, *History of Shell*, 475–91.

with the following offer.[233] The companies would accumulate one million tons in new storage facilities they would construct. This oil would become a "national reserve" for emergency use only. The companies would also maintain a working surplus of four months' consumption. Finally, the companies would not repatriate their earnings in Germany for five years. Instead, these profits would be loaned to German banks or invested in the German economy. In exchange, the Third Reich would not expand its support for the synthetic fuel industry and would allow the companies to maintain their market share.[234] Since the market was suffering from oversupply, Shell considered the proposal "desirable." As the group explained to the British Petroleum Department, Shell had considerable resources invested in Germany and was concerned about "autarchic elements" urging an expansion of synthetic fuel. By accepting the offer, Shell believed it could undermine support for autarky within Germany.[235] Shell could handle currency controls, and since it would soon have excess supplies from Iraq, the fact that Germany wished to increase its stockpile was a welcome development.[236]

Whitehall was unimpressed, and its chief economic advisor ridiculed Shell's position.[237] Jersey's main objection was the requirement to leave its profits in Germany, with no guarantees concerning the exchange rate when the five-year waiting period expired.[238] Despite having considerable assets stranded in Germany, Anglo-Persian was unwilling to cross the Foreign Office. Shell meanwhile modified its position—it was willing to abide by a three-year waiting period on repatriation of profits if Berlin guaranteed a fixed exchange rate and paid interest.[239] Jersey complained to the U.S. Embassy about Shell's position, especially the stipulation that Shell's support was conditional upon the other two companies agreeing.[240] Jersey suspected that Shell had engineered the whole scheme because of Deterding's pro-German sympathies.[241] Shell would probably gain disproportionately—under the German plan, companies wishing to import oil into Germany would have to do so using hard currency earned through exports or re-exports, and Shell's German

[233] The following discussion draws from: Zetzsche, "Logistik," 80–2; Mejcher, *Politik*, i: 160–70; Neil Forbes, *Doing Business with the Nazis: Britain's Economic and Financial Relations with Germany, 1931–1939* (London: Cass, 2000), 147–57.

[234] "Technische [...]," no date or author, enclosed with: Pasel to Foreign Office, "Mineralölversorgungsplan," July 25, 1934, PAAA, 118179; Brown to Runciman, "Supplies [...]," November 8, 1934, attached to: Runciman, "Supplies [...]," November 9, 1934, C.P. 252 (34), BNA, CAB 24/251.

[235] This opinion was widespread in Germany. "Royal Dutch als Weltmacht," *Die deutsche Volkswirtschaft* 21 (1936), T-77/425 (Wi/IF 5.3444). Hitler suspected as much—he later complained that Deterding was conspiring with Schacht. *Table Talk*: No. 128 (January 27, 1942).

[236] "Memorandum [...]," July 30, 1934, attached to: Starling to Leith-Ross, July 31, 1934, BNA, T 160/602.

[237] Leith-Ross to Starling, September 4, 1934, BNA, T 160/602.

[238] Jersey estimated the total figure for all three companies at $250 million. Phillips, memorandum of conversation, July 17, 1934, NARA, RG 59, 862.6363/152.

[239] Feis, memorandum of conversation, July 25, 1934, NARA, RG 59, 862.6363/153.

[240] White to Hull, "Certain Problems [...]," August 18, 1934, NARA, RG 59, 862.6363/154.

[241] Erhardt to Dodd, "Reported Negotiations [...]," August 21, 1934, NARA, RG 59, 862.6363/155.

subsidiary, Rhenania-Ossag, dominated the German export market (primarily lubricants).[242] Rumors also circulated that the War Ministry would place preferential orders with the compliant companies, which the Americans interpreted as a bid to favor Shell.[243]

Shell could not overcome resistance from Jersey, Anglo-Persian, and Whitehall, which all opposed the deal.[244] Berlin reacted in October 1934 by introducing a plan (the formation of BRABAG) to boost gasoline production to two million tons within two years, thereby eliminating any imports. Neither Whitehall nor the major oil companies took such threats seriously. For a savings of 60 million RM of hard currency, Germany would be foregoing 200 million RM in import duties and would have to scrape together one billion RM of financing.[245] Hitler's efforts to sway Deterding during his visit to Berchtesgaden that month by dangling the prospect of a Shell monopoly in Germany also fell flat.[246]

In November, the Germans tried to broker a new deal, this time using IG Farben as an intermediary. Schacht and War Minister Werner von Blomberg asked Krauch "to procure for them a rather great stock of oil" to tide Germany over until it had completed the construction of new synthetic plants. Krauch traveled to London with Fischer, who spearheaded the negotiations. IG Farben would purchase £4 million of petroleum (one million tons at current prices) over the following five years with a 10 percent down payment. The remaining compensation would come within five years of the completion of the contract, and IG Farben would have the option of demanding delivery of the petroleum within two years.[247] The British opposed the clause allowing IG Farben to purchase the fuel in two rather than five years, fearing it would accelerate rearmament. British officials by now grasped that the Germans were not putting all of their eggs in the synthetic basket.[248] The British were hamstrung, however, by a lack of diplomatic support—Washington refused to support Whitehall and the companies if Berlin retaliated.[249] IG Farben then modified the proposal to include full payment upon delivery.[250] Anglo-Persian backed out, but Shell and Jersey delivered $14–15 million of petroleum.[251] The former

242 Erhardt to Hull, "Foreign Interests [...]," February 15, 1935, NARA, RG 59, 862.6363/165. Rhenania evaded the exchange controls by exporting finished goods, but operations became difficult after the Four-Year Plan began, and the regime placed the company in receivership during the war. Jonker and van Zanden, *History of Shell*, 464–74; Howarth and Jonker, *History of Shell*, 20–1, 78.

243 Erhardt to Hull, "American Interests [...]," August 15, 1935, NARA, RG 59, 862.6363/169.

244 The Germans were convinced that "political" as well as commercial factors had motivated Anglo-Persian's opposition. Mulert, "Mineralölversorgungsplan," August 25, 1934, PAAA, R 118179.

245 "[...] Supply of Petrol [...]," October 9, 1934, Enclosure No. 1 to: Phipps to Simon, October 12, 1934, BNA, T 160/602.

246 "Reich Oil Monopoly [...]," *New York Times* (October 25, 1934).

247 Bingham to Hull, November 20, 1934, NARA, RG 59, 862.6363/158; "Extracts... 25 August 1945 [...]," *NMT*, vii: 1202–9.

248 Mejcher, *Politik*, i: 168.

249 Feis, memorandum, November 21, 1934, attached to: NARA, RG 59, 862.6363/158.

250 Phillips, memorandum of telephone conversation, NARA, RG 59, 862.6363/159; Department of State to the U.S. Embassy, London, December 4, 1934, NARA, RG 59, 862.6363/158.

251 "Extracts," *NMT*, vii: 1207; Kockel, *Von Krieg*, 143–5.

paid heavily, as the regime retaliated against Anglo-Persian's German subsidiary by denying it foreign exchange to import crude oil from its parent company.[252]

Although Shell and Jersey also chafed at German restrictions, they did not complain too much since they earned high profits in Germany and found loopholes to evade capital controls by constructing physical assets such as tankers or refineries.[253] Their various partnerships with IG Farben also provided them with a measure of political security. As for Whitehall, it feared applying excessive pressure. While there is no evidence that Britain pursued a conscious strategy "to keep both Germany and Italy dependent on foreign controlled oil," it did have an incentive to ensure that Germany could import sufficient quantities.[254] Three years before Chamberlain began appeasement, Whitehall—with Washington's support—was dangling economic inducements to Berlin to modify its threatening behavior.[255] If it discouraged autarky by supplying Germany with oil, this would serve British interests by limiting Berlin's capacity to wage war. Time was running out, however, as British scientists and businessmen warned that Germany could be "sanctions proof" as early as 1937.[256]

Whitehall was overstating Germany's capabilities, but it had reason to worry. The Third Reich had no desire to relinquish its synthetic fuel industry for greater access to overseas imports. To do so would have flown in the face of the Germans' experience during World War I, when they had confronted the extent of their vulnerability to a British blockade. Henceforth, Germany would seek continental solutions to satisfy its geopolitical ambitions. In terms of petroleum, that meant embracing the possibilities of synthetic fuel, despite its high cost relative to crude oil. For Weimar-era governments and business interests, synthetic fuel was a possible means of stabilizing the economy by creating additional demand for coal and steel, strengthening Germany's finances, and promoting motorization. The military was obviously concerned with waging a credible defense of Germany's borders against hostile neighbors. The Third Reich shared these concerns and referenced them when justifying its policy of self-sufficiency, but unlike its predecessors, it would not allow Germany to remain the victim of forces beyond its control.

[252] See DBA, P-1701.

[253] Larson, *New Horizons*, 333–4; Wilkins, *Maturing*, 237–9; Zetzsche, "Logistik," 73–82.

[254] Contra Helmut Mejcher, "The International Petroleum Cartel (1928), Arab and Turkish Oil Aspirations and German Oil Policy towards the Middle East on the Eve of the Second World War," in: *Oil, the Middle East, North Africa and the Industrial States*, ed. Klaus Jürgen Gantzel and Helmut Mejcher (Paderborn: F. Schöningh, 1984), 47.

[255] C.A. MacDonald, "Economic Appeasement and the German 'Moderates', 1937–1939," *Past & Present* 56 (1972): 105–35. Washington believed that a deal was "very much in the interest of" the oil companies if it reduced German support for self-sufficiency. Feis, memorandum, December 3, 1934, NARA, RG 59, 862/6363/160.

[256] Clarke, "Steps [...]," January 13, 1936, O.B. 154, BNA, CAB 50/5.

6

Fueling the War to Come

1935–1939

Despite significant investments during the first two years of the Third Reich, Germany made little progress toward energy independence. Modest increases in production could not keep pace with the higher consumption once the economy recovered and rearmament began. Germany also could not afford to continue importing oil while rearming, which limited the industrial surplus available for export, which in turn reduced the supply of foreign exchange. There were only two ways to resolve this impasse: accelerate the expansion of synthetic fuel production, which risked overheating the economy, or throttle back on rearmament and export the industrial surplus. The Third Reich's response to this problem was typical of its response to other challenges—it rejected the caution espoused by conservatives such as Schacht and embraced escalation in the hope that a short-term surge could pay long-term dividends. The result was the Four-Year Plan, which dictated that the aim of the German economy would not be to sustain the fragile recovery but rather to gird the country for a war of revenge and conquest.

The centerpiece of the Four-Year Plan was self-sufficiency, but not as an end in and of itself. Rather, synthetic fuel, along with higher production and stockpiling, would enable the military to achieve an enduring solution to Germany's resource deficiencies through conquest.[1] The New Defense Economy Supply Plan of 1938, in turn, represented a further radicalization of Germany's economic policy, which mirrored the regime's increasingly belligerent foreign policy as Hitler sped up his timetable to take advantage of a favorable geopolitical environment. The results were impressive but insufficient. The problem was fundamentally one of opportunity costs. The synthetic fuel industry's requirements for coal, steel, labor, and capital could only be satisfied at the expense of other sectors of the economy, including munitions production. There was no way to boost synthetic output quickly enough without disrupting the overall rearmament effort, so, starting in 1938, Berlin's gaze returned to Romania, whose oil once again became indispensable to meeting the stubborn gap between German consumption and production.

[1] Richard Overy, "The Four Year Plan," European Yearbook of Business History (2000): 93–5; Norman Rich, *Hitler's War Aims: Ideology, the Nazi State, and the Course of Expansion* (New York: W.W. Norton & Company, 1973), 64–7.

THE BATTLE TO CONTROL POLICY, 1935–1937

By the summer of 1935, domestic production would cover only three months of consumption.[2] While the military judged Germany's immediate prospects in a conflict as "unhealthy," IG Farben was more optimistic thanks to its estimates of future supply and demand during both peacetime and wartime.[3] According to the final assessment from May 1935, peacetime consumption would rise by 300,000 tons between 1934 and 1938. Internal production would increase by 75 percent, thereby allowing Germany to cover 65 percent of its peacetime consumption by 1938, against 28.1 percent in 1935.[4] Germany would, however, still have to import 7.73 million tons between 1935 and 1938 (half of consumption), at a cost of 511 million RM in foreign exchange. The situation during mobilization was even trickier. Against an expected demand of 5.68 million tons, only 1.66 million tons of production and 792,000 tons of stockpiles would be available by 1938—43 percent of total demand. To satisfy its requirements for an entire year, Germany would have to import 3.15 million tons by April 1937 for 242 million RM. Further, Germany would require an additional 2.97 million tons of storage capacity—a further expenditure of 387 million RM.[5]

IG Farben was pleased with Germany's progress. Whereas production only covered one-fifth to one-third of consumption between 1930 and 1934, the situation would soon improve dramatically. By 1937, IG Farben estimated, Germany would only need to import 35 percent of its consumption, and once the planned expansion of Leuna and construction of the new facilities was complete, Germany would be almost self-sufficient (85–90 percent) in gasoline. Even where major shortfalls did exist, such as for diesel fuel or lubricants, IG Farben was confident that a synthetic substitute would soon be available.[6]

The military was less than sanguine. In March 1935, the Defense Ministry warned that Germany's survival during a future conflict would depend upon not just open sea lanes but also the availability of foreign exchange. Germany's shortage of the latter and the impossibility of securing foreign credits put it in an impossible position, especially as it was already depleting its stockpile to satisfy current requirements.[7] No one was more concerned than Blomberg, who repeatedly raised the matter with

[2] "Beitrag [. . .]," August 1934, enclosed with: Wa Wi II to Wi, August 14, 1934, T-77/300 (Wi/IF 5.1638).

[3] Note, December 16, 1934, T-77/205 (Wi/IF 5.1029); Ritter to Krauch, December 12, 1935, NARA, RG 238, T-301/58 (NI-7295). Kockel claims that IG Farben sought cooperation with the military because it worried that the Economics Ministry was hostile to its synthetic fuel program. Kockel, *Von Krieg*, 143.

[4] Public estimates from this period assumed an increase in production from 850,000 tons in 1933 to 2.03 million tons by 1936–7. Steinberger, "Kriegswirtschaft und Selbstversorgung," *Deutsche Wehr* (January 31, 1935).

[5] "Zur Deckung [. . .]," May 22, 1935, no author, NARA, RG 238, T-301/58 (NI-7295).

[6] "Die deutsche Mineralölversorgung," June 1935, no author, attached to: Vollard-Bockelberg to Ritter, December 4, 1935 and March 10, 1936, NARA, RG 238, T-301/58 (NI-7295).

[7] Defense Ministry, "Die Wehrwirtschaftliche Lage [. . .]," March 11, 1935, *Akten*, ii: 416–17.

Table 6.1. Petroleum consumption, 1938

Type of product	% Increase
Aviation fuel	160%
Gasoline	120%
Diesel fuel	320%
Fuel oil	20%
Aircraft lubricants	160%
Automobile lubricants	200%

Table 6.2. Production as a share of consumption, 1935

Type of product	% of consumption
Aviation fuel	52%
Gasoline	62%
Diesel fuel	42%
Aircraft lubricants	0%
Automobile lubricants	43%

Table 6.3. Production as a share of consumption, 1938

Type of product	% of consumption
Aviation fuel	40%
Gasoline	45%
Diesel fuel	11%
Fuel oil	40%
Aircraft lubricants	0%
Automobile lubricants	14%

Schacht starting in August 1934.[8] As he explained in June 1935, "[the] expansion of supply is not keeping pace with rise in demand, such that the supply position, rather than improving, is actually deteriorating."[9] By 1938, petroleum consumption would have increased significantly, as can be seen in Table 6.1.

As of October 1935, domestic output only covered the percentage of consumption shown in Table 6.2.

By 1938, the situation had deteriorated even further, as can be seen in Table 6.3.

According to the War Ministry's estimate of Germany's supply position, by April 1, 1937, production would cover only two to four months of consumption. Blomberg urged the appointment of a "Fuel Commissar" to exercise leadership over the relevant military and civilian agencies.[10]

[8] See *Akten*, ii: 194, n. 1. Besides the documents cited, the following discussion is based on Kockel, *Von Krieg*, 134–70; Petzina, *Autarkiepolitik*, 36–9.

[9] The data for the following three tables comes from Blomberg to Schacht, "Förderung [...]," June 3, 1935, *Akten*, ii: 611–13.

[10] Blomberg to Schacht, "Förderung [...]," June 3, 1935, *Akten*, ii: 611–13. The estimate for 1937 was not reprinted—it is attached to a copy of Blomberg's letter in: T-77/341 (Wi/IF 5.2164, 2687).

The military was already thinking of something like the Four-Year Plan.[11] During a meeting in July 1935, the Economics Ministry resisted any diminution of Schacht's powers, as he had become Plenipotentiary for the War Economy in May. Military planners such as now-Colonel Thomas, head of the fledgling Defense Economy Office, were unconvinced. They criticized the ministry's handling of petroleum policy, which they believed had "led to strategic considerations being entirely ignored."[12] Blomberg advised Schacht that the Fuel Commissar's "tasks lay to a great extent within the practical realm and beyond the competence" of his ministry.[13] Schacht reiterated his objections at the end of July.[14] He pointed out that Germany still required the assistance of the oil companies, which were unlikely to cooperate with any commissar. The military resigned itself to the situation, but one of Blomberg's deputies reminded everyone that, while Germany could satisfy 51 percent of its mobilization demand in 1935, by 1938 this would decrease to 40 percent.[15]

The War and Economics ministries tried to reconcile their positions. During a meeting in August 1935, the Economics Ministry expressed doubts about the potential of the synthetic fuel industry, worrying that "we have overextended ourselves... with regard to hydrogenation." The new Fischer–Tropsch plants, however, could boost Germany's meager diesel fuel production, which only covered 10 percent of the navy's mobilization demand of two million tons.[16] Following the meeting, Blomberg wrote to Schacht to restate his wishes, including "ensuring" that all hydrogenation plants were capable of producing aviation fuel and using only soft coal as a feedstock, as well as boosting crude oil production.[17]

As 1935 ended, the War Ministry remained concerned by the supply situation. As of January 1, 1936, Germany's supply outlook was terrible, as can be seen in Table 6.4.

When combined with existing stocks, "full supply" of gasoline and aviation fuel was possible for only six months, and that of other products for only three. Despite the anticipated production increases by 1938 (800,000 tons), Germany's overall position would actually deteriorate due to higher consumption. The only recourse was to stockpile, but this required foreign exchange. A German industrial consortium hoped one day to exchange Iraqi crude oil for German industrial goods, but this was years away.[18] There were no easy options, and the military was aghast that

[11] IG Farben was also in favor of the installation of a fuel commissar, but it is a stretch to claim the whole matter "amounted to a privatization of oil policy." Kockel, *Von Krieg*, 150–1.

[12] Wi II, "Aktenvermerk [...]," July 11, 1935, T-77/341 (Wi/IF 5.2164, 2687).

[13] Blomberg to Schacht, July 11, 1935, T-77/341 (Wi/IF 5.2164, 2687). The military expected that any commissar should be "an expert" recruited from the industry. "Gründe [...]," no date or author, T-77/101 (Wi/IF 5.433). Supposedly, Blomberg's choice was Fischer. Birkenfeld, *Synthetische Treibstoff*, 54, n. 10.

[14] No record of an earlier meeting between Blomberg and Schacht (on July 19 or 20, 1935) survives. Kockel, *Von Krieg*, 154, n. 165.

[15] "Aktenvermerk... 27.7.35 [...]," *Akten*, ii: 693–4.

[16] Wi II, "Aktenvermerk [...]," August 9, 1935, T-77/341 (Wi/IF 5.2164, 2687).

[17] Blomberg to Schacht, "Programme [...]," September 6, 1935, T-77/341 (Wi/IF 5.2164, 2687).

[18] Between 1932 and 1936, German firms tried to develop Iraq as a source of oil that Germany could import without expending hard currency. In 1934 and 1936, however, Hitler rejected this

Table 6.4. Production and reserves, January 1, 1936[19]

Type of product	% of demand
Gasoline	45%
Aviation fuel	45%
Diesel fuel	17.5%
Fuel oil	33.5%
Aircraft lubricants	0%
Automobile lubricants	18%

Germany would be able to meet less than half of its wartime consumption through internal production by 1938.[20]

Blomberg met with Hitler on October 17, 1935 to make the case for a Fuel Commissar.[21] As Blomberg explained to Schacht a week later, he could not see how else to improve the situation. The War Minister beseeched his colleague to reconsider his opposition.[22] Schacht stalled and, in November, circulated an update on the status of the petroleum industry. Under a government-sponsored program, German oil companies had received five million RM in public assistance to offset their costs, starting in 1934, with another nine million RM earmarked for 1935 and 1936 plus a similar quantity of private investment, but the results were disappointing. Although five new fields had come online, with one exception their average production was a meager 100 tons per month. On the bright side, expansion of hydrogenation production at Leuna was complete, and seven new facilities would come into operation in 1936. There was a stockpile of 450,000 tons, and an additional 95,000 tons of aviation fuel would soon be added. Finally, negotiations were under way to stockpile additional imports from Romania and the United States without expenditure of hard currency.[23]

On November 18, 1935, Thomas, now a colonel, met with Schacht to discuss the points raised in the recent exchange of letters, and the Economics Ministry again refused to back down.[24] The dispute had already taken an unexpected twist

option for fear of "[incurring] the enmity of powerful international oil interests." The navy then unsuccessfully sought a concession in or long-term supply contract with Mexico. For Iraq, see: Fritz Grobba, *Männer und Mächte im Orient: 25 Jahre diplomatischer Tätigkeit im Orient* (Göttingen: Musterschidt-Verlag, 1967), 85–94; Mejcher, *Politik*, i: 102–45. For Mexico, see: Volland, *Mexiko*, 83–174. For the navy's perspective, see: Meier-Dörnberg, *Mineralölversorgung*, 31–44; Zetzsche, "Logistik," 46–72. For summaries, see: Kockel, *Von Krieg*, 151–60 and 179–92; Anand Toprani, "'The Navy's success speaks for itself'? The German Navy's Independent Energy Security Strategy, 1932–1940," *Naval War College Review* 68: 3 (2015): 91–118.

19 Wi II, "Vortragsnotiz [...]," August 23, 1935, T-77/101 (Wi/IF 5.433).

20 Wi II, "Vortragsnotiz [...]," August 23, 1935, T-77/101 (Wi/IF 5.433).

21 "Gründe [...]," no date or author, T-77/101 (Wi/IF 5.433).

22 Blomberg to Schacht, "Einheitliche Treibstoffpolitik," October 25, 1935, T-77/341 (Wi/IF 5.2164, 2687). This letter was based on: Thomas, "Entwurf [...]," no date, T-77/228 (Wi/IF 5.1171). See also: Thomas to Wa A, "Betriebstoffversorgung," November 2, 1935, T-77/341 (Wi/IF 5.2164, 2687).

23 "Übersicht [...]," no date, attached to: Schacht to Blomberg, November 12, 1935, T-77/101 (Wi/IF 5.433). The British Ambassador informed Whitehall in December 1935 that Germany had signed an agreement to stockpile a war reserve of 200,000 tons, possibly with a U.S. supplier. Mejcher, *Politik*, i: 169–70.

24 Wi II, "[...] Treibstoffkommissar [...]," T-77/101 (Wi/IF 5.433).

after the head of the air force, Hermann Göring, entered the picture. Göring had no expertise in economic affairs, but he was a skilled empire builder who saw an opening to solidify his position as Hitler's successor. Göring was also committed to enhancing the status of the fledgling air force but had no wish to provoke any feuds with the much larger army—yet. He assured Blomberg that the War Ministry had his full support.[25] Göring's intervention heartened Blomberg, who wrote to Schacht again following Italy's invasion of Abyssinia.[26]

The dispute between the Economics and War ministries was not a product of any dovishness on Schacht's part—his feud with Blomberg and, later, Göring was a turf war.[27] Schacht's primary demand was that the regime should not increase its support for synthetic fuel to the point that it harmed the rest of the economy.[28] By the end of 1935, Schacht had realized he was fighting a losing battle, and announced that he was appointing one of his deputies as the "Plenipotentiary for Fuel Questions."[29] Blomberg acceded to Schacht's suggestion that the military detail a representative to the plenipotentiary's staff.[30] The following month, a draft decree was presented to Hitler using guidelines prepared by the military. The decree tasked the plenipotentiary with "organizing the entire petroleum industry along strategic and defense-economic lines. The aim is securing the foundation for the mobilization of the armed forces and for the conduct of war."[31] For reasons that are unclear, Hitler never signed the decree; in any event, the position would have become superfluous with the advent of the Four-Year Plan the following year.[32]

By the spring of 1936, the military's frustration with Schacht had reached boiling point. Specifically, it felt he had not done enough to curb civilian consumption.[33] The outlook was mixed. On the one hand, the Chancellery forecast 1.47 million tons of production in 1936, against two million tons of demand. On the other, the supply of fuel oil and diesel was "significantly less favorable."[34] As the Defense Economy Office concluded in March, Germany could not expect any "appreciable

[25] Milch to Blomberg, "Treibstoffkommissar," November 15, 1935, T-77/341 (Wi/IF 5.2164, 2687).

[26] Blomberg to Schacht, November 30, 1935; reprinted in: Dietrich Eichholtz, *Deutsche Ölpolitik im Zeitalter der Weltkriege* (Leipzig: Leipziger Universitätsverlag, 2010), 265. See also: W Wi II, "[...] Treibstoffkommissar," December 17, 1935, T-77/341 (Wi/IF 5.2164, 2687).

[27] Karlsch, *Faktor Öl*, 185–6.

[28] Dieter Petzina, "Vierjahresplan und Rüstungspolitik," in: *Wirtschaft und Rüstung am Vorabend des Zweiten Weltkriegs*, ed. Friedrich Forstmeier and Hans-Erich Volkmann (Düsseldorf: Droste, 1975), 69–70.

[29] Schacht to Keitel et al., December 11, 1935, T-77/101 (Wi/IF 5.433); Schacht to Blomberg, "Einheitliche Treibstoffpolitik," December 13, 1935, T-77/341 (Wi/IF 5.2164, 2687).

[30] Blomberg to Schacht, December 24, 1935, T-77/341 (Wi/IF 5.2164, 2687). Blomberg settled upon an army officer to serve as the War Ministry's liaison: "Dienstanweisung [...]," no date, attached to: Blomberg to Schacht, "2.Verbindungsoffizier [...]," January 21, 1936, T-77/341 (Wi/IF 5.2164, 2687). Meanwhile, the Defense Economy Office would handle "the particular interests of the armed forces" and coordinate with the military liaison on the staff of the Fuel Commissar, as it was already doing with the Aviation Ministry. W Wi II, "Vortragsnotiz [...]," January 9, 1936, T-77/101(Wi/IF 5.433).

[31] Decree dated January 1936, T-77/341 (Wi/IF 5.2164, 2687). Earlier drafts are reprinted in: *Akten*, iii: 101–2.

[32] See the commentary included in: *Akten*, iii: 102, n. 5.

[33] Richard Overy, *War and Economy in the Third Reich* (Oxford: Oxford University Press, 1994), 183ff.

[34] Willuhn to Lammers, January 21, 1936, *Akten*, iii: 62–3. Hitler apparently took "notice with the greatest of interest." *Akten*, iii: 808.

relief" until 1938. Even then, only the supply of gasoline and aviation fuel would improve significantly—production would cover 61 percent of mobilization demand. The supply of diesel fuel relative to consumption would double, but it would only be 22 percent of demand. The ratio with regard to fuel oil would actually decrease, from 31 percent to 22 percent. One bright spot was the availability of aviation lubricants. In March 1936, Germany produced none of its requirements, but by 1938, 22 percent would come from domestic sources. Two crises the previous year, however, threatened to "shake the very foundations" of German strategy. The first was the Soviet Union's halting of petroleum exports.[35] The second was Romania's refusal to continue trading unlimited amounts of oil on a clearing basis.[36] The only options were giving in to Romanian demands, boosting imports, or increasing domestic production. By 1935, Romania's share of German imports had reached 40 percent, which was unsustainable. Under the terms of a new agreement signed that year, Romania limited its oil exports under clearing to 25 percent of the value of all exports to Germany.[37] The second cited option was out of the question since it required the diversion of foreign exchange. Finally, there were no means of quickly boosting internal production.[38]

Germany had to choose between solvency and self-sufficiency. During its first three years, the regime could increase public expenditure with minimal risk of inflation because the economy still had slack capacity.[39] Once Germany approached full employment, however, the danger of overheating became acute. Germany's current account deficit between 1933 and 1935 exceeded 900 million RM, and its foreign exchange reserves dwindled from 400 million RM to 72 million RM.[40] By March 1936, raw materials stockpiles had declined by two-thirds against the previous year, and munitions firms had to reduce output by 30 percent. Only two options were available—either cool down the economy through interest-rate hikes and risk a recession, or continue spending but buy time before a crisis hit by imposing further controls on the economy. Not surprisingly, Hitler chose the latter, which meant that he had to replace Schacht. Although the latter had used the tools of economic nationalism to promote recovery and aid rearmament, he rejected autarky as impractical and sought to reintegrate Germany into the global

[35] Moscow was putting pressure on Berlin in advance of negotiations concerning the 1935 Credit Agreement. Mejcher, *Politik*, i: 169. Contra Yergin, *Prize*, 332, exports resumed, but they declined by half between 1935 and 1937. Geological Research Office, *Erdöl in Rußland* (Berlin, 1941), Library of Congress (LOC).

[36] Germany used three methods of financing foreign trade to avoid expenditure of foreign exchange: clearing agreements, barter agreements, and payment in blocked currency. Clearing agreements are pools of money, usually controlled by the central bank, into which a nation's exporters deposit the proceeds of their sales. Importers then draw from these accounts to finance their purchases. Larry Neal, "The Economics and Finance of Bilateral Clearing Agreements: Germany, 1934–1938," *Economic History Review* 32: 3 (1979): 391–404.

[37] Dietrich Eichholtz, *Deutsche Politik und rumänisches Öl, 1938–1941* (Leipzig: Leipziger Universitätsverlag, 2005), 17.

[38] "Memorandum [...]," March 9, 1936, reprinted in: Office of United States Chief of Counsel for Prosecution of Axis Criminality, *Nazi Conspiracy and Aggression* (Washington, DC: GPO, 1946), iii: 871–4 (hereafter cited as *NCA*).

[39] Petzina, "Vierjahresplan," 65ff.

[40] Barkai, *Nazi Economics*, 255.

economy.[41] In the short run, with Germany approaching full employment, he worried about overheating and wanted to cut back on defense spending. The regime was equally concerned with inflation, but it had different ambitions than Schacht and could not afford to let rearmament slacken.[42] In April, Hitler took the first step by granting Göring extensive powers to tackle the foreign exchange problem.[43]

Göring and Schacht clashed repeatedly during a series of interagency conferences over the period May–June 1936. On May 12, 1936, Schacht observed that Germany was trapped between the need to import raw materials and its inability to pay for them.[44] Göring was uninterested in importing oil: "If we have war tomorrow, we must help ourselves by substitutes."[45] That meant finding a replacement for the 1.77 million tons German would have to import if it went to war immediately.[46] During a subsequent meeting, Schacht agreed that autarky was "theoretically" possible but risked inflation. Finance Minister Johann Ludwig von Krosigk suggested limited public support, but Göring overruled him.[47] Germany would eliminate its outstanding mobilization deficit, as of 1938, of three million tons of petroleum by constructing ten new synthetic plants, at a cost of 1.15 billion RM.[48] In June, officials explained that Germany would likely be self-sufficient in gasoline in peacetime by 1937, but petroleum imports in 1936 would cost 155 million RM and more in future. Göring demanded progress—Germany's fate depended upon boosting domestic production, for in wartime, "we won't get a drop of oil from abroad."[49]

Hitler ended any debate concerning the future of rearmament in his secret memorandum of August 1936.[50] The paper was a reaction to Germany's escalating financial and economic crisis, and to calls by conservatives to cut spending. Hitler had already articulated many of the underlying ideas in his unpublished "Second Book" of 1928, in which he argued that expansion eastwards was necessary to provide Germans with "a life comparable to that of the American people."[51] He made only one mention of petroleum, in the context of the Soviet Union being "a possessor of oilfields, which have the same significance today as iron and coal mines had in the past century."[52]

[41] "Notizen [...]," November 12, 1935, no author, T-77/300 (Wi/IF 5.1638).

[42] Tooze, Wages, 207ff.

[43] Lammers to Hess et al., April 6, 1936, BA-B, R 26 I/35.

[44] This was the lament of the Economics Ministry official responsible for petroleum distribution, who asked Schacht for greater freedom to conduct barter or third party trade agreements. "Besprechung... 13 August 1936," *Akten*, iii: 457–8.

[45] "Record... 12 May 1936," *NCA*, iii: 878–84.

[46] Wi II to Wi Ia, "Vorbereitung [...]," February 1, 1936, T-77/205 (Wi/IF 5.1029).

[47] "Copy [...]," attached to: Löb to Blomberg, May 30, 1936, *NCA*, iii: 886–90.

[48] "Fuel Supply Situation," May 27, 1936, *NCA*, iii: 891.

[49] "Niederschrift...25.6.36," NARA, RG 238, T-301/39 (NI-5380).

[50] Wilhelm Treue, "Hitlers Denkschrift zum Vierjahresplan, 1936," *Vierteljahrshefte für Zeitgeschichte*, 3: 2 (1955): 188–202; Petzina, *Autarkiepolitik*, 30–53; Tooze, *Wages*, 203–43.

[51] For Hitler's view of the United States, see: Tooze, *Wages*, 9–11 and 658.

[52] Gerhard Weinberg, ed., *Hitler's Second Book: The Unpublished Sequel to Mein Kampf* (New York: Enigma, 2003), 105–7, 158–9, and 172. Although Hitler appreciated the significance of petroleum, he also supported the development of natural gas and alternative energy sources. See: *Table Talk*: No. 1 (July 5, 1941), No. 16 (August 2, 1941), No. 104 (January 13, 1942), and No. 269 (July 26, 1942).

Six years later, Hitler took an important step toward realizing these objectives. Specifically, he ordered that both the military and economy be capable of "action" and "war," respectively, within four years. His aim was not autarky as the concept is commonly understood. The National Socialists never believed that Germany could be self-sufficient within its prewar borders and said as much both before and after taking power.[53] While they rejected participation in a Jewish-dominated "global economy," they did not rule out that Germany might benefit from international trade, only that its prosperity should not depend upon it.[54] The point of measures that reduced Germany's dependence on foreign trade was "to secure Germany's political and economic freedom."[55] Such freedom was not an end but rather a means. The aim, Hitler believed, was to prepare Germany to accomplish its long-term objective: "the extension of our living space, that is, an extension of the raw materials and food basis of our nation."[56] As one of Göring's deputies later explained, the whole point was that the Four-Year Plan was to enable a "large sphere solution"—seizing the resources of Germany's neighbors.[57]

Within the context of 1936, therefore, Hitler's instruction that "German fuel production must now be developed with the utmost speed and brought to a definitive completion within 18 months" was less a call for autarky than a demand to conserve foreign exchange.[58] This deadline was not an empty boast—Göring apparently informed Hitler using information provided by Krauch that it was possible for Germany to meet 80 percent of its estimated petroleum consumption by 1938.[59] Schacht was appalled by Hitler's decision and urged Blomberg to intercede, but it was too late.[60] On September 4, 1936, Göring announced the new policy before a ministerial council including Schacht and Blomberg. Rearmament would "be sped up" because war with the Soviet Union was "inevitable." Germany must achieve "autonomy in all those fields in which it is technically possible," as this would save 600 million RM of foreign exchange.[61]

[53] Ludolf Herbst, *Der Totale Krieg und die Ordnung der Wirtschaft: Die Kriegswirtschaft im Spannungsfeld von Politik, Ideologie und Propaganda* (Stuttgart: Deutsche Verlags-Anstalt, 1982), 65–6.

[54] Barkai, "Grundanschauungen," 355–85; Wilhelm Deist et al., *Build-Up of German Aggression*, vol. 1 of *Germany and the Second World War*, ed. Militärgeschichtliches Forschungsamt (Oxford: Clarendon Press, 2003), 173–93.

[55] Georg Freiherr von Wrangel, *Deutschlands Kampf um Rohstoff-Freiheit* (Berlin: Propaganda-Verlag Paul Hochmuth, 1942), 6.

[56] "[...] Tasks of the Four Year Plan—1936," *NMT*, xii: 430–9. For Hitler's economic views, see: Barkai, Nazi Economics, 21ff; Berenice Carroll, *Design for Total War: Arms and Economics in the Third Reich* (The Hague: Mouton, 1968), 93–105; Henry Ashby Turner, "Hitlers Einstellung zu Wirtschaft und Gesellschaft vor 1933," *Geschichte und Gesellschaft* 2: 1 (1976): 89–117, and Big Business, 71–83.

[57] "Four Year Plan," April 29, 1941, *NMT*, xii: 538–42. See also: Lesser, *Resources*, 51; Petzina, "Vierjahresplan," 70–2; Steiner, *Triumph*, 252–8.

[58] "[...] Tasks of the Four Year Plan—1936," *NMT*, xii: 430–9.

[59] "[...] Bericht [...]," No. 46 in: Dietrich Eichholtz and Wolfgang Schumann, ed., *Anatomie des Krieges: Neue Dokumente über die Rolle des deutschen Monopolskapitals bei der Durchführung des zweiten Weltkrieges* (Berlin: VEB Deutscher Verlag der Wissenschaften, 1969), 139–42.

[60] Thomas, memorandum, September 2, 1936, *NCA*, iii: 894–5.

[61] "Minutes...4 September 1936," *NMT*, xii: 439–42.

When Hitler and leading officials made pronouncements on the Four-Year Plan, they claimed its aim was to save foreign exchange and stimulate employment.[62] There was no mention of general self-sufficiency. In his proclamation before the 1936 Party Congress, Hitler stressed that the plan was only supposed to make Germany "independent…with respect to those materials which we are capable of obtaining ourselves" through synthesis.[63] Some officials even claimed that Germany was contributing to world peace by providing an alternative energy supply once oil reserves gave out, as well as eliminating competition over oilfields.[64] Private speeches, while not denying that the plan was supposed to facilitate rearmament, made no mention of war against the Soviet Union.[65] Nevertheless, only three days after announcing the Four-Year Plan, Hitler publicly boasted that Germany could make better use of Russia's natural resources than its "Jewish-Bolshevist leadership." He clearly understood that the only way to reconcile Germany's demand for guns and butter was through conquest.[66]

Göring received a formal appointment to implement the Four-Year Plan in October 1936.[67] A week later, he unveiled the plan's executive structure. Among its agencies was the Office for German Raw and Basic Materials, which succeeded the Raw Materials and Foreign Exchange Staff created in April and would have the same director, Lieutenant-Colonel Fritz Löb. This office would be responsible for supervising the entire petroleum industry.[68] Schacht's reluctance to appoint to a Fuel Commissar in the summer of 1935 had now come back to haunt him. His status as Plenipotentiary for the War Economy appeared incompatible with Göring's new position. The War Ministry suggested that Blomberg should wield "executive power" in wartime, while in peacetime Schacht shared authority with Göring.[69] Blomberg was hopeful that Göring would cooperate with Schacht, since the former's mission "to clear all bottlenecks" would support the latter's task of preparing the economy for war.[70]

This could not have been further from Göring's actual objective of imposing his control over the economy.[71] Hitler, meanwhile, had little incentive to impose order, since overlapping authorities and bureaucratic infighting were "characteristic" of his leadership style.[72] Starting in November 1936, the Office for German Raw Materials would assume responsibility for supervision of the civilian petroleum

[62] Koerner Document 140, *NMT*, xii, 442–6.

[63] Max Domarus, *Complete Hitler: A Digital Desktop Reference to His Speeches and Proclamations, 1932–1945* (Wauconda: Bolchazy-Carducci, 2007), 828–31.

[64] Koppenberg, "Mineralölgewinnung aus Kohle," *Vierjahresplan*, 1937/V; Werlin, "Rohstoffplan und Motorisierung," *Vierjahresplan*, 1937/I.

[65] "[…] Execution of the Four Year Plan…17 December 1936," *NMT*, vii: 814–17; Thomas, "Vortrag…1. XI. 37," *IMT*, xxxvi: 110–11.

[66] Domarus, *Complete Hitler*, 835–6.

[67] "Ordinance […]," *NCA*, vi: 499.

[68] Göring, "Decree […]," October 22, 1936, *NMT*, xii: 447–52. Göring announced that his staff would exercise oversight of the petroleum industry a month before issuing his decree concerning the Four-Year Plan: "[…] Aktenvermerk von 23.9.36," no author, T-77/101 (Wi/IF 5.433).

[69] Division of National Defense, "Report […]," December 30, 1936, *NCA*, vii: 465–6.

[70] Blomberg to Hitler, February 22, 1937, *NCA*, vii: 342.

[71] Koerner, "Leadership and Economy," February 1938, *NMT*, xii: 489–92.

[72] Petzina, "Vierjahresplan," 72–5 (quotation from p. 72).

market—previously the responsibility of the Economics Ministry.[73] The Defense Economy Office retained its prerogatives, specifically "the supervision of all work that is necessary for securing the supply of the armed forces and the armaments industry."[74] The navy also carved out its own niche with imports for naval consumption.[75]

Under the initial planning for the Four-Year Plan, of the 1.4 billion RM to be spent, 42 percent (574 million RM) was earmarked for synthetic fuel production, nearly all of it (98 percent) raised from private capital.[76] It would take some time before the plan would have any tangible effects—two years to double synthetic output, from 630,000 tons in 1936 to 1.5 million tons in 1938, and another two years before it peaked at 3.43 million tons (split 4:1 between hydrogenation and the Fischer–Tropsch process).[77] The growth in imports decreased significantly starting in 1937, but that had less to do with larger domestic supplies than it did with a shortage of hard currency.[78] Germany's foreign exchange position also deteriorated as the Four-Year Plan and rearmament consumed larger portions of German industrial output. In April 1937, Schacht had observed that Göring's assurance that higher synthetic fuel production would improve Germany's financial position was untrue, since the increase in gasoline production was eclipsed by consumption growth.[79]

The following month, the first round of planning for the Four-Year Plan was complete. Between 1937 and 1940, 8.8 billion RM of government and private funds would be invested across the German economy, with the aim of self-sufficiency in gasoline and aviation fuel within eighteen months and complete independence by 1940.[80] Not everyone was pleased. The imbalance between expected supplies of gasoline and those of fuel oil and diesel prompted Blomberg to ask Göring to reallocate resources since, under existing planning, by the end of the Four-Year Plan Germany would only produce 55 percent of its diesel requirements and 30 percent of its fuel oil consumption.[81] The navy quickly

[73] Göring, "Jurisdiction [...]," November 26, 1936, *NCA*, vii: 338–41. Division II was responsible for petroleum, while Division III, led by Krauch, would handle R&D. Heemskerck, "Richtlinien [...]," November 27, 1936, T-77/341 (Wi/IF 5.2164, 2687).

[74] Chief W/Ro, "Vorschlag [...]," December 3, 1936, T-77/101 (Wi/IF 5.433). Göring's office soon muscled into matters under the military's jurisdiction, such as mobilization requirements. Heemskerck to Schacht, "Vorbereitungen [...]," January 12, 1937, T-77/101 (Wi/IF 5.433).

[75] Navy High Command to Göring, "Zusammenarbeit [...]," January 6, 1937, T-77/101 (Wi/IF 5.433).

[76] The regime's initial contribution came from a hefty increase in the duty on imported gasoline—14.3¢ to 30¢ per gallon—but by 1939, it was also providing half of the necessary funding once private capital became scarce. Stranges, "Fuel Industry," 162.

[77] USSBS, *German Oil Industry, Ministerial Report Team 78* (Washington, DC: GPO, 1945), 21.

[78] The growth between 1937 and 1939 was 245,562 tons. Planning Office, "Mineralöl-Einführ [...]," October 1944, T-77/341 (Wi/IF 5.2164, 2687).

[79] Schacht to Göring, April 2, 1937, attached to: Blessing to Thomas, April 2, 1937, *NCA*, vii: 380–8. Schacht reiterated many of these points in his letter of resignation. Schacht to Göring, August 5, 1937, *NCA*, vii: 567–75.

[80] Karlsch, *Faktor Öl*, 188–9.

[81] Blomberg to Göring, January 26, 1937, reprinted in: Meier-Dörnberg, *Mineralölversorgung*, 89–90. Blomberg's letter was based on: Defense Economy Office, "Forderungen [...]," January 25, 1937, T-77/229 (Wi/IF 5.1171).

Table 6.5. Mobilization consumption, 1937–8

Type of product	% increase
Aviation fuel	10%
Gasoline	80%
Diesel fuel	100%
Fuel oil	40%
Aircraft lubricants	110%
Automobile lubricants	90%

realized that it would gain little from the Four-Year Plan and, using its relative autonomy, put greater emphasis on stockpiling.[82]

Until the investments of the Four-Year Plan started to pay dividends, Germany would be woefully unprepared for a military contingency. Between 1928 and 1937, various policies had enabled a quadrupling of production, from 651,000 tons to 2.26 million tons. Unfortunately, imports had also doubled during that time, from 1.83 million tons to 3.37 million tons.[83] At the beginning of 1937, the Defense Economy Office estimated that Germany's mobilization demand in 1937/8 (5.3 million tons) would be 55 percent higher than the previous year's figure, as shown in Table 6.5.

Regarding supplies, there was good news and bad news. On the one hand, overall production had doubled over the past year. The regime was especially pleased with the new refinery at Misburg, which would double production of lubricants to 55,000 tons.[84] On the other, the percentage of consumption covered by internal production had barely changed, at 36 percent versus 35 percent in 1936–7. Whereas Germany's production of gasoline and aviation fuel was improving (40 percent and 60 percent of demand, respectively), supplies of diesel (8 percent), fuel oil (32 percent), and aircraft and automobile lubricants (18 percent and 25 percent, respectively) were grossly insufficient. The results of stockpiling were mixed. An additional 250,000 tons had been added during the past year alone, but there was little progress in accumulating diesel and fuel oil due to the shortage of foreign exchange and storage. Combining production with reserves would only cover six months of full consumption of gasoline and lubricants, and two months of diesel and fuel oil consumption. The situation would improve by 1940, when Germany would produce 60 percent of the estimated mobilization demand—3.3 million tons out of seven million tons. Setting aside the navy's requirements, Germany would then be capable of fueling its war for nine to twelve months.[85] Unfortunately, Germany's political leadership soon adopted policies that shrank the time horizon before war commenced.

82 W Ro III, "Aktenvermerk [...]," June 16, 1937, T-77/175 (Wi/IF 5.694).

83 Supervisory Office for Petroleum, "Deutschlands Mineralölbilanz [...]," May 3, 1938, T-77/341 (Wi/IF 5.2164, 2687).

84 See: Gönningen, "Verstärkte Schmierölversorgung [...]," *Vierjahresplan*, 1937/XI.

85 W/Ro, "Vortragsnotiz [...]," January 11, 1937, T-77/123 (Wi/IF 5.533). See also: Thomas, memorandum, January 15, 1937, T-77/341 (Wi/IF 5.2164, 2687).

ESCALATION AND CRISIS, 1937–1939

Since 1933, German officials had operated on the assumption that the regime would afford them enough time to prepare for a great power conflict. Hitler's rash decision in the winter of 1935/6 to exploit international preoccupation with Abyssinia in order to remilitarize the Rhineland, followed a few months later by the decision to support the Spanish Nationalists, disturbed officials but also appeared to be aberrations borne of opportunism. They were mistaken.[86] In November 1937, Hitler unveiled a new strategy to his military and diplomatic leaders. According to the minutes of the meeting, the so-called "Hossbach Memorandum," Hitler declared that self-sufficiency for the Third Reich was only partially possible in raw materials and impossible for food. Germany could neither reduce living standards by reducing food consumption nor import the additional amounts required. Hitler ruled out greater foreign trade for economic and strategic reasons. Germany therefore had no option but to follow Japan and Italy's example and expand by force. Hitler still aimed at achieving a military solution by 1945, but he stunned his audience by suggesting that Germany might take advantage of Anglo-French preoccupation with the Mediterranean to overrun Austria and Czechoslovakia in 1938.[87] He followed up on this the next month, modifying an earlier directive to focus on defensive preparations by ordering the military to commence operational planning against Czechoslovakia.[88] This sharpening of the Third Reich's strategic posture had profound repercussions. Besides being the catalyst for a purge of the military and civilian leadership, the possibility that war was imminent provided the impetus for a renewed bid for self-sufficiency.[89]

In a January 1938 speech, Löb claimed that Germany's fuel problem was now "solvable." Demand in 1936 was 4.22 million tons, 42 percent of which came from domestic sources. By the end of the Four-Year Plan, peacetime consumption would increase by 33 percent for gasoline, 50 percent for diesel, and 25 percent for lubricants—or 6.3 million tons in total. Löb expected that Germany could then eliminate its imports of gasoline and lubricants, and he anticipated the total savings in foreign exchange to reach 350 million RM. Germany would continue importing crude oil to be refined in synthetic facilities that could use either oil or coal as a feedstock.[90] Löb did not provide specific production figures, but according to General Hermann von Hanneken, the Plenipotentiary for Iron and Steel Management, as of January 1, 1938, "production capacity" stood at 4.39 million tons, with an

[86] Tobias Jersak, "A Matter of Foreign Policy: 'Final Solution' and 'Final Victory' in Nazi Germany," *German Studies Review* 21: 3 (2003): 371–6.

[87] Hossbach, "Notes [...]," November 10, 1937, *NCA*, iii: 295–305.

[88] Blomberg, "Weisung [...]," June 26, 1937, and Blomberg, "Nachtrag [...]," December 7, 1937, both in: *IMT*, xxxiv: 733–47; Blomberg, "Directive [...]," December 21, 1937, in: Auswärtiges Amt, *Documents on German Foreign Policy, 1918–1945, Series D* (Washington, DC: GPO, 1949–1964), vii: 635–8 (hereafter cited as *DGFP* (D)).

[89] Michael Geyer, "The Crisis of Military Leadership in the 1930s," *Journal of Strategic Studies* 14: 4 (Dec. 1991), 448–62.

[90] "Die Mineralölwirtschaft [...]," January 12, 1938, T-77/107 (Wi/IF 5.469). See also: *Vierjahresplan* 1938/II.

additional 624,000 tons coming online during the year and another 1.63 million tons in 1939. This would leave Germany with 6.64 million tons of capacity by January 1, 1940, plus another 850,000 tons of crude oil.[91]

These figures turned out to be wildly optimistic, but we have a good overview of Germany's actual position at the beginning of 1938 thanks to an intercepted British estimate, which the Defense Economy Office contrasted with the actual figures. British estimates for German peacetime demand between 1936 and 1939 coincided with internal German figures, but the estimate of mobilization demand in 1939 was significantly higher—8.6 million versus 7.8 million tons. Another major discrepancy existed for German supplies in 1939—whereas the British estimated production plus reserves at 6.3 million tons, the Germans only expected 5.2 million tons. There was only a 300,000-ton difference between both sides' figures for Germany's import requirement: 2.3 million tons in the case of the British, and 2.6 million tons according to the Defense Economy Office. British and German estimates of Romanian supplies were identical: 2.5 million tons. When combined with Polish exports (100,000 tons), it was clear by the beginning of 1938 that Germany had enough oil to fight a war in 1939—but only with large imports from Romania.[92]

Additional petroleum production during the first year of the Four-Year Plan (October 1937) was worth 39.7 million RM. For the plan as a whole, the value of greater exports and import savings reached 367.3 million RM. Higher consumption, however, negated any positive effect on Germany's trade balance.[93] In February 1938, Göring decided to reorganize the Four-Year Plan by integrating its constituent offices within the Ministry of Economics. The Office for German Raw Materials now became the Office for Economic Development and functioned in effect as the executive arm of the Four-Year Plan.[94] It was composed of four departments, the most important being the R&D division under Krauch, who was effectively in charge of the agency and took control formally in December 1939. The Geological Research Office, led by Wilhelm Keppler (one of Hitler's economic advisors), would handle geological affairs, with Bentz serving as the Director of the Crude Oil Division.[95] Finally, the Economics Ministry acquired nominal oversight over raw materials, and the ubiquitous Fischer finally took over the ministry's Petroleum Division in 1939.

Göring also made greater use of plenipotentiaries for particularly important tasks.[96] This process was an outgrowth of Hitler's effective ban on cabinet government the

[91] Hanneken, "Bericht [...]," no date, attached to: Hanneken to the Office for Economic Development, January 21, 1939, BA-B, 3112/36.

[92] Defense Economy Office to Military Intelligence, "Mineralölversorgung Deutschlands," January 29, 1938, T-77/427 (Wi/IF 5.3481). A subsequent British assessment overstated Germany's wartime consumption by 1.3 million tons and understated its production in 1937 by 700,000 tons. Griebel to W Wi, "Stellungnahme [...]," November 28, 1938, T-77/123 (Wi/IF 5.533).

[93] Löb, "Ergebnis [...]," October 30, 1937, RG 238, T-301/70 (NI-8590).

[94] "Über die Organisation [...]," January 27, 1938, *Akten*, v: 84–92; Göring, "Decree [...]," February 5, 1938, *NMT*, xii: 482–8; Goering, "Reorganization [...]," December 5, 1939, *NMT*, vii: 963–4; Donner, "Organizational Questions [...]," August 27 (1940), *NMT*, xii: 543–8.

[95] Bentz, "Supplement [...]," February 11, 1946, BNA, FO 1039/496.

[96] Göring, Decree, July 16, 1938, *NMT*, vii: 900–1.

previous year. To centralize decision-making, policymakers empowered lower-level officials—usually plenipotentiaries or experts within the bureaucracy—to execute high-level decisions.[97] One beneficiary was Krauch, who in early 1938 was still a second-tier bureaucrat.[98] During the period August–December 1938, he received two appointments from Göring as plenipotentiary for the chemical industry.[99] Another winner was Bentz, whom Göring appointed as Plenipotentiary for Crude Oil in July 1938, extending his brief to include Czechoslovakia and Poland the following year—a dizzying ascent for someone who had been a low-ranking bureaucrat until 1934.[100] Overall, Krauch would oversee the synthetic fuel industry, while Fischer and Bentz would share responsibility for the oil industry. Although all three were technically subordinate to the new Economics Minister, Walther Funk, in practice they reported to Göring. This quadrumvirate of Krauch, Bentz, Fischer, and now-General Thomas, who became chief of an expanded Defense Economy Office within the Armed Forces High Command in the summer of 1939, would shape petroleum policy until 1942, when Albert Speer's Central Planning Board sidelined Göring.[101] The fact that two of these men were IG Farben executives suggests that the conglomerate gained a great deal from the reorganization even if there is no proof it engineered it.

Following the annexation of Austria in March 1938, Hitler turned his attention to liquidating Czechoslovakia, which carried the risk of a general European conflagration. Germany had to be ready for war well before its rearmament programs were complete, which was impossible.[102] As officials concluded during a conference with Göring on July 16, 1938, the progress of the Four-Year Plan had heretofore been "unsatisfactory."[103] The previous day, the Defense Economy Office had estimated that existing production and reserves of gasoline and diesel would last for only three months in wartime.[104] By October, Göring had admitted that Germany was broke and demanded that the Four-Year Plan focus on those industries that were central to the war effort or saved foreign exchange.[105]

97 Alex Kay, "Germany's Staatssekretäre, Mass Starvation and the Meeting of 2 May 1941," *Journal of Contemporary History* 41: 4 (2006): 688 and 695–7.

98 The unprofitability of hydrogenation ($40 million in losses by 1932) temporarily cost Krauch his chance to replace Carl Bosch as the head of IG Farben, thus encouraging him to seek opportunities within the regime. Hayes, "Bosch," 358–9.

99 Göring to Krauch, August 22, 1938, NARA, RG 238, T-301/113 (NI-13901); Neumann to Krauch, August 26, 1938, NARA, RG 238, T-301/113 (NI-13900). Krauch's tasks, as well as the mission of the Office for Economic Development, are spelled out in Kehrl to Krauch, March 30, 1943, *NMT*, vii: 971–7.

100 Göring to Bentz, July 28, 1938, reprinted in: Eichholtz, *Deutsche Ölpolitik*, 268–9; Göring to Bentz, October 4, 1939, Bundesanstalt für Geowissenschaften und Rohstoffe (BGR), 86534.

101 Fischer retained his position following Göring's eclipse, although starting in 1943 he reported to Hans Kehrl, chief of the Raw Materials Office within the Economics Ministry. Kehrl, *Krisenmanager im Dritten Reich: 6 Jahre Frieden—6 Jahre Krieg, Erinnerungen* (Düsseldorf: Droste Verlag, 1973), 203 and 227. For a schematic of the post-1942 administrative structure for petroleum policy, see: Stokes, "Oil Industry," 262–4.

102 Germany's position on the eve of the Sudeten Crisis is summarized in Murray, *Balance*, 3–49.

103 "Business... 16 July 1938," *NMT*, vii: 895–7.

104 Defense Economy Office to General Staff, "Betriebsstoff-Versorgungslage [...]," July 15, 1938, T-77/123 (Wi/IF 5.533).

105 "Conference... 14 Oct. 38[...]," *NCA*, iii: 901–4.

Table 6.6. European motorization, 1937

Country	# of automobiles (per capita)
Britain	688,000 (21)
Germany	621,000 (47)
France	455,000 (19)

One industry that satisfied both criteria was petroleum. Production had kept pace with consumption, but the foreign exchange burden had swollen from 145 million RM to 230 million RM between 1933 and 1937 due to the larger volume of imports—2.48 million tons to 4.31 million tons—even if the share of imports relative to demand was unchanged (3:1).[106] The increase in imports was distributed unevenly, with demand for diesel—90 percent of which was imported—and crude oil tripling, and that for fuel oil doubling. Germany had also gone from the least motorized of the great powers in absolute terms to near parity with Britain, while vaulting past France, although it still lagged behind both in per-capita figures, as seen in Table 6.6.

Likewise, per-capita consumption was unimpressive—only 40 liters in Germany, compared to 83 liters in France, 144 liters in Britain, and 609 liters in the United States.[107] Nevertheless, there had been a noticeable drop in Germany's gasoline imports, from 1.09 million tons in 1932 to 1.06 million tons five years later. This was only possible because production of lighter varieties of petroleum had quadrupled between 1933 and 1937, from 386,000 tons to 1.48 million tons. Nevertheless, by the beginning of 1938, 60 percent of Germany's gasoline supply came from domestic sources. Crude oil production had doubled from 230,000 tons in 1932 to 453,000 tons in 1937, while the number of producing oilfields had risen from five to seventeen.[108] Just four oilfields near Hannover accounted for 91 percent of crude oil production, and while the results elsewhere in Germany were unimpressive, recent discoveries in Austria, where output quadrupled from 7,500 tons to 33,000 tons in 1936–7, appeared promising. Moreover, synthetic fuel was becoming less expensive. At first glance, the price differential between synthetic and natural petroleum—roughly 16 Pfennigs per liter of synthetic fuel versus 3.45 Pfennigs per liter of gasoline sold at the U.S. Gulf Coast—was immense. The gap narrowed, however, if the cost of shipping to Hamburg was included, which raised the price of U.S. gasoline to 8–9 Pfennigs per liter.

In the summer of 1938, the Reichsbank was pleased with the increase in foreign exchange earnings since 1934 (from six to 7.74 billion RM), but the outlook for

[106] Except where noted, all of the information in this paragraph is drawn from: Reichs-Kredit-Gesellschaft, "Treibstoffwirtschaft [...]," April 1938, T-84/51 (EAP 66-c-2-10/22).

[107] "Bestand [...]," *Wehrtechnische Monatshefte* 2 (1936), T-77/425 (Wi/IF 5.3444).

[108] All told, the regime sponsored 643 drillings between 1934 and 1945, 505 of which yielded some oil, at a cost of forty-four million RM. The government also passed a new law encouraging unit operation of oilfields. Berlin recouped its expenses through royalties and taxes, but although it helped develop nineteen oilfields, German firms found little oil in pre-1937 Germany. Karlsch, *Faktor Öl*, 171–5 and 177–8; Stokes, "Oil Industry," 259–64.

1938 was unfavorable, as it forecast a decrease in exports.[109] Until the expansion of synthetic production was completed, the situation during a protracted war would be dire. A Defense Economy Office evaluation from the spring of 1938 determined that Germany could cover its full requirements for only four months. Monthly consumption would run to 636,000 tons—84,000 tons for the army, 162,000 tons for the navy, 175,000 tons for the air force, and 216,000 tons for the civilian economy. Once stockpiles had evaporated, Germany could only satisfy one-quarter of its consumption. If the Germans wished to have enough fuel to cover six months of consumption, they would need to import 934,000 tons, at a cost of 65 million RM. The cost of one year's consumption was even larger—300 million RM, largely in foreign exchange.[110]

Germany's petroleum position would not have been dire had war not been imminent, but that was no longer the case following Hitler's diplomatic offensive in 1938. If the regime was to make good on its boasts concerning energy independence, it had to make some hard choices concerning the short-term allocation of Germany's scarce resources. Assuming that self-sufficiency was even technologically feasible, there was always the question of the opportunity cost. Supporters of synthetic fuel argued that if there was one item in which Germany needed to be self-sufficient, it was petroleum. Others resources such as coal, ore, and food could be bought or taken from neighbors. With the exception of Romania, however, Continental Europe had no significant oil producers. Wartime imports from the Western Hemisphere were naturally out of the question, and Soviet exports were drying up. Moreover, it was not enough to have an ample supply of oil. As planners understood, specific varieties of refined petroleum had to be available in sufficient quantities at the start of the hostilities. The volume required, and the cost and vulnerability of storage facilities, made it impractical for any country to stockpile enough for more than a few months' consumption. These and other resource constraints inspired and then circumscribed the most ambitious synthetic fuel project ever attempted—the New Defense Economy Supply Plan of 1938, otherwise known as the Krauch Plan.

This plan marked the start of the second phase of the Four-Year Plan—the shift in focus from broad management of the economy to a narrow emphasis on increasing output of several war-essential intermediate goods.[111] Krauch's plan was ready in June 1938 and commenced a month later. It required a near doubling of the steel allocation to the synthetic fuel industry after December 1938—from 60,000 tons to 110,000 tons through 1942. Krauch also wanted to expand the capacity of existing "standby plants" and stockpile any surplus rather than constructing new facilities.[112] He estimated peacetime consumption in 1942 at 7.2 million tons and

[109] Reichsbank, "Entwicklung [...]," July 9, 1938, *Akten*, v: 515–17.

[110] W/Ro, "Mineralöl-Versorgung [...]," June 8, 1938, T-77/341 (Wi/IF 5.2164, 2687). See also: W/Ro, "Bemerkungen [...]," May 3, 1938, and "Schluss [...]," May 17, 1938, both in: T-77/123 (Wi/IF 5.533).

[111] Petzina, "Vierjahresplan," 75ff; Karlsch, *Faktor Öl*, 197–8.

[112] For modifications to existing plants, see: Krauch to Economics Ministry, "Sofortprogramm," September 29, 1938, attached to: Krauch to Armed Forces High Command, September 29, 1938, NARA, RG 238, T-301/60 (NI-7471).

consumption in the event of mobilization at 13.84 million tons. Krauch ruled out meeting 100 percent of the latter—instead, Germany would seek self-sufficiency only where it was feasible (diesel and lubricants). Meanwhile, it would produce gasoline equivalent to 85 percent of demand, aviation fuel to 75 percent, and fuel oil to 60 percent. In total, Krauch estimated output during mobilization at 11.09 million tons (79 percent of consumption).[113] This gap could be filled using the surplus from peacetime, when production was 9.25 million tons (128 percent of demand). The scale of the investment necessary was, however, staggering. The new facilities would consume an additional 28.8 million tons of soft coal and seventeen million tons of hard coal, plus a further 1.25 million kilowatts of electrical power. The plan consumed 4.46 million tons of steel and the labor of 77,500 workers, at a cost of 4.35 billion RM.[114] This quantity of steel would have sufficed to build a fleet 3.5 times the size of the Royal Navy in January 1940.[115]

Although the figures the Krauch Plan envisaged—eleven million tons—were tiny compared to U.S. and Soviet oil production, they are significant considering that Germany's total supply peaked in 1943 at 11.3 million tons including imports, withdrawals from stockpiles, and booty from the occupied territories.[116] Considering Germany's limited means and requirements, the Krauch Plan was certainly an ambitious objective—but it was not autarky, since Germany would still have to import oil for the foreseeable future.

Romania was the only way to square the circle, as the Office for Economic Development concluded in August 1938. It surveyed Germany's situation following mobilization under three scenarios: (1) overseas imports cut off but imports from Romania continuing at existing levels; (2) all imports ceasing; and (3) Germany monopolizing Romanian exports. In the first case, Germany would only be able to meet 61 percent of its estimated demand and would have to stockpile 2.43 million tons. In the second case, it would supply only 57 percent of its needs and require an additional 233,000 tons of stockpiles. In the third case, however, there was no need whatsoever to stockpile. In short, if Germany was to have a sufficient supply of petroleum in wartime, it had to secure at least half of Romania's current exports to other nations (4.9 million tons). This was even clearer when using the new 1942 mobilization requirements—13.84 tons—as a benchmark. In the case that it lost access to imports, Germany could meet 26 percent of its requirements in 1939 and 79 percent by 1943. Again, the only way of coping with any shortfall was with a share of Romania's exports to other countries—from

[113] This figure would satisfy only 25 percent of estimated British and French requirements. Office for Economic Development, "Verzögerung [...]," November 29, 1938, NARA, RG 238, T-301/73 (NI-8838).

[114] Office for Economic Development, "Neuer Mineralölplan," July 12, 1938, T-77/423 (Wi/IF 5.3398). Additional statistics may be found in: BA-B, R 3112/89. The Krauch Plan also included increases in the production of synthetic rubber, light metals, gunpowder, and explosives. Reich Office for Economic Development, "New Military-Economic Production Plan [...]," July 12, 1938, *NMT*, vii: 890–3.

[115] USSBS, *Oil Industry*, 22; Stranges, "Fuel Industry," 162–3.

[116] USSBS, *Effects*, 75.

98 percent in 1939 to 39 percent by 1943. Even if it secured the additional oil, Germany lacked the refinery or railway capacity to exploit it.[117]

In the meantime, the Krauch Plan was running into difficulties. The annexation of Austria, followed by that of the Sudeten territory, could only provide temporary respite from Germany's mounting economic difficulties. The Mining Division of the Economics Ministry determined in August 1938 that Krauch's expectations concerning additional coal production were unrealistic.[118] There was also the perennial shortage of steel, over whose allocation Hitler exercised personal scrutiny after 1937.[119] According to Krauch's original projections, the synthetic fuel industry would have a temporary cut to its steel allocation until April 1, 1939, when it would increase to more than 110,000 tons. Once the regime decided in late 1938 that it needed to boost exports to avoid a financial crisis, however, it turned out there was insufficient steel to go around.[120] In November, after Krauch learned that his steel allocation would not be raised until April 1, 1940 at the earliest, he warned this would delay fulfillment of his plan by eighteen months. Krauch could not understand why this was happening, considering Germany would save 144 million RM in foreign exchange by 1943 under his plan, while it would lose 165 million RM if the steel allocation only went up in April 1940 and 379 million RM if the allocation stayed at 42,000 tons.[121]

The following month, Krauch explained to General Wilhelm Keitel (Chief of the Armed Forces High Command) that a number of firms had already committed themselves to providing the necessary equipment for his plan in 1939. Forcing them to restructure their operations would add another six months to the completion date. In that case, Germany could not begin stockpiling excess synthetic production until 1942. Krauch also pointed out that any delays would jeopardize the production of irreplaceable high-octane aviation fuel. The overall shortfall would reach seven million tons by 1944, valued at 420 million RM. Meanwhile, the 120,000-ton steel allocation he required was only 8.7 percent of Germany's output. While diverting steel to munitions production would increase

[117] Office for Economic Development, "Mob-Bedarf [...]," August 5, 1938, T-77/500 (Wi/I.54). The office subsequently produced detailed plans for stockpiling in case Romania proved unreliable. The necessary fuel (5.45 million tons) would cost 245 million RM, not including the provision of eight million cubic meters of storage, 18,000 railway tank cars, and twelve tankers. Office for Economic Development, "Ergänzung [...]," October 21, 1938, T-77/427 (Wi/IF 5.3481).

[118] Economics Ministry, "Neuer wehrwirtschaftlicher Erzeugungsplan," August 31, 1938, *Akten*, v: 643–8.

[119] Tooze, *Wages*, 243.

[120] For background, see: Tooze, *Wages*, 300ff.

[121] Office for Economic Development, "Verzögerung [...]," November 29, 1938, and Office for Economic Development, "Auswirkungen [...]," December 14, 1938, both in: NARA, RG 238, T-301/73 (NI-8838). The petroleum industry was not the only sector hammered by shortages. The steel allocation for the first quarter of 1939 was the same as that in the last quarter of 1938 (306,600 tons), but much less than the required 613,500 tons. The military saw a small reduction (from 584,333 tons to 573,113 tons), which was only 53 percent of its requirements (1,088,300 tons). Overall, the entire economy had to cope with a shortfall of one million tons at the start of 1939. W Wi Id, "Material [...]," October 27, 1938, *NCA*, iii: 904–6.

weapons output, what was the point if there was insufficient fuel?[122] If necessary, Krauch's office argued, the Third Reich should cut back on exports and allocations of capital and raw materials to other sectors of the economy to satisfy the demands of the Krauch Plan.[123]

The Defense Economy Office backed Krauch. With a steel allocation of 42,000 tons, the Krauch Plan would not be complete until 1947, while only 38 percent of the average mobilization requirements between 1941 and 1943 could be satisfied internally. Raising the allocation from April 1940 onwards would shave three years off the waiting period, but Germany would only produce 54 percent of its mobilization requirements as of 1944.[124] The Defense Economy Office was mortified that a two-thirds reduction in the allocation to the petroleum industry would limit production by 1943 to 6.5 million tons—well short of the expected peacetime demand of 8.3 million tons and nowhere near the revised mobilization figure of twenty-one million tons.[125] To make matters worse, a scarcity of credit threatened even the 6.5-million-ton figure. The Defense Economy Office warned that bond markets were unwilling to make further loans and urged the government to intervene. Except for the navy, which grumbled at Krauch's indifference to its needs, stockpiling was not a viable alternative.[126] Therefore, it recommended a one-year emphasis on expanding fuel versus armaments output.[127]

The newly established Armed Forces High Command took its case to Göring directly, reminding him that Hitler's promise of 1936 regarding self-sufficiency within eighteen months was "nowhere near" reality. The High Command offered a less ambitious program than that recommended by the Defense Economy Office: increasing production to eight million tons by 1942. Even this limited objective required Göring to put the petroleum industry "in first place" for access to resources. The regime would also have to provide 1.5 billion RM per year and ensure that the steel allocation increased to 120,000 tons per month.[128]

The armed forces eventually endured reductions of between 20 percent and 50 percent in their allocations of steel, copper, aluminum, rubber, and cement.[129]

[122] Office for Economic Development, "Verzögerung [...]" and "Zur Auswirkung [...]," December 18, 1938, both attached to: Krauch to Keitel, December 23, 1938, NARA, RG 238, T-301/60 (NI-7471).

[123] Office for Economic Development, "Lagebericht [...]," March 1, 1939, T-77/331 (Wi/IF 5.2015).

[124] W/Ro, "Entwicklung [...]," January 20, 1939, T-77/231 (Wi/IF 5.1180).

[125] The Defense Economy Office's upward revision of the 1943 mobilization figures (from 13,835,000 to 21 million tons) was due to a 250 percent increase in estimated air force consumption after 1941—from four million tons to ten million tons. "Vorausgeschätzter Gesamt-Mineralölbedarf [...]," attached to: Armed Forces High Command, "Vortragsnotiz [...]," January 31, 1939, NARA, RG 238, T-301/60 (NI-7471).

[126] Grossman, "Vermerk [...]," October 12, 1938, T-77/228 (Wi/IF 5.1171). In February 1939, the navy estimated it required 180 million RM worth of imports (six million tons) to establish a reserve equal to three years of wartime consumption. "Sicherstellung [...]," February 15, 1939, no author, attached to: W/Ro III to Ro V, March 22, 1939, NARA, RG 238, T-301/60 (NI-7471).

[127] W/Ro, "Vortragsnotiz [...]," January 10, 1939, T-77/341 (Wi/IF 5.2164, 2687).

[128] Armed Forces High Command, "Vortragsnotiz [...]," January 31, 1939, NARA, RG 238, T-301/60 (NI-7471).

[129] Williamson Murray, *German Military Effectiveness* (Baltimore: Nautical & Aviation, 1992), 218–19.

Table 6.7. Germany's supply position, 1939[130]

Type of product	% covered
Gasoline	42%
Aviation fuel	30.8%
Diesel fuel	19.2%
Fuel oil	20.5%

The synthetic fuel industry, by contrast, endured only a temporary delay. On February 9, 1939, Göring restored the Krauch Plan's steel allocation as of the third quarter of 1939, thus delaying its completion by only six months.[131] Nevertheless, synthetic fuel was not going to make Germany energy independent even temporarily. As of March 1939, Germany could use domestic sources to cover only the shares of its wartime needs shown in Table 6.7.

This worked out to between three to six months of consumption.[132] A month before the start of World War II, the Office for Defense-Economic Planning, responsible for collecting statistical data for rearmament, estimated that Germany's annual mobilization deficit would be 117 percent of its 1937 imports (4.29 million tons), 70 percent of which came from unreliable nations.[133]

The various crises confronting Germany by the latter half of 1938 lent a new urgency to resolving its petroleum shortfall. Germany had triumphed during the Sudeten crisis of September–October 1938 without the use of force, successfully coercing Britain and France into overseeing the dismemberment of Czechoslovakia, even though it was still unprepared for war. The expansion of synthetic production was nowhere near complete, and Germany still imported about five million tons (two-thirds of demand) in 1938, with three of its four largest suppliers in the Western Hemisphere: Venezuela/Dutch West Indies (first place), the United States (second), and Mexico (fourth). Third-placed Romania was no substitute—imports from that country in 1938 totaled only 451,000 tons.[134] Moreover, Bucharest had warned Berlin on October 3, 1938 that it would cut off oil exports in wartime.[135] If that had happened, German reserves would have lasted no more than five months.[136]

130 W/Ro, "Mineralöl-Versorgungslage [...]," February 22, 1939, T-77/341 (Wi/IF 5.2164, 2687).

131 Office for Economic Development, "Wehrwirtschaftlicher neuer Erzeugungsplan [...]," February 14, 1939, T-77/423 (Wi/IF 5.3394, [illegible], 3398, 3400).

132 W/Ro, "Mineralöl-Versorgungslage [...]," February 22, 1939, T-77/341 (Wi/IF 5.2164, 2687). Such figures fluctuated—an evaluation the following month had slightly different results. W/Ro, "Mineralöl-Bilanz," May 8, 1939, T-77/341 (Wi/IF 5.2164, 2687). One can compare five different assessments from between 1937 and 1939 in Armed Forces High Command, "Vergleichende Zusammenstellung [...]," September 1944, T-77/341 (Wi/IF 5.2164, 2687).

133 Office for Defense-Economic Planning, "Die Einfuhrabhänigigkeit [...]," August 1939, T-84/72 (EAP 66-c-12/32). For a breakdown by petroleum product, see: Office for Defense-Economic Planning, "Der deutsche Außenhandel [...]," May 1939, T-84/195 (EAP 66-c-12-62/160).

134 "Deutschlands Mineralöl-Einfuhr [...]," no date or author, T-77/295 (Wi/IF 5.1589).

135 Williamson Murray, "Munich, 1938: The Military Confrontation," *Journal of Strategic Studies* 2: 3 (1979), 282–302.

136 W Ro Chef, "Kurze Beurteilung [...]," August 10, 1939, T-77/167 (Wi/IF 5.665).

At the start of the New Year, the Defense Economy Office concluded that self-sufficiency had failed. Imports were still almost double production (4.4 million tons versus 2.4 million tons), and the latter only covered 27 percent of mobilization demand (eight million tons). As of January 1939, Germany could only expect production to rise to 6.5 million tons by 1943—well short of the target of 11.3 million tons—while mobilization requirements would probably be much greater due to higher air force requirements. As for external suppliers, Romania was the obvious choice, but only one-fifth (800,000 tons) of Germany's petroleum imports from Romania in 1936 traveled along the Danube, while the capacity of Germany's Danubian fleet on war's eve was at most 1.4 million tons.[137] The remainder came by rail (1.4 million tons) or by sea via the Black Sea and the Mediterranean (1.8 million tons)—the latter requiring a journey of more than 4,000 miles. Germany's requirements were still so small that Romania's total exports plus domestic production would suffice in wartime, but since foreign capital controlled two-thirds of Romanian production, Germany could not afford to rely solely upon "Romania's good will."[138]

Synthetics might one day render Germany independent, but the military had to be prepared for conflict right away. In April 1939, the Defense Economy Office assessed the best way to meet Germany's requirements during a protracted war. Under the worst-case scenario, Germany would be at war with both the Western Allies and the Soviet Union. It would require 673,000 tons of petroleum per month. As of March 1939, it was only capable of producing 193,000 tons during mobilization, leaving a shortfall of 480,000 tons. Existing production and stockpiles would keep the German war machine running for four months. There were few viable options for filling the gap of five million tons. Three-quarters of Germany's imports in 1938 came from just three countries—the United States, Mexico, and Venezuela—none of whom were reliable suppliers in wartime. That left only one option: "Romania!"[139]

BUILDING BRIDGES TO BUCHAREST, 1938–1939

Despite its name, the vaunted prewar "Reichsmark bloc"—Bulgaria, Greece, Hungary, Romania, Turkey, and Yugoslavia—was never at Berlin's mercy. If anything, its members took advantage of Germany, which paid dearly for the privilege of buying imports with blocked currency rather than foreign exchange.[140] German food imports from the region cost 40 percent more than the world price.[141]

[137] Haβmann, "[...] Ausbau der Donau-Flotte," October 21, 1938, *Akten*, v: 748–54; First Danubian Steamship Company, "Memorandum [...]," May 3, 1939, BA-B, R 901/112770.

[138] Griebel, "Mineralölversorgung [...]," January 19, 1938 (*sic*; 1939), T-77/526 (Wi/I. 164). See also: "Aktennotiz [...]," January 19, 1939, T-77/649 (Wi/VI. 39).

[139] Defense Economy Office, "Die Mineralölversorgung [...]," April 1939, T-77/500 (Wi/I.37). See also: Defense Economy Office, "Mineralöl Rumänien," January 17, 1939, T-77/645 (Wi/Vi. 13a, b).

[140] Milward, "Reichsmark Bloc," 331–59; Steiner, *Triumph*, 1046.

[141] Neal, "Economics," 391–404.

Table 6.8. Largest oil companies in Romania, 1938

Company	Production	Nationality
Astra Romana (Shell)	1,461,000 tons	Majority British–Dutch
Romano-Americana (Jersey)	899,000 tons	U.S.
Concordia	864,000 tons	Romanian–Belgian–French
Steaua	740,000 tons	French–British–Romanian
Unirea (Phoenix Oil and Transport)	671,000 tons	British
Creditul Minier	381,000 tons	Romanian
Colombia	377,000 tons	French
Prahova	335,000 tons	Italian
Total for top seven companies	5,728,000 tons	
Total Romanian production	6,603,000 tons	

Romania enjoyed an especially strong position because it had many customers for its oil who, unlike Germany, were willing to pay in hard currency.[142] On the eve of World War II, Romania's proven reserves stood at sixty million tons and its probable reserves were perhaps twice as great. Refining capacity was twelve million tons, while consumption (1.67 million tons in 1938) was low.[143]

It is clear why the Third Reich considered Romania the answer to its petroleum problems—at least in the short run. Before World War II, the German stake in Romania's oil industry was miniscule. Fifty-six companies were active there, but seven had 80 percent of production and the top three more than 50 percent. Moreover, seven of the eight largest by output were foreign-owned.

Furthermore, British, Dutch, French, Belgian, and U.S. capital contributed 54 percent of investment, while German capital claimed a paltry 0.2 percent.[144]

Romania did not regain prewar levels of production until 1923 due to material shortages, as can be seen in Table 6.9.

Production in 1919 was 920,000 tons, of which only 4 percent was exported. Two years later, it had increased to 1.17 million tons, 24 percent of which went abroad. Nonetheless, Germany evinced little interest in Romanian oil, whose exports went all over Europe, as well as to Turkey and Egypt.[145] The Romanians were actually the first to suggest larger exports to Germany, since they feared losing market share there to the Soviet Union. The Romanian minister raised the matter in 1934 but received an equivocal response—the Germans were happy to purchase more oil from Romania, but only under less stringent financial terms.[146]

The following year, the two countries signed an agreement limiting the amount of Romanian oil exports paid for on a clearing basis to 25 percent of Romania's

[142] Milward, "Reichsmark Bloc," 341–3 and 354–5.

[143] IG Farben, "Das rumänische Erdöl," April 11, 1939, T-77/611 (Wi/IC 4.16).

[144] Eichholtz, *Rumänisches Öl*, 13–15.

[145] Exports to Germany did not resume until 1921—27,749 tons (7 percent of total exports). Division X to Wiedenfeld, et al., October 14, 1920, PAAA, R 97745; Freytag, "Rumänische Petroleumindustrie," December 31, 1922, PAAA, R 89383.

[146] Romanian Legation to Foreign Office, May 11, 1934, Ritter to Division II, et al., "File Note," May 14, 1934, and Foreign Office to Romanian Legation, June 26, 1934, all in: PAAA, R 118179.

Table 6.9. Romanian Oil Production, 1914–1945[147]

Year	Quantity	Year	Quantity
1914	1,784,000 tons	1930	41,624,000 bbl
1915	1,673,000 tons	1931	49,127,000 bbl
1916	1,244,000 tons	1932	53,815,000 bbl
1917	517,000 tons	1933	53,020,000 bbl
1918	1,214,000 tons	1934	62,063,000 bbl
1919	6,618,000 bbl	1935	61,310,000 bbl
1920	7,435,000 bbl	1936	63,659,000 bbl
1921	8,368,000 bbl	1937	52,452,000 bbl
1922	9,843,000 bbl	1938	48,487,000 bbl
1923	10,867,000 bbl	1939	45,648,000 bbl
1924	13,369,000 bbl	1940	43,168,000 bbl
1925	16,650,000 bbl	1941	40,517,000 bbl
1926	23,314,000 bbl	1942	42,094,000 bbl
1927	26,368,000 bbl	1943	39,182,000 bbl
1928	30,773,000 bbl	1944	26,191,000 bbl
1929	34,758,000 bbl	1945	34,772,000 bbl

total exports to Germany.[148] Between 1933 and 1936, Romania's share of German oil imports rose from 7 percent to 20 percent, and oil comprised most of the 41.7 million RM increase in German–Romanian trade.[149] By 1936, Romania was Germany's second-largest oil supplier, at 901,000 tons, behind only Venezuela/Dutch West Indies, at 1.48 million tons. That year, however, a dispute over payment resulted in a 50 percent drop over two years, and Germany paid exorbitant rates in Reichsmarks to save foreign exchange.[150] The value of German exports to Romania in 1937 and 1938 was double that of 1936, even though the volume of oil imports plummeted.[151] By 1937, only 15.5 percent of Romania's oil exports went to Germany and Austria, compared to 26.8 percent to Britain and France.[152]

Relations did not improve until after the Sudeten crisis, when Romania, still a French ally, reached out to Germany out of fear of the Soviet Union and Hungary. Bucharest was now willing to cease privileging exports to hard currency nations, since this policy had failed to promote the modernization of the Romanian oil industry, such that Romanian gasoline was no longer acceptable for import into Britain by 1938.[153] Hitler and Göring met with King Carol in Germany in November of that year to mend fences in exchange for economic cooperation.[154]

[147] Friedensburg, *Erdöl*, 26; DeGolyer, *Historical Data*, 7.

[148] Eichholtz, *Rumänisches Öl*, 17; Pearton, *Oil*, 194.

[149] David Kaiser, *Economic Diplomacy and the Origins of the Second World War: Germany, Britain, France, and Eastern Europe, 1930–1939* (Princeton: Princeton University Press, 1980), 137–8.

[150] Germany imported 900,500 tons from Romania in 1936 for 54.12 million RM. Two years later, it imported only 450,000 tons for 36.02 million RM. "Deutschlands Mineralöleinfuhr," April 21, 1939, T-77/295 (Wi/IF 5.1589).

[151] Kaiser, *Economic Diplomacy*, 318.

[152] "Mineralöl-Statistik," no date, T-77/658 (Wi/VI.109).

[153] Fabricius to Foreign Office, "Rumänische Petroleumpolitik," August 1, 1938, BA-B, R 901/112769.

[154] *DGFP* (D), v: 309–11, 338–42, 344–7, 392–4, and 394–5.

The following month the two nations signed a trade agreement, which retained the 25 percent limit on oil exports under clearing but allowed Berlin to pay for additional imports with weapons.[155] The conclusion of the Molotov–Ribbentrop Pact in August 1939 finally convinced the Romanians to tilt to the Axis.[156] As evidence of their good faith, that same month they suddenly made a large quantity of oil available to Germany.[157]

Romanian oil production dropped again after June 1936—from 25,330 tons per day to 18,500 tons a year later. German diplomats initially paid little attention to this development.[158] That changed in 1938, following the annexation of Austria and the Sudeten crisis. German officials now paid much closer attention to Romania's potential utility as a raw materials supplier.[159] Of course, the most pressing task was financial—if Germany was to boost imports from Romania, it had to find some alternative method of financing due to the shortage of foreign exchange. Arms sales were the obvious choice.[160]

Romania was clearly worth the trouble. In January 1939, the Office for Defense-Economic Planning determined that Romania could satisfy Germany's entire peacetime consumption and stressed that its refining industry was equally important.[161] Another assessment from March 1939 raised concerns about the oilfields' vulnerability to Soviet attack. The Red Army had easy access to the oilfields, most of which were clustered around Ploieşti, about thirty-five miles north of Bucharest. Meanwhile, the Carpathian Mountains blocked the path of the German Army. Although it dismissed fears of the imminent exhaustion of Romania's oilfields, the Office for Defense-Economic Planning recommended continuing investment in synthetic production.[162] IG Farben took a more optimistic view. According to its economic intelligence division, the main culprits for the decline in production were inefficient drilling practices and Bucharest's nationalistic policy.[163] The drop in production after 1936 coincided with a decrease in drilling

[155] Kaiser, *Economic Diplomacy*, 263ff.

[156] Eichholtz, *Rumänisches Öl*, 17–32; Steiner, *Triumph*, 946–9. See also: Christian Leitz, "Arms as Levers: 'Matériel' and Raw Materials in Germany's Trade with Romania in the 1930s," *International History Review* 19: 2 (1997): 312–32.

[157] German Legation, Bucharest, "Erdölausfuhr [...]," August 25, 1939, BA-B, R 901/112770.

[158] Opinions differed over the causes. The German minister considered it evidence of the oilfields' impending exhaustion. See Fabricius, "Rückgang [...]," March 24, 1937, BA-B, R 901/112769; Fabricius, "Rückgang [...]," July 8, 1937, T-77/599 (Wi/IC 4.41). See also: note to Military Intelligence, "Rückgang [...]," January 25, 1938, no author, T-77/599 (Wi/IC 4.41). Other argued that the decline was due to a lack of investment and the unfavorable regulatory climate. Gensch, "Rumänien/Erdöl," November 1939, BA-B, R 901/112772; "Rumäniens Öl [...]," *Militär-Wochenblatt* (December 27, 1940); W.F., "Neuordnung [...]," *Deutsche Wehr* (November 28, 1941).

[159] Kaiser, *Economic Diplomacy*, 277–8.

[160] Memorandum, April 28, 1938, no author, BA-B, R 901/112769.

[161] Office for Defense-Economic Planning, "Rumäniens Rohstoffwirtschaft [...]," January 1939, T-84/109 (EAP 66-c-12-42/34). See also: Ruprecht, "Rumäniens Rohstoffbesitz," *Deutsche Wehr* (May 16, 1941).

[162] Office for Defense-Economic Planning, "Die rohstoffwirtschaftliche Bedeutung [...]," March 1939, T-84/80 (EAP 66-c-12-4/24). See also: IG Farben, "Der deutsche Rohstoffbezug [...]," September 4, 1939, BA-B, R 8128/579; Becker, "Der Kriegsaussenhandel [...]," June 26, 1939, and "Die Zufuhren [...]," February 15, 1940, both in: T-77/1060 (Wi/ID. 37).

[163] IG Farben, "Das rumänische Erdöl," April 11, 1939, T-77/611 (Wi/IC 4.16).

activity—395,000 meters in 1937 but only 152,000 meters during the first seven months of 1939.[164] Romania was not running out of oil—production recovered after World War II and did not, in fact, peak until the mid-1970s, at 110 million bbl per year—but after 1936 output ebbed until the 1950s, due to inefficient drilling practices and a regulatory framework that discouraged exploration.[165]

The Armed Forces High Command lauded the German–Romanian Economic Treaty of March 23, 1939.[166] This treaty was the first step in Romania's conversion into a German satellite.[167] Britain and France did not stand by idly. France secured a doubling of oil imports (600,000 tons) between March 1939 and March 1940, and on April 13, 1939, both countries extended credits to Romania to purchase war materiel and guarantees for its security, but the tide was already turning in Germany's favor politically.[168] In September 1939, Berlin and Bucharest signed a one-year agreement worth 300 million RM, whereby Germany would exchange captured Polish war materiel for Romanian foodstuffs and oil.[169] The Romanian Minister-President (George Tatarescu) committed his country to selling to Germany up to 130,000 tons of oil per month. The two sides bickered, however, over the exchange rate.[170] They finally concluded a supplementary agreement on December 21, 1939, which finalized monthly oil deliveries for the first year at 130,000 tons.[171] Another complication arose in November, when the army refused to make any booty from Poland available. This matter was not resolved until March 1940.[172] The Germans now agreed to exchange weapons for a credit of 600 million lei to finance purchases of Romanian oil at a cost of 78.1 RM per ton exclusive of freight costs from Constanța.[173]

Only one obstacle remained—inadequate transportation capacity. Most analyses of the German–Romanian oil trade tend to focus on disputes over compensation.[174] Although Romania forced Germany to pay a premium—three times more

[164] Busch-Zantner, "Die Zukunft [...]," *Vierjahresplan*, 1939/XXIV.

[165] DeGolyer and MacNaughton, *Twentieth Century Petroleum Statistics* (Dallas: DeGolyer and MacNaughton, 2012), 7, and Historical Data, 7; James Gadea, "Ulei Română" During World War II and Beyond: Development of the Romanian Oil Industry," *Yale Review of International Studies* 5: 1 (2014): 75–85; Pearton, *Oil*, 174–5 and 202.

[166] "Vertrag [...]," March 23, 1940, *1940 League of Nations Treaty Series*, accessed August 15, 2018, <http://www.worldlii.org/int/other/LNTSer/1940/9.html>. Although the treaty was supposed to promote economic integration, the Germans initially failed to expand their presence in the Romanian oil industry. Eichholtz, *Rumänisches Öl*, 23–7.

[167] Armed Forces High Command, "Übersicht [...]," March 31, 1939, T-77/611 (Wi/IC 4.37). See also a follow-up study completed in mid-1940: Armed Forces High Command, "Die Wehrwirtschaft Rumäniens [...]," no date, T-77/611 (Wi/IC 4.33). This is shared by: Steiner, *Triumph*, 729.

[168] Eichholtz, *Rumänisches Öl*, 29–30; Steiner, *Triumph*, 744–8. Part of the problem, at least in France, was political opposition to greater imports of Romanian oil. Steiner, *Triumph*, 402 and 715–16.

[169] Clodius and Fabricius to Foreign Office, September 29, 1939, *DGFP* (D), viii: 172–3.

[170] Clodius and Fabricius to Foreign Office, December 6 and 14, 1939, *DGFP* (D), viii: 493–4, and 530–2.

[171] Wiehl, memorandum, January 3, 1940; Ribbentrop to Göring, March 16, 1940; *DGFP* (D), viii: 598–603 and 925–6.

[172] Wi VII, "Verhandlungen [...]," March 4, 1940, T-77/400 (Wi/IF 5.3063).

[173] *DGFP* (D), xiii: 73–4, 111, 401–2, 435, and 868–9.

[174] Tooze, *Wages*, 307–8.

than customers offering hard currency even after the war began—the financial differences were reconcilable.[175] Logistics was another matter. Romanian oil went abroad three ways: across the high seas, along the Danube, and by rail. The cheapest option was by sea. The distance from Constanța to Hamburg was 3,610 nautical miles (nineteen days when traveling at eight knots) and cost one-third less than transporting by rail.[176] Accordingly, before the war, most of Romania's exports had traveled by sea from Constanța—3.32 million of 4.5 million tons in 1938, and 58 percent of Germany's imports during the first eight months of 1939. After the war began, exports from Constanța thereafter fell to only 1.39 million tons in 1940 and 224,000 tons the following year, as traffic was rerouted to rail or the Danube.[177] Germany had to find some way of spreading the burden between rail and river. Both were necessary, since it was impossible to meet Germany's requirements solely by rail, which could deliver at most 130,000 tons per month even under ideal circumstances.[178]

Romania could theoretically export five million tons after subtracting its consumption from production, but Germany could only have 3.5 million after subtracting the requirements of Italy (1.15 million tons) and Slovakia, Hungary, and Yugoslavia (350,000 tons). In 1939, however, the existing transportation network could only deliver one million tons annually—60,000 tons per month via the Danube on 133 tankers and 30,000 tons per month by rail on 1,500 railway tank cars. The completion of sixty-four tankers under construction in April–June 1940 would raise the export ceiling to 1.5 million tons. Getting a further two million tons was trickier. The Germans could choose between a combination of expanding rail, river, or pipeline throughput. Building 10,000 new railway tank cars, plus expanding the rail lines from Ploiești to the Danube, could move an additional 1.5 million tons annually. Construction of the tank cars would consume 120,000 tons of steel, require eighteen to twenty-four months to complete, and cost 40 million RM. The job could also be done by 420 riverine tankers, at a cost of 80,000 tons of iron and 63 million RM—this cost would not include, however, the expense of building small pipelines to and from the river. The costliest option, a pipeline between Romanian and Germany, required 300,000 tons of steel, twenty to twenty-four months for construction, and 400 million RM. Railway cars were a cheaper option, but they increased strain on the rail network. As many as 5,000 railway cars were needed every day to move the throughput of a single pipeline with an hourly carrying capacity of 2,000 tons.[179] Pipelines were cheaper over the

175 Gensch, "Rumänien/Erdöl," November 1939, BA-B, R 901/112772.

176 German Legation, Bucharest, to Foreign Office, "Transportweg [...]," September 26, 1935, PAAA, R 118180; <https://sea-distances.org/>.

177 Fabricius, "Erdölausfuhr [...]," January 5, 1940, BA-B, R 901/112771; Neubacher, "Rückgang [...]," October 5, 1940, BA-B, R 901/112772; Foreign Office, "Die rumänische Mineralölwirtschaft," no date (circa 1942), T-120/2618.

178 "Erdöltransportmöglichkeiten [...]," October 4, 1940, author unclear, BA-B, R 901/112772.

179 Ruprecht, "Erdölbeförderung," *Militär-Wochenblatt* (October 23, 1942).

long run but also vulnerable to sabotage and could only handle one kind of petroleum product at a time. Realistically, the cost of a pipeline was only justified if Germany could also use it to transport Caucasian oil.[180]

GERMANY'S POSITION ON THE EVE OF WAR, 1939

By the spring of 1939, Germany could boast of considerable success in expanding its petroleum output. Between 1933 and 1938, consumption had more than doubled, from 2.92 million tons to 6.25 million. This led to an 80 percent increase in imports, from 2.44 million tons to 4.42 million.[181] Nevertheless, between 1933 and 1938, German production had almost tripled, going from 835,000 tons to 2.34 million tons, thanks largely to synthesis, which had provided an eightfold increase in motor and aviation fuel production (from 110,000 tons to 810,000 tons).[182] The acquisition of Austria and the Sudetenland also promised significant dividends. The latter's output of brown coal was 10 percent of Germany's production.[183] As for Austria, oil output there jumped after 1938, rising from 57,000 tons to 145,000 tons between 1938 and 1939 and reaching a wartime peak of 1.21 million tons in 1944. It may not have been much oil in global terms, but it could not have come at a better time for Germany, where crude production peaked in 1940 at 1.06 million tons.[184] The Third Reich squandered its Austrian bonanza by encouraging all-out production, which depleted the pressure of most fields, but it had no alternative.[185] As 1939 dawned, gap between supply and demand—480,000 tons per month during mobilization—was still immense.[186] A dip in imports during the first third of 1939 also forced Germany to dip into its reserves.[187]

Krauch was still proud of Germany's success in expanding its petroleum supply. Skeptics had criticized his plans, but he had proven them wrong. The major challenge was the lower steel allocation, but Göring's intervention had prevented catastrophe.[188] In April 1939, Krauch predicted that Germany would reach a

[180] "Erdöl-Transportfrage [...]," July 6, 1939, no author, T-77/612 (Wi/IC 4.55); W Wi III, "Betrachtungen [...]," October 18, 1939, T-77/612 (Wi/IC 4.71).

[181] Institute for the World Economy, "Die Versorgung [...]," February 1940, T-84/72 (EAP 66-c-12/33).

[182] Supervisory Office for Petroleum, "Deutschlands Mineralölbilanz [...]," May 3, 1938, T-77/341(Wi/IF 5.2164, 2687).

[183] Krauch, "Weg und Erfolg [...]," *Vierjahresplan*, 1939/I–II.

[184] German Institute for Economic Research, "Erdölgewinnung [...]," October 1944, T-77/341 (Wi/IF 5.2164, 2687); Bentz to Armed Forces High Command, "Entwicklung [...]," January 29, 1945, Imperial War Museum, Duxford (IWM), Foreign Documents Collection (FD) 4809/45; Dietrich Eichholtz, *Krieg um Öl: Ein Erdölimperium als deutsches Kriegsziel, 1938–1943* (Leipzig: Leigziger Universitätsverlag, 2006), 15–18.

[185] Karlsch, *Faktor Öl*, 178–9 and 211–13.

[186] W Wi VI d to W/Ro, "Sicherstellung [...]," February 20, 1939, and Griebel to W Wi VI, "Sicherstellung [...]," February 24, 1939, both in: T-77/649 (Wi/VI. 39).

[187] "Mineralöl-Situation," June 9, 1939, no author, T-77/123 (Wi/IF 5.533).

[188] At the start of the war, the monthly allocation was only 90,000 tons, but the Office for Economic Development believed that this was enough to reach an output of 8.35 million tons by 1943. "Treibstoffversorgung," January 15, 1940, no author, T-84/216 (EAP 66-c-12-62/29).

Table 6.10. Expected German supplies, 1944

Type of product	% of mobilization demand
Aviation fuel	100%
Gasoline	75% (in 1943)
Diesel fuel	60%
Fuel oil	40%

production capacity of 8.3 million tons by 1942–3 and a further three million tons by 1944, which corresponded with the Economics Ministry's estimate for peacetime consumption. By 1944, he expected that Germany would be able to meet the shares of its mobilization demand set out in Table 6.10.

Even if the Krauch Plan did not guarantee independence, it would still cover aviation fuel requirements and allow the country to accumulate larger stockpiles.[189]

Krauch's ambitions went far beyond self-sufficiency—he and his subordinates were already planning to harness the economies of Eastern Europe under German supervision.[190] By the spring of 1939, the Office for Economic Development estimated that the Krauch Plan would save Germany 2.26 billion RM of foreign exchange just for petroleum and rubber. This was still insufficient.[191] Just weeks before the invasion of Poland, the office studied whether "blockade security" was possible within a German sphere encompassing Eastern and Southern Europe. Peacetime petroleum consumption within this "Greater Economic Area" in 1938 was 11.5 million tons, half of which was supplied internally. By 1942, it would rise to 14.95 million tons in peacetime and 23.85 million tons in war. At best, the peacetime deficit in 1942 would be 1.5 million tons and 7.9 million tons during mobilization, and the situation would become catastrophic if Italy entered the war. Italian petroleum consumption in 1938 was 2.85 million tons, of which 2.63 million tons was imported, mainly from the United States (37 percent), Romania (20 percent), Latin America (16 percent), and the Middle East (10 percent).[192] In June, the Defense Economy Office estimated that Germany and Italy required 9.4 million tons of imports to cover one year's wartime consumption. Romania's export surplus would shrink from five to four million tons between 1938 and 1939, while the transportation network could only deliver 2.2 million tons to Germany and one million tons to Italy.[193] "Blockade security" was therefore impossible within the

189 Office for Economic Development, "Arbeitsbericht [...]," April 20/21, 1939, T-77/430 (Wi/IF 5.3594).

190 Tooze, *Wages*, 307–8.

191 Office for Economic Development, "Notizen [...]," May 16, 1939, reprinted in: Dietrich Eichholtz, "Die 'Groβraumwehrwirtschaft' für den großen Krieg: Zwei geheime Memoranden der Reichstelle für Wirtschaftsausbau vom Früjahr/Sommer 1939," *Bulletin des Arbeitskreises "Zweiter Weltkrieg"* (1986): 82–5.

192 "Mineralöl-Statistik," no date, T-77/658 (Wi/VI.109). For details on Italy's unpreparedness, see: Steiner, *Triumph*, 850–5.

193 W Wi IV d, "Transportmöglichkeiten [...]," June 21, 1939, T-77/526 (Wi/I. 156). This assessment was based on a longer study, which concluded that Romania, Hungary, Yugoslavia, and Estonia could supply at best 6.15 million tons of oil to Germany and Italy. Defense Economy Office, "Vortrag...20. 6. 1939 [...]," T-77/516 (Wi/I. 84).

"Greater Economic Area." The only way to make it a reality was through "a close economic association" with the Soviet Union.[194]

Depending on one's perspective, the immediate outlook was either manageable or grim. Only days before the outbreak of hostilities, the Office for Economic Development estimated that existing production and reserves could cover between five to six months of mobilization consumption of aviation fuel, gasoline, and fuel oil, but only three months for diesel. Thereafter, Germany could meet 60 percent of its gasoline demand and 29 percent of its demand for fuel oil and diesel. The completion of several synthetic facilities would improve the situation by 1940.[195] As Thomas explained to German diplomats in May, Germany also had to consider Italy's needs—in which case Romania was insufficient.[196] He was more candid with his superiors in August—Germany must have "complete domination" of the Southeastern Europe, he argued, even though it could only satisfy 60 percent of the Axis' import requirements during mobilization.[197]

Following the end of World War II, British intelligence officers came upon German estimates of the Third Reich's fuel position from October 1939. They found the similarities with their own evaluations "little short of staggering." The only difference concerned the size of Germany's reserves. British intelligence had put forward a figure of three million tons, but the rest of Whitehall found this implausible—surely no country would go to war with such meager stocks—and they revised the final estimate upwards by two million tons. The analysts had in fact been correct, and all wartime British assessments of Germany's petroleum position had overstated reserves by two million tons. After discovering this fact, one British official concluded that there was "much deliberate bluff" behind the Germans' decision for war, and that their early successes had created "a false sense of security."[198] Under the stress of war, the shortcomings of a system that depended on efficient means of producing and transporting petroleum, the generosity of ambivalent partners such as Romania, and the spoils of conquest soon became apparent.

[194] "Möglichkeiten [...]," August 1939, no author, reprinted in: Eichholtz, "Großraumwirtschaft," 86–160. See also: Institute for the World Economy, "Die Versorgung [...]," February 1940, T-84/72 (EAP 66-c-12/33).

[195] Memorandum, August 23, 1939, attached to: Office for Economic Development to Defense Economy Office, August 24, 1939, T-77/123 (Wi/IF 5.533).

[196] "Vortrag...24 May 1939 [...]," *IMT*, xxxvi: 112–32.

[197] Thomas, "Stand [...]," August 9, 1939, T-77/312 (Wi/IF 5.1788).

[198] Watson to Morton, September 5, 1945, BNA, PREM 7/14.

7

From Crisis to Opportunity

1939–1941

Germany had once again bet on war, with only the slimmest margin of error. Its survival would depend, as it had during the previous war, upon superior organization and greater ruthlessness to overcome "the superiority in raw materials and manpower of the enemy states."[1] For a time, the gamble seemed to pay off. The Third Reich's early successes would stun even its most ardent supporters, while leaving it with challenges Germany was utterly unequipped to surmount.

Germany's supply position began to deteriorate in the winter of 1939/40. Thanks to a mixture of luck and skill, these shortages did not prevent Germany from overrunning France and the Low Countries in six weeks in the period May–June 1940. Ironically, Germany's greatest triumph planted the seeds for a crippling energy crisis—a fact that is absent from most historical narratives.[2] By defeating France but not Britain, Germany's victory remained incomplete, while its liabilities had increased sharply. In 1938, the Continent had consumed 26.8 million tons of petroleum but produced only 7.95 million tons, not including German synthetic fuel.[3] Now, two years later, Germany was responsible for meeting this entire deficit, without recourse to either the Western Hemisphere or the Middle East.

German planners had never considered how they would supply their occupied territories with energy before World War II—a curious omission considering the scope of their ambitions. The system they developed was concerned only with supplying the Third Reich and was wholly inadequate to supply the rest of Europe.[4] The latter would require not only acquiring new sources of oil but also transforming the European market under German leadership. It soon became clear that the solution to Germany's problems lay in the east. Unless Germany took control of the oilfields of either the Soviet Union or the Middle East—ideally, both—fuel shortages would soon derail the entire war effort.[5]

[1] Thomas "File Note [...]," December 13, 1939, T-77/331 (Wi/IF 5.2014).

[2] Contra Paul Kennedy, *Rise and Fall of the Great Powers: Economic Change and Military Conflict from 1500 to 2000* (New York: Random House, 1987), 341, see: Tooze, *Wages*, 411.

[3] Continental Oil, "Förderung [...]," May 1942, T-84/51 (EAP 66-c-2-10/21); Mejcher, *Politik*, ii: 31–6.

[4] Walter Levy, *Oil Strategy and Politics, 1941–1981* (Boulder: Westview Press, 1982), 9–23. This fact eluded many otherwise shrewd observers. See Friedensburg, "Die deutsche Roh- und Treibstofflage," October 3, 1940, T-77/344 (Wi/IF 5.2199).

[5] Military Intelligence, "Schätzung [...]," February 20, 1941, T-77/86 (Wi/IF 5.384); "Meldung [...]," January 20, 1942 T-77/1375 (Wi/IVB. 21); "Europäischer Krieg [...]," June 27, 1942, attached to: Continental Oil to Thomas, [illegible] August 1942, T-77/551 (Wi/I. 315).

THE FRUITS OF VICTORY, 1939–1941

On the eve of World War II, Thomas warned that "Germany finds itself still in a position of economic weakness." The European Axis had an annual petroleum deficit of four million tons but increasing imports was difficult due to the transportation bottlenecks and Turkey's pro-Allied attitude, since Italy imported Romanian oil through the Turkish Straits.[6] A week into the Polish campaign (Case White), the Defense Economy Office stated that, over the first year of hostilities, Germany would have to import 3.12 million tons to cover the gap between supply and demand after its reserves evaporated.[7]

On the bright side, a new supplier became available after August 1939.[8] The Soviet Union sent more than 600,000 tons of petroleum to Germany in 1940 under the German–Soviet Commercial Agreement of February 1940.[9] According to the lead German negotiator, Karl Schnurre, by February 1941 Germany would receive 900,000 tons, valued at 115 million RM.[10] If the quantities covered under the August 1939 credit agreement and Germany's share of Galician production under the Molotov–Ribbentrop Pact are included, the figure would increase to 1.1 million tons of petroleum. As of October 1, 1940, however, only 532,000 tons of the 909,000 tons Moscow agreed to deliver had arrived.[11] In January 1941, the two sides concluded a new commercial agreement, under which the Soviets would deliver almost one million tons between February 1941 and August 1942.[12] The Petroleum Import Company, a public–private venture, distributed the petroleum under the guidance of the Economics Ministry.[13] More than half of Soviet exports (510,000 tons) arrived by rail across the German–Soviet border, but a significant quantity traveled via the Black Sea through Romania (384,000 tons—primarily fuel oil) or the Baltic (68,207 tons).[14] Ultimately, Germany received 876,000 tons of petroleum from the Soviet Union during their partnership—a trifling amount compared to the 8.82 million tons it imported from Romania between 1940 and 1944.[15] The gap between Soviet and Romanian exports was, however,

[6] Thomas, "Stand [...]," August 9, 1939, no author, T-77/312 (Wi/IF 5.1788). For Germany's petroleum position at the start of the war, see: Joel Hayward, "Hitler's Quest for Oil: The Impact of Economic Considerations on Military Strategy," *Journal of Strategic Studies* 18: 4 (1995): 100; Williamson Murray, "Change in the European Balance of Power, 1938–1939" (Ph.D. Dissertation, Yale University, 1975), 128–36; Tooze, *Wages*, 412. For Turkey's tilt toward the Allies, see: Steiner, *Triumph*, 961–5.

[7] W/Ro, "Einfuhrbedarf Deutschland [...]," September 9, 1939, IWM, FD 4809/45.

[8] The following discussion draws from: Hartmut Schustereit, "Die Mineralöllieferungen der Sowjetunion an das Deutsche Reich 1940/41," *Vierteljahrschrift für Sozial- und Wirtschaftsgeschichte*, 67: 3 (1980): 334–53.

[9] Defense Economy Office, "Mineralöl-Einfuhr [...]," December 22, 1941, IWM 4809/45. For the text of the agreement, see: *DGFP* (D), viii: 762–9.

[10] Schnurre, "German–Soviet Agreement [...]," February 26, 1940, *DGFP* (D), viii: 814–17.

[11] "Mineralölbezüge Deutschlands [...]," October 3, 1940, author unclear, PAAA, R 105990. See also: Defense Economy Office, "Aktenvermerk [...]," December 23, 1939, T-77/671 (Wi/IV. 236).

[12] Hedler, "Die Mineralöle [...]," August 31, 1941, T-77/438 (Wi/IF 5.2726).

[13] Griebel to Thomas, January 5, 1940, IWM, FD 4809/45.

[14] Schustereit, "Mineralöllieferungen," 338–47.

[15] Tomberg, "Wehrwirtschaftliche Erkenntnisse [...]," November 1944, T-77/429 (Wi/IF 5.3517).

much narrower between January 1940 and August 1941: 911,000 tons, versus 2.48 million tons.[16] Moreover, to maintain deliveries to Germany and increase its own stockpile, the Soviet Union may have even become a net oil importer, purchasing its deficit from the United States.[17]

Besides logistical difficulties, disputes over compensation caused delays and kept exports erratic, fluctuating, for instance, from 19,513 tons in April 1940 to 60,895 tons in July before dropping to 19,105 tons in November.[18] Prior to the fall of France, German diplomats worried that the Soviet Union might not wish to provoke the Allies by being too helpful to Germany.[19] The Germans' failure to deliver promised military hardware and raw materials (coal) also prompted the Soviets to throttle oil and grain exports in February 1940. Göring convened a meeting at which he reiterated that Hitler had made "an explicit decision" to satisfy obligations to the Soviet Union even at the expense of the German military.[20] Exports of Soviet grain and oil resumed in April 1940 after Göring reassured Moscow.[21]

The two sides also signed a sales contract in May 1940 (renewed in March/April 1941) stipulating the terms of future deliveries and guaranteeing the Germans a better price for oil than they had paid the Romanians in 1939. Soviet deliveries then spiked to the point of overloading the German railway system after the collapse of France, and again in the spring of 1941 following the German occupation of the Balkans, when reports of the German military buildup along the German–Soviet border reached Moscow. In both cases, the Soviets may have increased their exports to Germany to soften their hard line on political matters, such as the Soviet Union's annexation of the Baltic Republics and Bessarabia in June 1940 and Foreign Minister Vyacheslav Molotov's territorial demands during his visit to Berlin in November 1940. Moscow was probably trying to mollify Berlin by demonstrating the Soviet Union's value to Germany as a trading partner, but if so, its manipulation of exports backfired by eroding German trust and patience.[22] Overall, it appears that Germany got the better end of the economic partnership. Although the Soviets received advanced German hardware, it quickly became obsolete. The Germans, meanwhile, used Soviet raw materials to top up their

[16] Defense Economy Office, "Mineralöl-Einfuhr [...]," December 22, 1941, IWM, FD 4809/45.

[17] Felix Rehschuh, "From Crisis to Plenty: The Soviet 'Oil Campaign' Under Stalin," in: *Cold War Energy*, ed. Jeronim Perović (Cham: Springer International Publishing, 2017), 55.

[18] Schustereit, "Mineralöllieferungen," 339. For 1940, the Economics Ministry planned on 60,800 tons of imports per month. Ro V, "Mineralöl-Einfuhr aus Rußland 1941 [...]," January 11, 1941, and Defense Economy Office, "Mineralöl-Einfuhr [...]," December 22, 1941, both in: IWM 4809/45. Manfred Zeidler concludes that the Germans came out ahead by 146 million RM (the German deficit vis-à-vis Soviet deliveries). The overall financial balance was, however, unfavorable to Germany if one included the 1935 and 1939 credit agreements (a loss of 300 million RM): "German-Soviet Economic Relations during the Hitler-Stalin Pact," in: From Peace to War: Germany, Soviet Russia and the World, 1939–1941, ed. Bernd Wegner (Providence: Berghahn Books, 1997), 95–111. For problems with oil exports, see: Horst Boog et al., *Attack on the Soviet Union*, vol. 4 of *Germany and the Second World War*, ed. Militärgeschichtliches Forschungsamt (New York: Oxford University Press, 1998), 118–36.

[19] Schustereit, "Mineralöllieferungen," 340–2.

[20] Gramsch, memorandum, April 1, 1940, *DGFP* (D), ix: 59–61.

[21] *DGFP* (D), ix: 82–3, and 106.

[22] Schustereit, "Mineralöllieferungen," 345ff and 351–3.

stockpiles in advance of Operation Barbarossa.[23] Germany also imported 250 million RM of goods via the Trans-Siberian Railway.[24]

The major threat to Soviet exports actually came from the Allies, who were planning to bomb the Caucasian oilfields, pipelines, and refineries via airbases in Iraq and Syria before the German invasion of France and the Low Countries.[25] The Germans were terrified at the prospect—in January 1940, the Defense Economy Office estimated the loss of Soviet imports following an attack on Baku (700,000 tons in 1940) would cause a reduction of 44 percent in Germany's imports and 14 percent of its total supply.[26] In any event, Soviet imports could not compensate for disappointing imports from Romania and high consumption following the invasion of Poland. In October 1939, Thomas assessed the situation as critical. Although Germany signed a new economic agreement with Romania in September 1939, Thomas judged the prospect of boosting imports as "unfavorable," while increasing Soviet deliveries was difficult due to transportation bottlenecks.[27]

At home, by October 1939, Germany was still unable to satisfy Krauch's steel requirements.[28] Even before German forces attacked Western Europe in May 1940, the assumptions that underpinned German planning before September 1939 had proven unrealistic. During the first six months of the war, reserves of gasoline, diesel, and naval fuel oil dropped from 870,000 tons to 438,000 tons, although aviation fuel reserves increased from 442,000 tons to 480,000 tons.[29] The Polish campaign did not pose tremendous demands. Germany began the war with 2.4 million tons of reserves, and the consumption of aviation fuel in September 1939 was 54,000 tons, against 276,000 tons of gasoline and 155,000 tons of diesel.[30] As of October 1, 1939, stocks stood at 1.81 million tons, or 4.5 months of gasoline and aviation fuel consumption, three months of diesel, and 6.5 months of fuel oil. With these reserves, Germany could fight a war for twelve months after importing 3.12 million tons—half of Romania's production.[31]

The Germans only began to grasp the seriousness of their situation after Case White had ended.[32] At the end of September 1939, Plenipotentiary for Automotive

[23] Birkenfeld, "Stalin als Wirtschaftspartner Hitlers, 1939–1941," *Vierteljahrschrift für Sozial- und Wirtschaftsgeschichte* 53: 4 (1966): 477–510.

[24] Perkins, "Coins," 95; Heinrich Schwendemann, "German-Soviet Economic Relations at the Time of the Hitler–Stalin Pact, 1939–1941," *Cahiers du Monde russe* 36: 1/2 (1995): 162. See also: Karlsch, *Faktor Öl*, 207–8.

[25] Auswärtiges Amt, *Weißbuch Nr. 6: Die Geheimakten des französischen Generalstabs* (Berlin: 1941), 43, 48–50, 57–60, 64–8, 74–5, 77–8, 87–8; Ronald Cooke and Roy Nesbit, *Target, Hitler's Oil: Allied Attacks on German Oil Supplies, 1939–1945* (London: W. Kimber, 1985), 21–51; Charles Richardson, "French Plans for Allied Attacks on the Caucasus Oil Fields, January–April, 1940," *French Historical Studies* 8: 1 (1973): 130–56.

[26] Griebel, "Vortragsnotiz [...]," January 27, 1940, T-77/128 (Wi/IF 5.549).

[27] "Inspekteurbesprechung am 12. 10. 39 [...]," T-77/312 (Wi/IF 5.1788). These difficulties are summarized in Murray, *Balance*, 326–34.

[28] Only 95,000 tons were left after the Economics Ministry and military received their shares. Office for Economic Development, "[...] Stand der Durchführung [...]," October 2, 1939, T-77/310 (Wi/IF 5.1773–4).

[29] Defense Economy Office, "Entwicklung [...]," September 11, 1940, T-77/123 (Wi/IF 5.533).

[30] Tomberg, "Wehrwirtschaftliche Erkenntnisse [...]," November 1944, T-77/429 (Wi/IF 5.3517).

[31] W/Ro, "Einfuhrbedarf [...]," September 9, 1939, T-77/341 (Wi/IF 5.2164, 2687).

[32] The following discussion draws from: Defense Economy Office, "Die deutsche Treibstoffversorgung [...]," February 16, 1942, T-77/668 (Wi/VI. 216); the situation reports from

Affairs General Adolf von Schell assured the Chief of the General Staff, General Franz Halder, that the "fuel problem" had been "solved."[33] This proved premature. Between October and December, reserves of gasoline and diesel suffered a precipitous decline: 40 percent in the case of the former (from 454,000 tons to 280,000 tons) and 50 percent for the latter (from 298,000 tons to 150,000 tons). The shortage arose from the fact that gasoline production had been curtailed by 30 percent in favor of aviation fuel and diesel, but it could be rectified by shifting the emphasis back to gasoline. Krauch shot down this idea, suggesting instead that Germany import the required gasoline from Romania and the Soviet Union.[34] Later that month, the Office for Economic Development explained that air force consumption would increase in the second half of 1940, leaving Germany capable of producing only 58 percent of its requirements.[35] The situation was equally dire when it came to diesel fuel for the German civilian economy, although supplies of gasoline and naval fuel oil (production plus reserves) would suffice for another year.[36]

Among the other reasons for the shortage was higher than expected gasoline and diesel consumption during the lull following the Polish campaign. There was also the shock of losing overseas imports. Perhaps the most important reason, however, was difficulties in Romania. Deliveries along the Danube were erratic in winter, when the river often froze.[37] Already by November 1939, Romanian exports had dropped to 60,000 tons.[38] Worse was to come—while deliveries during the last quarter of 1939 totaled 234,000 tons, they collapsed to 94,000 tons in the first quarter of 1940, before recovering to 183,000 tons in the second quarter. On the eve of the invasion of Norway in April 1940, the Defense Economy Office forecast that, unless the situation improved, gasoline reserves would run out by the summer.[39] The Germans hoped that things would improve after April due to higher domestic production, the start of Soviet exports, and the resumption of full deliveries from Romania.[40]

1940 to May 1941 in: Defense Economy Office, "Vortragsnotizen [...]," [?] June 1941, and Defense Economy Office, "Schreiben [...]," June 6, 1941, both in: IWM, FD 4809/45; and the running commentary on Germany's petroleum position between September 1939 and November 1940 and from January 1941 to September 1942 in: Defense Economy Office, "Abschriften [...]," no date, T-77/668 (Wi/VI. 216) and "Auszug [...]," no date, T-77/668 (Wi/VI. 216).

[33] Diary entries for September 28–29, 1939: Arnold Lissance, ed., *War Journal of Franz Halder* (1950), retrieved from the Combined Arms Research Library, Digital Library (CARL).

[34] "Meeting of 10 January 1940," in: *NMT*, vii: 965–7.

[35] "Treibstoffversorgung," January 15, 1940, no author, T-84/216 (EAP 66-c-12-62/29).

[36] "Darstellung [...]," no date or author (circa late 1939), T-77/364 (Wi/IF 5.2507).

[37] The Germans should have expected this—they had already used the Danube to ship oil during their occupation of Romania in World War I. Sußdorf, "Feldkraftfahrwesen," viii: 363.

[38] Wiehl to German Legation, Bucharest, November 30, 1939, *DGFP* (D), viii: 467–8.

[39] Tomberg, "Abschrift [...]," September 1940, and Defense Economy Office, "Abschriften [...]," no date, both in: T-77/668 (Wi/VI. 216).

[40] Defense Economy Office, "Überblick [...]," December 16, 1939, T-77/341 (Wi/IF 5.2164, 2687). The Armed Forces High Command demanded that the army and air force provide daily supply reports, and that the Reich Petroleum Office (which oversaw the domestic oil industry) do the same for imports. They also demanded monthly reports on the overall situation. Keitel, "Vortragsnotiz [...]," January 16, 1940, T-77/341 (Wi/IF 5.2164, 2687).

Before May 1940, it was clear to both sides in the conflict that Germany depended on Romania and the Soviet Union.[41] In February, the Institute for the World Economy estimated Germany's annual wartime consumption at between eight and ten million tons, with five to seven million tons imported, which turned out to be close to the mark—in May 1940, Germany's mobilization demand including Denmark, Norway, and the former Czechoslovakia was eight million tons.[42] There was no way to satisfy this demand if Germany shared Romanian and Soviet exports with the rest of Europe. Even if all of their exports in 1938 (5.28 million tons) had remained in Europe, there would still have been a continental shortfall of 9.21 million tons.[43] This figure was greater than the quantities provided under Germany's commercial agreements with either country: 130,000 tons per month with Romania and 650,000 tons for the remainder of 1940 from the Soviet Union. It also exceeded the capacity of the transportation network—100,000 tons per month in the case of the former and 50,000 tons for the latter.[44] Fortunately, this was not Germany's problem—yet.

Germany survived 1939 through a combination of planning and luck. British and French passivity during the "Phony War," in particular, allowed Germany to husband its meager raw materials stockpiles. The Germans, however, could not expect their adversaries to remain so accommodating in future. Germany's economic weakness spurred Hitler to finish off France as quickly as possible.[45] As Göring informed Thomas in January 1940, Hitler was determined "to bring about a decision to the war in 1940." Accordingly, Germany would devote all available resources "without consideration" for its long-term position.[46] As planning for Case Yellow—the invasion of France and the Low Countries—progressed, Germany confronted not an overall shortage of petroleum but one of gasoline.[47] The Defense Economy Office assumed a spike in monthly army consumption for three months, starting in March 1940, to 110,000 tons; air force consumption peaking at 150,000 tons; naval consumption of fuel oil and diesel increasing to 95,000 tons and 26,000 tons, respectively; and civilian consumption declining to 64,000 tons. It did not believe there would be any problems with supplies of aviation fuel, or diesel and fuel oil for the navy, but worried that higher army gasoline and diesel consumption would create "difficulties within weeks."[48]

[41] Attachment 1 to: Chief of the Security Police to Foreign Office, "Beurteilung [...]," January 31, 1940, BA-B, R 901/116640.

[42] Ro V to Ro III, May 15, 1940, T-77/123 (Wi/IF 5.533).

[43] Institute for the World Economy, "Die Versorgung [...]," February 1940, T-84/72 (EAP 66-c-12/33). See also: Institute for Economic Research, "Das Gewerbe [...]," June 1940, T-77/201 (Wi/IF 5.991).

[44] Ro V, "Notiz [...]," February 14, 1940, T-77/123 (Wi/IF 5.533).

[45] Murray, *Balance*, 326–38.

[46] Thomas, "Besprechung [...]," January 30, 1940, Hoover Institution, Records of the International Military Tribunal at Nuremberg, Box 995 (PS-1456).

[47] Halder learned in April 1940 that the existing fuel supply would "[probably] be sufficient until fall," although the army's operational reserve would run out after three weeks. *Halder Diary*, April 12, 1940.

[48] The Defense Economy Office recommended the creation of special operational reserves of 50,000 tons of both diesel and aviation fuel. "Untersuchungen [...]," February 8, 1940, T-77/341(Wi/IF 5.2164, 2687).

Although the Armed Forces High Command established an operational reserve on March 8, 1940, the gasoline situation did not improve. Keitel informed Göring six days later that it had "worsened" to the point that it jeopardized operations. Consumption continued to outpace new supplies—between September 1, 1939 and March 1, 1940, gasoline reserves fell from 324,000 tons to 108,000 tons, while diesel reserves dropped from 230,000 tons to 55,000 tons. If Germany undertook major offensive operations, it would be able to meet its gasoline requirements for only "4–6 weeks," after which it could provide supplies equal to 30 percent of consumption. Following Thomas' recommendations, Keitel offered a list of corrective measures, including withdrawals from air force and naval reserves.[49]

Göring obliged ten days later, authorizing not only the expansion of the operational reserves, but also increased production of gasoline at the expense of aviation fuel and diesel. He also freed essential personnel within the petroleum industry from military service, directed the Labor Ministry to grant the industry "preferential" treatment, and requested that Krauch, Bentz, and Funk to devise a plan for assuring a "steady supply" for both the military and the essential sectors of the civilian economy.[50] Finally, he instructed the navy to make available some of its considerable diesel reserves. The navy complied for the sake of the war effort, but its Commander-in-Chief, Admiral Erich Raeder, did not miss the opportunity to complain to Hitler that none of this would have been necessary had the Economics Ministry not "hindered" his efforts to secure greater quantities of oil from Mexico before the war.[51]

Germany's first wartime energy crisis was the subject of a high-level conference chaired by Göring on March 27, 1940. Thomas explained that the existing shortage had arisen primarily due to poor weather and higher military consumption, now at 230,000 tons of gasoline per month instead of the usual 150,000 tons. Fischer then gave a briefing on imports and production. The maximum throughput from Romania was 130,000 tons per month, but imports stood at only 50,000 tons. Fischer expected to increase the capacity of the Danubian tanker fleet starting in May 1940, but for the time being it would be impossible to boost railway carrying capacity (45,000 tons per month using 2,200 tanker cars).[52] As for the Soviet Union, it had delivered only 80,000 tons during the first quarter of 1940, but Fischer expected that Soviet exports would rise to 84,000 tons per month by May. Poor weather had also delayed the expansion of crude oil production, while hydrogenation plants were operating at 113,000 tons per month.[53]

[49] Keitel to Göring, March 14, 1940, T-77/377 (Wi/IF 5.2694). Thomas briefed Keitel on the depletion of Germany's reserves over the winter on March 9, 1940, and Keitel ordered the Defense Economy Office's recommendations sent to Göring. Defense Economy Office, "Vortrag [...]," March 9, 1940, IWM, FD 4809/45.

[50] Göring to Keitel, "Maβnahmen [...]," March 24, 1940, T-77/606 (Wi/IC 4.40).

[51] "Vortrag [...]," March 29, 1940, T-1022 (PG 32184). For background, see: Toprani, "German Navy," 103ff.

[52] The problem of disappointing deliveries from Romania had already been a topic at a conference in January 1940. The former transportation chief in Romania during World War I suggested that the lack of unified leadership over railway deliveries was holding back exports. Wi VII, "Sitzung [...]," January 29, 1940, T-77/400 (Wi/IF 5.3063).

[53] Göring, "Sitzung [...]," March 27, 1940, T-77/435 (Wi/IF 5.3654).

The navy's representative claimed the Romanians were not doing enough to procure oil from the foreign companies. He suggested that Germany should, instead, sell captured Polish weapons to third parties for hard currency and use the proceeds to buy the oil from Romania.[54] The Foreign Office disagreed—its officials pointed out that the largest U.S. oil company in Romania, Romano-Americana (Jersey's subsidiary), was "one of our principal suppliers."[55] Besides, there was no point in trying to get additional oil while the Danube was frozen. Finally, the navy's suggestion to buy Romanian oil rather than trade for it was "very bad business"—prices had increased by 150 percent since the war began. Arms sales were also indispensable for wooing the Romanians, and Bucharest would not take kindly to Berlin reneging on its commitments.[56] Nor was there any reason to doubt Romania's goodwill.[57]

Göring accepted most of the proposals made at the conference and, after securing Hitler's assent, reissued his letter of March 24, 1940 the following month as a formal directive.[58] He also wrote to Thomas and Bentz requesting their advice.[59] Thomas was pleased to report that the military-supply position had improved despite operations against Denmark and Norway. Air force consumption would rise slightly, while naval consumption would decrease due to the navy's losses in Norway.[60] The import situation showed little improvement, with deliveries from Romania and the Soviet Union "sluggish." In April, they had delivered only 32,000 tons of gasoline out of the expected 46,000 tons, and 35,000 tons of diesel instead of 53,000 tons.[61] While Moscow was "suspending" petroleum exports due to disagreements over counter-deliveries from Germany, in Romania foreign-owned refineries were holding back exports. High water levels were also impairing transit along the Danube, but deliveries would climb to 107,000 tons per month by May/June.[62]

Any concerns soon dissolved into euphoria. The German campaign against France and the Low Countries in May–June 1940 ranks among the most dazzling victories in history. We can reevaluate, however, the events of 1940 in a new

[54] Wiehl to German Legation, Bucharest, March 27, 1940, *DGFP* (D), ix: 30–1.

[55] One-quarter of Romana-Americana's exports went to Germany between January 1939 and January 1940. Berle to Roosevelt, "Romanian Deliveries [...]," February 9, 1940, FDRL, President's Secretary's File, Box 72.

[56] Clodius and Fabricius to Foreign Office, March 29, 1940; Neubacher and Fabricius to Foreign Office, March 29, 1940; *DGFP* (D), ix: 41–3 and 44.

[57] Clodius and Fabricius to Foreign Office, March 30, 1940 and April 1, 1940; *DGFP* (D), ix: 49–53 and 61–2.

[58] Göring to Keitel, "Maβnahmen [...]," April 2, 1940, T-77/377 (Wi/IF 5.2694).

[59] Göring to Thomas, April 3, 1940, T-401/5 (RBF 96).

[60] The Office for Economic Development had previously warned that Germany would need to import an addition 1.9 million tons of fuel oil between January 1940 and July 1, 1941 to cover the navy's requirements. "Treibstoffversorgung," January 15, 1940, no author, T-84/216 (EAP 66-c-12-62/29).

[61] The sticking point was the price. Berlin was offering a 20 percent premium on the world price and was willing to settle for 80 percent, while Moscow was demanding 138 percent. In fact, Moscow was seeking a lower price for a cruiser promised under the February 1940 agreement. *DGFP* (D), ix: 317–18 and 454.

[62] Thomas to Göring, "Massnahmen [...]," April 25, 1940, T-77/606 (Wi/IC 4.40).

light—one that illuminates the limits of German power.[63] Much as was the case in Napoleonic France following its victories over the Third and Fourth Coalition in the period 1803–7, instead of improving Germany's strategic position, operational successes had worsened it. After its defeat at Trafalgar, France lacked the means to coerce Britain and found itself at the mercy of British blockade that eroded its control of Europe. Similarly, after July 1940, Germany lacked the air and naval assets to invade Britain, which would continue to draw support from the empire. Meanwhile, in the spring and summer of 1940, the United States began extending assistance to Britain while commencing an ambitious military and naval buildup.

Just as troubling for Germany was that the energy balance also tipped against it. By the time of the Norwegian campaign, Germany was in danger of burning through its remaining reserves of gasoline in the summer of 1940 and thereafter depending entirely on domestic production and imports. It only escaped a crippling shortage thanks to increased deliveries from Romania and the Soviet Union, which more than doubled from 156,000 tons to 350,000 tons between the first and second quarters of 1940. These supplies, along with one-off windfalls such as unexpectedly low air force consumption and booty, enabled Germany to survive the summer of 1940.[64]

Stocks captured during Case Yellow exceeded consumption and fueled the Battle of Britain in the autumn.[65] The émigré petroleum economist Walter Levy assessed Germany's consumption between September 1939 and June 1941 at 13.5 million bbl—one-quarter of the annual prewar figure. Levy surmised that Germany may even have "came out ahead of the game," since he estimated that German forces captured twenty million bbl in France and the Low Countries.[66] The actual haul was, in fact, much smaller. According to Whitehall, the Germans acquired 550,000 tons of oil in Norway, Denmark, Holland, and Belgium.[67] As for France, contemporaries assumed it had large reserves.[68] This was true for nickel—the Germans made off with a one-year supply—but not oil.[69] Germany only secured 800,000 tons from Norway, Holland, Belgium, and France, not including supplies captured in the field.[70] In fact, considerable quantities of fuel

[63] For discussion of the 1940 campaign as a strategic turning point, see: Hillgruber, Hitlers Strategie, 273ff; Michael Geyer, "German Strategy in the Age of Machine Warfare," in: *Makers of Modern Strategy: From Machiavelli to the Nuclear Age*, ed. Peter Paret (Princeton: Princeton University Press, 1986), 572ff; Tooze, Wages, 380ff.

[64] Tomberg, "Die Probleme [...]," September 1940, T-77/167 (Wi/IF 5.662).

[65] Eichholtz, *Krieg um Öl*, 23–5; Hayward, "Hitler's Quest," 100–1.

[66] Levy, "Oil in This War's Strategy," no date (circa 1940/41), "Petroleum's Part in Axis Strategy," *World Petroleum* 12: 7 (July 1941), 39–43, and "Oil Drives Hitler's Army," March 1942, all in: American Heritage Center, Papers of Walter Levy, Box 1; Levy, *Oil*, 14–15.

[67] Greenwood, "Economic Consequences [...]," June 17, 1940, W.P. (G) (40) 155, BNA, CAB 67/7/5.

[68] G.F., "Treibstoffe auf Vorrat," *Vierjahresplan*, 1937/VII; "Aufgaben [...]," *Vierjahresplan*, 1939/IV. The Italians were astonished at the dearth of French reserves in 1940, as the major oil companies opposed accumulating more than a three-month reserve. Nowell, *Cartel*, 252–62 and 275–6.

[69] Perkins, "Coins," 96.

[70] Defense Economy Office, "Die deutsche Treibstoffversorgung [...]," February 16, 1942, T-77/668 (Wi/VI. 216).

were destroyed during the fighting.[71] One million tons—one-fourth of German production—was destroyed before capture.[72] This is not to say the Germans left empty-handed—they captured many of France's refineries intact—only that their haul would not provide much relief.[73]

Germany's petroleum position was definitely better by the end of the 1940 campaigns than it was before they began, but German supplies were utterly inadequate for waging an intensive, protracted war. To make matters worse, Germany's conquests did little to improve its energy balance. Its new empire in Europe was worthless without new supplies of oil, coal, and foodstuffs to replace overseas imports cut off by the British blockade.[74] Contrary to the prevailing historical wisdom, the victories of 1940 did not "put the Third Reich in a more secure economic position."[75] Europe had rich economic potential, but there was a "vast gulf" between it and the Anglo-American sphere in terms of energy.[76] Europe was almost self-sufficient in coal (80 percent of its energy consumption) but distributing it "required a quite heroic effort of production, organization, and logistics." Agricultural output was suffering. Prewar Europe imported up to thirteen million tons of food annually—enough to feed twenty-five million people—and the situation was deteriorating in wartime.[77] One problem was the lack of fertilizer, since synthetic nitrates were required for munitions production. Shortages of horses, labor, and petroleum also reduced output, such that by 1941 the Germans feared they would have to impose further rationing, with dire political consequences.[78]

The petroleum situation was most troubling. European production beyond Romania and the Soviet Union was negligible, while consumption was growing rapidly. In 1913, Britain, France, Germany, and Italy had imported 4.49 million tons of petroleum. By 1931, even in the depths of the Great Depression, the figure was 16.81 million tons.[79] By 1938, European consumption including Britain and the Soviet Union was 65.2 million tons, but output was only 36.8 million tons.[80] Excluding the Soviet Union, daily consumption was 925,000 bbl against 229,800 bbl of local crude and synthetic production. To meet the difference, Europe had relied on overseas imports, three-quarters of which came from the United State and Venezuela, with the Middle East accounting for most of the remainder (see Map 6). Germany could barely fuel its war machine even with imports from Romania and the Soviet Union, but after June 1940, it acquired allies and dependencies that

[71] *Halder Diary*, May 22, 1940.

[72] Wi VIIIa2, "Wirtschafts-Pressekonferenz [...]," November 13, 1940, T-77/220 (Wi/IF 5.1143).

[73] DiNardo, *Panzer Arm*, 14–15.

[74] Edward Ericson, *Feeding the German Eagle: Soviet Economic Aid to Nazi Germany, 1933–1941* (Westport: Praeger, 1999), 127.

[75] Contra Murray, *War, Strategy, and Military Effectiveness* (Cambridge: Cambridge University Press, 2011), 212.

[76] Fritz, *Ostkrieg*, 28–9; Tooze, *Wages*, 411.

[77] Götz Aly and Susanne Heim, *Architects of Annihilation: Auschwitz and the Logic of Destruction* (Princeton: Princeton University Press, 2002), 234–5.

[78] Tooze, *Wages*, 411–25.

[79] Heberlein, "Deutscher Kraftstoff," *Deutsche Wehr* (February 7, 1934).

[80] Ferdinand Friedensburg, *Die Rohstoffe und Energiequellen im neuen Europa* (Berlin: G. Stalling, 1943), 140–50.

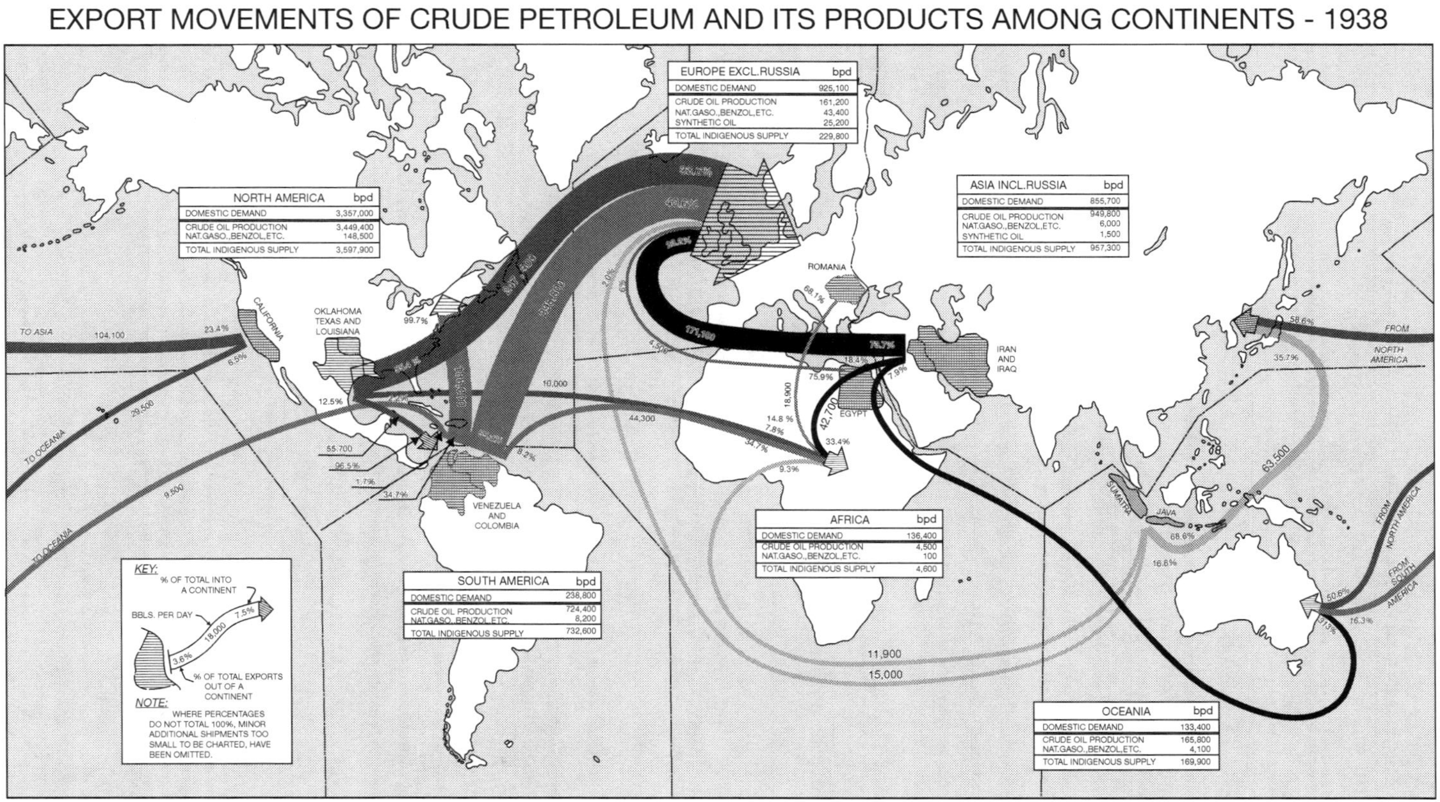

Map 6. "Export Movements of Crude Petroleum and its Products among Continents—1938 [in barrels per day]," December 1942
Source: NARA, RG 59, Lot File 77D141, Records of the Petroleum Division (PED), Box 3.

placed further strain on German petroleum supplies, while simultaneously losing access to Europe's primary suppliers. As Krauch explained in January 1941, existing production was sufficient to cover Germany's requirements with only a 25 percent reduction in reserves. If one included the occupied territories and Italy, however, Germany had a deficit of 1.6 million tons for 1941 alone.[81]

The Third Reich could hardly allow its new empire to starve for lack of fuel. Without fuel, Germany could not hope to exploit Europe's industrial and agricultural resources to defeat Britain or the Soviet Union. There was also the matter of maintaining its fragile alliance, since Italy's energy position was precarious. Soon after entering the war, Rome demanded a share of French reserves on the mainland and North Africa, but this was impossible since Germany had to keep France's industry and fleet supplied for use against Britain.[82] As a result, the Italians warned that their reserves would run out by May 1941, and the Foreign Office advised the Armed Forces High Command in February 1941 that the Italians might leave the war unless Germany rushed 214,000 tons of petroleum to them for the first six months of the year (as against the 574,000 tons Rome had requested). The Germans begged the Italians to make further economies to avoid shortages that could create military difficulties.[83]

Germany had no option but to acquire more oil. The situation would not be serious for the remainder of 1940, but thereafter, Germany's position would become critical.[84] As Thomas explained to Göring in June 1940, now that Germany was responsible for supplying most of Europe, its long-term outlook had worsened significantly. Thomas estimated that even after ruthlessly curbing consumption, existing supplies could satisfy Axis Europe (not including Italy) only until the end of 1940. The following year, assuming that Germany resumed offensive operations, there would be a monthly deficit of 118,000 tons—768,000 tons of demand against 650,000 tons of supply.[85]

According to the Defense Economy Office's assessment of Germany's raw materials position after the first year of fighting, petroleum reserves had increased from 200,000 to 2.1 million tons. The only immediate concern was the decline in diesel supplies, from 850,000 to 470,000 tons, since demand would rise with increased U-boat construction.[86] By September 1940, the office took a relatively sanguine view of Germany's ability to meet the requirements of the occupied territories, whose demand (besides France) was small.[87] This was also the opinion of the navy's economic staff, which concluded that Germany enjoyed a "not

[81] Office for Economic Development, "Vierjahresplan…30. 1. 1941," BA-B, R 3112/173.

[82] Thomas, "Aktennotiz […]," November 8, 1940, Nuremberg Records, Box 995 (PS-1456).

[83] Thomas, file notes, January 6 and February 6, 1941, and Thomas to Rintelen, February 8, 1941, all in: Hoover Institution, Nuremberg Records, Box 995 (PS-1456).

[84] Thomas to Göring, "Betriebstoffversorgungslage," August 29, 1940, T-77/377 (Wi/IF 5.2694).

[85] Thomas to Göring, "Maβnahmen […]," June 28, 1940, T-77/347 (Wi/IF 5.2233).

[86] Göring consequently ordered a boosting of diesel production. He was, however, "satisfied" with the overall situation. Thomas, "Besprechung […]," August 29, 1940, Hoover Institution, Nuremberg Records, Box 995 (PS-1456).

[87] Defense Economy Office, "Entwicklung […]," September 11, 1940, T-77/123 (Wi/IF 5.533). See also: Tomberg, "Abschrift […]," September 1940, T-77/668 (Wi/VI. 216).

inconsiderable" surplus except for diesel.[88] The long-term outlook was less promising. Neutral consumption in Europe in 1938, excluding Romania, the Soviet Union, and the Iberian Peninsula, was 9.65 million tons.[89] Even if French consumption was cut to the bone, it would still impose an additional drain of 100,000 tons per month in 1941, rising to 149,000 tons the following year. Even leaving aside the invasion of the Soviet Union, military consumption would still double from 305,000 tons per month in 1940 to 605,000 tons by 1942. Consumption in Axis Europe would therefore swell from 754,000 tons per month in 1940 to 1.25 million tons two years later.[90]

Rather than making progress toward energy independence, Germany had to rely even more on Romania and the Soviet Union. During a conference in October 1940, Göring insisted that Germany had "to counterbalance this dependence upon imports through stockpiling as much as possible."[91] It also needed to exercise greater responsibility for the protection of production in Romania. Since Operation Sea Lion (the invasion of Britain) had been postponed, 2,000 railway tank cars could be rerouted to Romania. Göring also ordered the construction of two pipelines in Romania with a monthly throughput of 350,000 tons to different ports along the Danube.[92]

Unexpectedly high consumption sparked a nasty exchange between the army and the Armed Forces High Command in the winter of 1940/1. After castigating the army for providing inaccurate consumption reports—by as much as 100,000 tons—the High Command instructed the army to accept a monthly allocation of 85,000 tons, which was double the figure of March 1940. The Commander-in-Chief of the Army, Field Marshal Walther von Brauchitsch, complained that the March 1940 baseline was unreasonable—back then the army had only ten armored and seven motorized divisions, whereas it was now increasing to nineteen and thirteen, respectively. The army was also preparing for operations "in the East" and against Gibraltar and the Balkans. Brauchitsch therefore insisted on 117,000 tons per month. In December, the High Command lost patience and imposed rationing, since, if nothing was done, the military would have to dip into its strategic reserve in spring 1941.[93]

Luckily, bombing was not yet a problem—the Allies began targeting the synthetic fuel plants in May 1940, but these attacks were only a nuisance due to Allied technological inadequacy and stiff German resistance.[94] As of January 1941, 120

[88] Borne, "Die allgemeine Rohstofflage [...]," September 23, 1940, T-1022/1772 (PG 32209).

[89] "Mineralöl-Statistik," no date, T-77/658 (Wi/VI.109).

[90] Ro V, "Mineralöl-Bedarf [...]," November 20, 1940, T-77/704 (Wi/IE 4.40).

[91] The previous August, Göring had instructed Thomas to ensure that deliveries from the Soviet Union took place as scheduled until the spring of 1941. Thomas, "Aktennotiz [...]," August 14, 1940, Hoover Institution, Nuremberg Records, Box 995 (PS-1456).

[92] Four-Year Plan, "Besprechung...4 Oktober 1940 [...]," BA-B, R 26 I/15. Only a fragment of the protocol survived, but the contents are summarized in: Thomas, "Ergebnis [...]," October 5, 1940, and Ro I, "Besprechung [...]," October 5, 1940, both in: T-77/606 (Wi/IC 4.40).

[93] The exchange may be reviewed in T-77/341 (Wi/IF 5.2164, 2687); Keitel to Brauchitsch et al., December 23, 1940, T-77/377 (Wi/IF 5.2694).

[94] Defense Economy Office, "Abschriften [...]," no date, T-77/668 (Wi/VI. 216).

raids had destroyed only 65,000 tons of petroleum.[95] Until 1943, total losses equaled only ten days of production.[96] Nevertheless, as Krauch explained in December 1940, the situation was still grim. Due to disappointing imports from Romania and the Soviet Union, Krauch had to downgrade his estimate of German reserves from 2,428,000 to 2.06 million tons. In 1941, production would increase to 4.47 million tons, but Romanian imports would be 300,000 tons fewer than planned, while those from the Soviet Union would remain at the 1940 level (740,000 tons). Military consumption would also rise by 1.07 million tons for the expansion of the air force and navy, which would eliminate Britain as a potential U.S. ally.[97] By January 1, 1942, Krauch estimated that Germany would have on hand 1.19 million tons of reserves plus 9.23 million tons of supplies (production plus imports), against 9.92 million tons of consumption. Germany could scrape through 1942, but thereafter the situation would be hopeless unless "it became possible to import petroleum from Romania and Russia through the Mediterranean, and most of all if oil from Iraq from either Tripoli or Haifa was available."[98]

The situation would deteriorate even faster if Germany invaded the Soviet Union. Besides eliminating a supplier, monthly military consumption would skyrocket to 200,000 tons of gasoline and 55,000 tons of diesel for the army, and 150,000 tons of aviation fuel for the air force.[99] In January 1941, the Armed Forces High Command estimated that Germany would begin Operation Barbarossa with a monthly deficit of 150,000 tons.[100] Following a meeting with Thomas that month, Halder noted that the fuel position was "serious." The situation had not improved by March, with Soviet exports "down to a dribble." Without them, stocks would last for less than three months—thereafter, Germany would have to depend on itself and Romania.[101] If the invasion began in June 1941, existing production, imports, and stocks would not cover Axis Europe's gasoline and diesel requirements by the end of August.[102] Two weeks later, Krauch laid out the situation for Göring. Expected supplies in 1941 would fall below expectations by 540,000 tons due to disappointing production in Romania and a shortage of skilled labor in Germany. The High Command also revised its requirements upward by 530,000 tons. Once one included the requirements of France and Italy, Germany would end 1941 with a deficit of 1.18 million tons.[103]

[95] Office for Economic Development, "Vierjahresplan…30. 1. 1941," BA-B, R 3112/173.

[96] Kehrl "sighed with relief" whenever he learned that Allied bombers were targeting Berlin rather than the synthetic plants. Kehrl, *Krisenmanager*, 366–7.

[97] Dr. R./vB., "Zur Entwicklung […]," January 28, 1941, NARA, RG 238, T-301/73 (NI-8835).

[98] Krauch to Körner, "Treibstoffversorgung," December 14, 1940, T-77/704 (Wi/IE 4.40). See also: Defense Economy Office, "Übersicht […]," December 12, 1940, T-77/704 (Wi/IE 4.40), which urged that the expansion of petroleum production should enjoy the same official "level of urgency" as that of tanks, planes, and U-boats.

[99] Griebel, "Untersuchung […]," February 8, 1941, and Raw Materials Division to Thomas, "Übersicht […]," February 25, 1941, both in: T-77/704 (Wi/IE 4.40).

[100] L IV/Qu, "Treibstofflage […]," January 24, 1941, IWM, FD 356/46.

[101] *Halder Diary*, January 28 and March 13–14, 1941.

[102] Raw Materials Division, "Kraftstoffversorgungslage […]," March 13, 1941, attached to: "Treibstoff-Versorgung," March 27, 1941, no author, T-77/704 (Wi/IE 4.40).

[103] Krauch to Göring, "Treibstoffversorgungslage […]," March 28, 1941, T-77/594 (Wi/I. 273).

By the end of 1940, all the signs were pointing to a crisis. Between August and December 1940, military consumption was higher than at any point prior to the invasion of France. This trend continued into 1941—consumption that year was two-thirds greater than in 1940: 5.15 million tons versus 3.01 million tons.[104] The resulting shortage was the product of a variety of systemic and contingent factors. Germany's obligations after June 1940 vastly exceeded the means available in Europe. In 1941, Germany would have to reconcile 2.3 million additional tons of consumption with only 200,000 tons of new supplies.[105] Moreover, railway cars were reassigned from moving petroleum to deploying troops and equipment for Operation Barbarossa.[106] The Germans reckoned on having enough petroleum to last the year provided they throttled non-military consumption, and if the war in the east ended quickly. Otherwise, there was no hope of supporting the expansion of the air force and navy. Whereas the gap between supply and demand in 1941 was estimated at 1.3 million tons, it would rise to 3.5 million tons the following year—11.5 million tons of supplies, including imports from the Soviet Union, vs. 15 million tons of demand. Even if consumption in 1943 remained unchanged while production expanded according to plan, Germany would still have a shortfall of 1.7 million tons. The only way to reduce this deficit was by constructing three synthetic plants, but this required one million tons of steel.[107]

The General Staff had drafted its plans on the assumption that the attacking forces would receive 6,710 tons of gasoline and diesel per day from twenty-two supply trains. The Defense Economy Office pointed out in May that this was impossible—Germany could provide sixteen trains per day, and then only for the first six weeks.[108] A lengthy campaign would be catastrophic, for Germany would have a 10 percent deficit as early as July 1941, while reserves would, as Thomas informed Halder nine days prior to Operation Barbarossa, "be exhausted in [the] fall." Halder scoffed at Thomas' concerns: "Operational plans cannot be tailored to suit economic planners."[109] Among the Chief of the General Staff's many flaws was his disregard for logistics. Such was "the arrogance of the men who had conquered France" but learned the wrong lessons.[110]

Other leaders were equally unsympathetic. In February 1941, Thomas warned Keitel and Alfred Jodl (chief of operations at the Armed Forces High Command)

[104] Four-Year Plan, "Treibstoffverbrauch [...]," August 3, 1944, IWM, FD 4809/45.

[105] Office for Economic Development, "Treibstoff-Versorgung [...]," June 21, 1941, T-84/217 (EAP 66-c-12-62/13).

[106] In May 1941, the Armed Forces High Command informed Halder that it could only send sixteen fuel trains to the forces assembling in the east, rather than the required twenty-two. *Halder Diary*, May 19, 1941. For the difficulties caused by the preparations for Operation Barbarossa, see: Hedler, "Die Mineralöle [...]," August 31, 1941, T-77/438 (Wi/IF 5.2726). After the invasion started, troop transports continued to receive higher priority than petroleum. The situation was only resolved at the end of July 1941. Ro V, "Aktennotiz [...]," July 31, 1941, T-77/612 (Wi/IC 4.75).

[107] "Mineralölerzeugung," no date or author, T-77/704 (Wi/IE 4.40).

[108] Defense Economy Office, "Untersuchung [...]," May 15, 1941, attached to: Griebel to Jansen, May 16, 1941, T-77/366 (Wi/IF 5.2528–34).

[109] *Halder Diary*, May 20 and June 13, 1941.

[110] Förster, "Hitler Turns East—German War Policy in 1940 and 1941," in: Wegner, *From Peace to War*, 115–33 (quotation from p. 129).

that, during an invasion of the Soviet Union, aviation fuel reserves would only last until the autumn, and gasoline and diesel until August. They responded that Hitler "would not allow his plans to be influenced by such economic difficulties."[111] The planners did their best to accommodate operational demands. In March, Thomas explained to Keitel and Göring that, with the exception of aviation fuel, Germany could fulfill its requirements for the first three months of campaigning. The situation would become either difficult (gasoline) or untenable (diesel) by September. Thomas suggested a series of conservation measures, including cutting the allocation to the civilian economy starting June 1, 1941 by a further 10 percent and reducing Romania's exports to third parties by 20,000 tons a month.[112] Even then, both the Defense Economy Office and the Economics Ministry estimated in May that Axis Europe would run a monthly deficit of 400,000 tons by October 1, 1941—one-third of expected demand. The only way to cope was "though the capture of a major oil-producing area" by no later than August.[113]

Krauch believed that Germany's only hope was the Caucasus. Two days after the start of Operation Barbarossa, he explained to the General Council of the Four-Year Plan that Germany could only meet its demands for the rest of the year by depleting its reserves. This would be impossible in 1942. Even assuming supplies increased as projected, Germany would still have a 600,000-ton gap that only Soviet imports could fill.[114] This left aside the requirements of the "Göring Plan" of June 23, 1941 to quadruple air force strength during the next two years. The only way to cope with the additional demand was to import four million tons of Soviet crude oil annually.[115]

By the autumn of 1940, it was clear that Axis Europe would soon face an energy crisis unless Germany took corrective action or ended the war. Even a successful invasion of the Soviet Union would not have solved all of Germany's problems. Postwar Europe would require imported oil, but a return to the status quo ante bellum was unacceptable—Germany had not fought the war to return Europe to a state of subservience to the United States and Britain. Even as planners coped with Germany's immediate shortage, they devoted some of their energies to the grander challenge of supplying their postwar empire.

POSTWAR STRATEGY, 1940–1941

Germany needed additional sources of oil after 1939 because synthetic fuel production had begun to plateau. Synthetics accounted for only 26.8 percent of Germany's

[111] Thomas, file note, February 8, 1941, IWM, FD 4809/45.

[112] Thomas to Göring et al., "Betriebstoffversorgung [...]," March 27, 1941, IWM, FD 356/46. See also: Hayward, "Hitler's Quest," 102–3. Göring was not keen on using aviation fuel for land warfare but relented. Thomas, "Vortrag [...]," June 26, 1941, IWM, FD 4809/45.

[113] Defense Economy Office, "Untersuchung [...]," May 15, 1941, attached to: Griebel to Jansen, May 16, 1941, T-77/366 (Wi/IF 5.2528–34); Landfried to Göring, "Anforderung [...]," May 24, 1941, T-77/704 (Wi/IE 4.40).

[114] Göring, "11. Sitzung...24. 6. 41 [...]," no date, T-77/430 (Wi/IF 5.3593). Krauch also believed that the existing transportation network could handle 400,000 tons of Soviet imports a month. Ritter, "Bericht [...]," June 25, 1941, T-84/217 (EAP 66-c-12-62/13).

[115] Eichholtz, *Krieg um Öl*, 90–1; Karlsch, *Faktor Öl*, 204.

petroleum supply between 1939 and 1943, compared to 32.6 percent from imports, 16.1 percent from German crude oil production, and the remainder from various alternative fuels such as benzol.[116] The establishment of the Continental Oil Company in 1941 (discussed later in the chapter) proves that control of oil reserves was at the heart of German strategy and postwar planning. This public–private, vertically integrated firm would operate oilfields across Europe, the Caucasus, and the Middle East after dispossessing their British, French, U.S., and Soviet owners, and reestablish the position Germany had lost in the international oil industry after 1918.

East German scholars were first to identify self-sufficiency and the acquisition of raw materials as key elements of German strategy. They also argued that Germany's oil ambitions extended included the Middle East.[117] Along with West German scholars, these demonstrated that the Third Reich had planned a massive pincer attack against Britain's empire in the Middle East following the defeat of the Soviet Union.[118] These campaigns were stages of a wider "global Blitzkrieg"—the first step was the defeat of the Soviet Union, followed by the destruction of the British position in the Middle East combined with a Japanese assault in Southeast Asia, and finally the Western Mediterranean would be closed through the capture of Gibraltar and occupation of French North Africa. In case the British continued to resist, Hitler ordered a feasibility study of operations against India in February 1941.[119] Underlying these operations, as Hitler informed Jodl in December 1940, was the need to "solve all Continental European problems" before the United States intervened.[120]

Immediately after the victory over France, German experts began planning for the "New Order" in Europe that would wage the generational struggle for global hegemony.[121] German schemes to absorb Southeastern Europe into a German-dominated "Greater Economic Area" long predated the Third Reich but consistently foundered. All of that changed with the victories of 1939–41. German officials began dusting off their old plans, which now expanded to cover all of Europe. No longer content with mere trade deals, they aimed at nothing less than redrawing Europe's borders and reorganizing its industries to satisfy German needs.[122] The question of who would formulate broad economic policy set off a feud between

[116] Plumpe, *I.G. Farbenindustrie*, 287–9.

[117] Dietrich Eichholtz, "Der Raubzug des faschistischen deutschen Imperialismus zu den Erdölquellen des Kaukasus, 1941–1943," *Jahrbuch für Geschichte* 14 (1976): 445–502; Hanns-Heinz Kasper, "Das Erdöl in den Raubplänen des deutschen Faschismus in Vorbereitung und bei der Durchführung des Zweiten Weltkrieges," *Jahrbuch für Wirtschaftsgeschichte* 3 (1976): 55–77.

[118] Eichholtz, *Krieg um Öl*, 80–121.

[119] Entry for February 17, 1941, *Kriegstagebuch des Oberkommandos der Wehrmacht*, ed. Helmuth Greiner and Percy Ernst Schramm (Frankfurt am Main: Bernard & Graefe, 1961–5), i: 328—hereafter cited as: OKW, *KTB*.

[120] Andreas Hillgruber, "Der Faktor Amerika in Hitlers Strategie, 1938–1941," in: *Nationalsozialistische Aussenpolitik*, ed. Wolfgang Michalka (Darmstadt: Wissenschaftliche Buchgesellschaft, 1978), 513–15. See also: Meir Michaelis, "World Power Status or World Dominion? A Survey of the Literature on Hitler's 'Plan of World Dominion' (1937-1940)," *Historical Journal* 15: 2 (1972): 355–6.

[121] Mark Mazower, *Hitler's Europe: How the Nazis Ruled Europe* (New York: Penguin, 2008), 102–36.

[122] Aly, *Architects*, 215–33.

the Foreign Office and the Four-Year Plan. Göring fired the first shot on June 20, 1940 when he claimed responsibility for "economic matters" and designated Funk to handle the formation of the "German-European economic sphere." Foreign Minister Joachim von Ribbentrop whined that his department had traditionally been responsible for foreign trade policy, and while he accepted that the Four-Year Plan was supreme in domestic economic policy, he insisted that the Foreign Office handle anything pertaining to Germany's external relations.[123] Within the Foreign Office, Carl Clodius (Deputy Director of the Commercial Policy Division) outlined a number of desiderata, including closer economic integration with Holland, Belgium, Luxembourg, and Norway. Germany would also have to reacquire its colonial empire, plus the Belgian Congo. Finally, he recommended purchasing "English and French interests in third countries," making specific reference to oil companies in Romania.[124] He then collaborated with Karl Ritter (Foreign Office liaison to the Armed Forces High Command), and on May 29, 1940 they presented Ribbentrop with their findings, which envisaged a European union of 200 million people, as well as a colonial empire in Africa.[125]

Various agencies also began considering how and where the Axis might address its petroleum deficiency, specifically. In Romania, the Geological Research Office estimated that sixty million tons of reserves remained. Romania's mining laws required that companies put 90 percent of any new oil they discovered at Bucharest's disposal, which naturally discouraged exploration and production.[126] Output had fallen from a peak of 8.7 million tons in 1936 to 6.24 million tons in 1939, and exports from 6.89 million tons to only 3.85 million. The major player in Europe was the Soviet Union, with possible reserves of one billion tons. The Soviets had started production in the Volga-Urals—the so-called "Second Baku"—but 90 percent of total production was still concentrated in the Caucasus. They had shown considerable ingenuity in increasing production from 3.14 million tons in 1921 to 29 million tons by 1939, but rising consumption had limited the exportable surplus to 1.5 million tons by the end of the 1930s. The Middle East was different—its production (16.7 million tons in 1939) was trivial compared to its reserves (960 million tons).[127] Two years later, the Geological Research Office determined that Middle Eastern output could increase to thirty-five million tons. There were complications—besides the problem of moving the oil to Europe, Middle Eastern crude oil was sour and difficult to refine.[128]

Interest in Middle Eastern oil peaked in the summer of 1940, when Germany's leading petroleum experts—Bentz and Fischer—took up the task of planning for

[123] Weizsäcker, memorandum, June 22, 1940, *DGFP* (D), ix: 683–4; *DGFP* (D), x: 24, 93, 115, and 170–3. For Funk's role, see: Funk to Lammers, July 30, 1940, *DGFP* (D), x: 369–70; Funk's testimony at Nuremberg: *IMT*, xiii: 129–30.

[124] Clodius, "Aspects [...]," May 30, 1940, *DGFP* (D), ix: 476–82.

[125] Ritter to Ribbentrop, June 1, 1940, *DGFP* (D), ix: 496–501.

[126] Keppler to the Foreign Minister et al., "Rumänisches Berggesetz," January 20, 1942, enclosed with: Wiehl to German Legation, Bucharest, January 28, 1942, BA-B, R 901/112777.

[127] Geological Research Office, *Erdöl in Europa und im nahen Orient* (Berlin, 1940), LOC.

[128] Geological Research Office, *Erdöl in Vorderasien* (Berlin, 1942), LOC.

Europe's energy future. The magnitude of their task was immense. Before the war, the German sphere in Central Europe had only produced half of its petroleum consumption, while the Continent minus the USSR imported two-thirds of its needs.[129] In July 1940, Bentz estimated that, excluding Britain, Ireland, the Soviet Union, and Turkey, annual European demand exceeded supply by nineteen million tons. If one added current Middle Eastern supplies and demand to those of Europe, the situation improved slightly—30.6 million tons of consumption versus 24.8 million tons of production. There were two ways to resolve this shortfall. The first was expanding synthetic fuel production, but the preferred option was developing the Middle East.[130] Other agencies reached similar conclusions.[131]

In August, Göring announced that "[one] goal of German economic policy is the increase of German influence in foreign enterprises."[132] Germany reserved the right to repossess assets seized by France after World War I, including the 23.75 percent stake of the Compagnie Française des Pétroles in the IPC and Steaua, both of which had previously belonged to Deutsche Bank.[133] The regime found an eager partner in Deutsche Bank, which never accepted the loss of its Romanian and Iraqi assets after the war and vigorously petitioned the Third Reich to assist in their return after July 1940.[134] Independent German oil companies also lined up to cooperate with the bank if it recovered its "old rights in Iraq."[135] The acquisition of raw materials was the subject of an interdepartmental conference on August 14, 1940. In his presentation on the oil industry, Bentz warned of an "extraordinary rise" in future petroleum consumption. Romania's declining output ruled it out as a "long-term" supplier, which made it "imperative" that Germany take control of Iraq and ideally Iran.[136]

Fischer had a similarly dim opinion of Romania, which he communicated to Deutsche Bank when it commenced negotiations to repurchase Steaua.[137] Since Germany had to look beyond Europe, Fischer drafted a long-term strategy to accomplish the ends Bentz had identified. Fischer calculated that immediate post-war demand within Germany's sphere would be 50 percent higher than that of 1938—forty-eight million tons including both Africa and the Middle East

[129] Statistical Office, "Die Mangelstoffe [...]," June 1940, T-84/70 (EAP 66-c-12/11), "Rohstoffversorgung [...]," July 1940, T-84/70 (EAP 66-c-12/10), and "Rohstoffversorgung [...]," October 1940, T-84/69 (EAP 66-c-12/8a).

[130] Bentz, "Sicherstellung [...]," July 24, 1940, BGR 49459.

[131] Office for Economic Development, "Rohstoff Bilanz [...]," June 26, 1940, no author, T-84/216 (EAP 66-c-12-62/29).

[132] Göring to Seyss-Inquart et al., August 2, 1940, *DGFP* (D), x: 401–3.

[133] "Vorschläge [...]," June 25, 1940, no author, PAAA, R 106289; note to Abs and Weigelt, July 4, 1940, author unclear, DBA, P-7068; Helmut Mejcher, *Die Politik und das Öl: II. Die Teilung der Welt, 1938–1950* (Stuttgart: Klett-Cotta, 1990), 29–30.

[134] See DBA, P-7068 (Romania) and P-8369 (Iraq); Baumgart, "Interessen," 60–5; Dietrich Eichholtz, *Deutsche Ölpolitik im Zeitalter der Weltkriege: Studien und Dokumente* (Leipzig: Leipziger Universitätsverlag, 2011), 96–167 and 404–5.

[135] Schirner to Weigelt, August 6, 1940, DBA, P-8369.

[136] "Protokoll... 14 August 1940," PAAA, R 106289.

[137] Note, October 1, 1940, DBA, P-7068.

but not Britain.[138] By 1943, annual synthetic production would reach 7.5 million tons, on top of two million tons of crude oil. This would cover two-thirds of Germany's requirements—fourteen million tons. Southeastern Europe could boost production to nine million tons. The Soviet Union was unlikely to have any large exportable surplus due to its high consumption. The only region capable of boosting exports considerably was the Middle East (see Map 7). Production could rise after the capacity of existing pipelines was expanded (Iraq), new territories for development were opened up (Iran), and production was allowed to rise beyond the capacity of local refineries (Bahrain). Excluding the Soviet Union, Europe and the Middle East produced twenty-three million tons of refined petroleum at this time. Provided current "transportation and refining bottlenecks" were eliminated, by 1945 production could rise to fifty-four million tons, against forty-eight million tons of consumption. As for existing concessions in the Middle East, Fischer recommended respecting those owned by U.S. companies. Anglo-Iranian and Shell would not be so fortunate.[139] Fischer wanted to replace them with a German firm, but a state-run company would lack the technical competence, while independent German companies lacked the capital and skilled personnel. Only a public–private partnership could "accomplish the intentions of the state most effectively."[140]

German oil experts all agreed that only Middle Eastern oil could free Axis Europe from its dependence on U.S. and British oil companies. Politicians and military planners concurred. In July 1941, the Minister for the Occupied Eastern Territories, Alfred Rosenberg, argued that Germany must acquire a "landbridge" to the Near East to secure enough oil to make it "independent of any maritime coalition for all time."[141] The Armed Forces High Command was also aware of the region's strategic value. In October 1940, it observed that Britain could not afford to lose control of Middle Eastern oil. For the Axis, Iraq was of particular importance because of its immense potential. The High Command recommended seizing Britain's share of Iraqi output and dividing the remainder between Italy and France. Germany had to act quickly. The British might sell Anglo-Iranian's shares in the IPC to their U.S. partners to frustrate the Axis from purchasing control.[142] The Iranians could also nationalize their oil industry, although this was preferable to Soviet control.[143]

The German Minister to Iran, Erwin Ettel, also focused on the Soviet threat. In the spring of 1940, he warned Berlin that the Soviet Union was trying to

[138] Much of the increase was due to the Autobahn. *Table Talk*: No. 2 (July 5–6, 1941), No. 101 (January 9–10, 1942), No. 161 (February 26, 1942), and No. 201 (July 18, 1942); Jochen Thies, "Hitler's European Building Programme," *Journal of Contemporary History* 13: 3 (1978): 415–16.

[139] Deutsche Bank preferred to gobble up all of Iraq, using its 1904 concession as a justification. Weigelt to Papen, November 11, 1940, author unclear, DBA, P-8369.

[140] Fischer, "Die Versorgung [. . .]," September 1940, reprinted in: Kockel, "Quelle," 198–208.

[141] Rosenberg, "Über [. . .]," July 27, 1941, *IMT*, vii: 326.

[142] Rumors abounded in the spring of 1941 that Anglo-Iranian was planning to sell out to the Texas Company. Note to Abs, April 21, 1941, author unclear, and "U.S.A. übernehmen [. . .]," June 6, 1941, enclosed with: Deutsche Bank to Grobba, June 6, 1941, DBA, P-8369.

[143] Armed Forces High Command, "Das Erdöl [. . .]," October 1, 1940, T-77/646 (Wi/VI. 19). A staff officer then gave a briefing to the high command on Britain's oil supplies on October 16, 1940. Entry for October 15–22, 1940, OKW, *KTB*, i: 122.

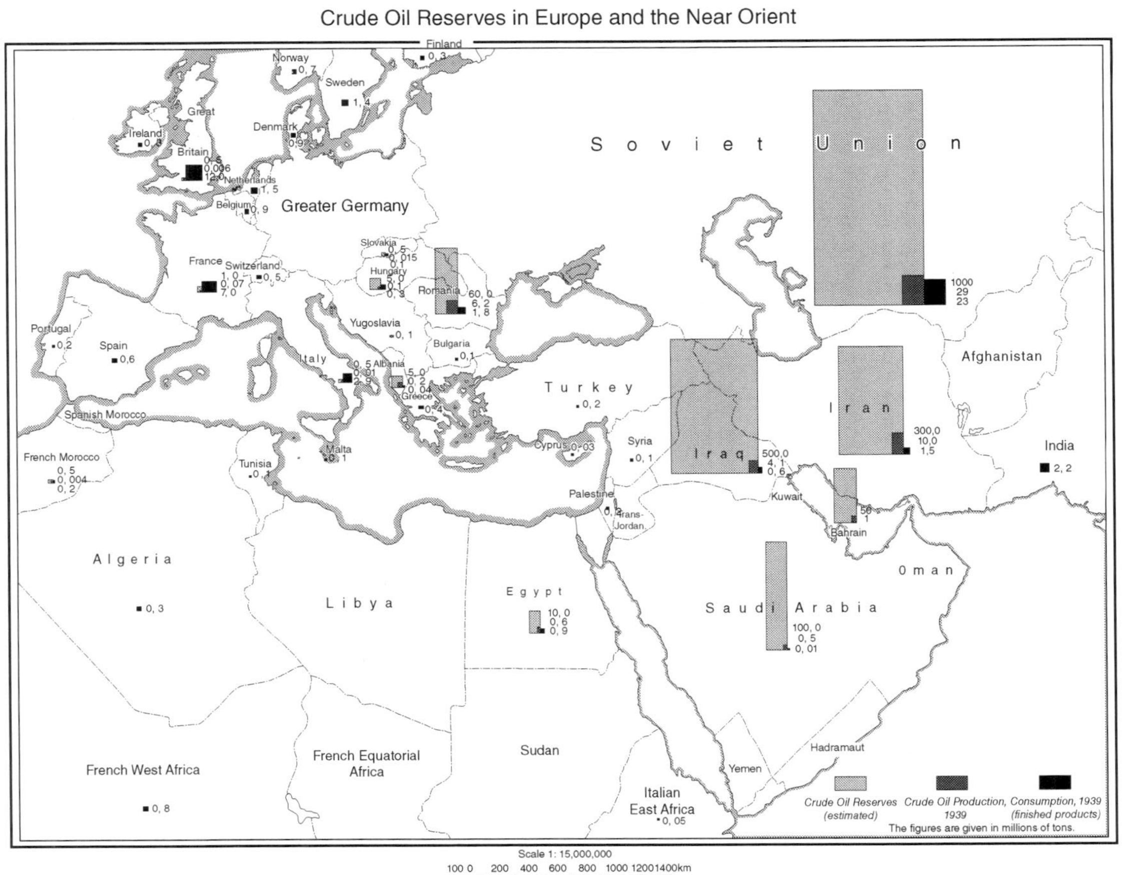

Map 7. "Crude Oil Deposits in Europe and the Near Orient," 1940

Source: Reichsstelle für Bodenforschung, *Erdöl in Europa und im nahen Orient*. Berlin: 1940. LOC.

exploit Britain's weakness in Iran.[144] He surmised that Moscow had exchanged its longstanding ambition of passage through the Turkish Straits for a landbridge through Iran. Ettel advised Germany to redirect Soviet energies "in another direction where there is less of a clash with the interests of the Reich," such as India.[145] The Foreign Office, however, was more concerned about Italy, which considered the Mediterranean its sphere of influence. German diplomats worried that competing Italian and Arab aspirations could complicate access to Middle Eastern oil and advised caution.[146]

Hitler agreed that Germany should occupy the Iraqi oilfields, but his aim was to undermine the Allied war effort. When it came to satisfying Germany's energy needs, however, he focused on Romania and the Caucasus.[147] Nevertheless, Bentz and Fischer's assessments demonstrate that high-ranking officials understood that occupying the Soviet Union was insufficient—the Third Reich would also need the Middle East to satisfy postwar Europe's energy demand. Informed observers of the international oil industry at the time reached a similar assessment. In the short run, however, the Caucasus was more valuable—not only was its production 66 percent greater than the Middle East's (200 million bbl per year versus 120 million bbl), but also its oil was easier to transport and refine.[148] The relative value of the Caucasus compared to the Middle East was, however, academic, since Germany controlled neither region. Until that changed, Germany could not afford to slacken its efforts in the one substantial oil producer under its influence.

ROMANIA CHOOSES SIDES, 1939–1941

Although Germany began pulling Romania into its orbit in March 1939, the results had not lived up to expectations.[149] Romania could satisfy Germany's entire wartime deficit, provided the Germans could solve the financial, political, geological, and logistical challenges. As the Defense Economy Office concluded at the end of 1939, Germany needed to import at least 190,000 tons a month to meet its minimum requirements. The 130,000 tons per month Berlin secured under the September 1939 agreement with Bucharest was insufficient.[150] The British and French oil companies were unwilling to sell anything to Germany unless Bucharest

[144] The Soviet Union and Iran had concluded a Treaty of Commerce and Navigation in March 1940. German analysts interpreted it as an attempt by Moscow to reduce the British threat to the Caucasian oilfields. "Die Anglo-Iranian [...]," *Vierjahresplan*, 1940/XXII.

[145] Ettel to Foreign Office, "Political Report," May 19, 1940, *DGFP* (D), ix: 379–83.

[146] "Mossul Oel," August 29, 1940, author unclear, DBA, P-8369.

[147] *Table Talk*: No. 269 (July 26, 1942).

[148] Levy, "Oil Drives Hitler's Army," March 1942, Levy Papers, Box 1. See also: Levy, "Petroleum's Part in Axis Strategy," *World Petroleum* 12: 7 (July 1941), 39–43, Levy Papers, Box 1, and *Oil Strategy*, 7–23 and 36–43.

[149] The following discussion is based on Pearton, Oil, 223ff; Dietrich Eichholtz, *Rumänisches Öl*, 17–68, and Krieg um Öl, 26–39.

[150] Griebel, "Bedeutung [...]," December 29, 1939, T-77/606 (Wi/IC 4.3a).

pressed them.[151] There was also the threat of Anglo-French sabotage.[152] German intelligence began cooperating with the Romanians to protect the oilfields in December 1939, but Göring complained the following month that German negotiators had devoted excessive attention to the "foreign exchange question" at the expense of security.[153] German intelligence had, however, developed "countermeasures" that would be in place later that month.[154]

The quickest way to resolve any concerns about Romania would be to occupy the oilfields, but this was out of the question. In May 1940, Romania supplied 1.64 million tons out of Germany's requirements of 8.22 million tons (20 percent). The Defense Economy Office advised that Germany should only undertake the occupation of Romania if it could forestall extensive sabotage of the oil industry and redirect third party exports to the Axis (1.49 million tons). German forces would have to occupy the oil-producing districts of Prohova, Dâmbovița, and Buzău, plus all of the territories north of Danube, including Bucharest.[155] Hitler understood the risks—he told Keitel in April 1940 that he was trying "to keep the Balkans quiet" and ordered the army to cease planning with Hungary to occupy Romania. He would only consider "defensive action" to protect Romania from either the Soviet Union or the Allies.[156] Hitler was in fact terrified of provoking the Soviet Union into a preemptive attack against Romania. In January 1941, he worried that Soviet bombers could turn Romania's oilfields "into an expanse of smoking debris."[157] The following year, he confessed to the Finnish Commander-in-Chief, Carl Mannerheim, that a Soviet invasion of Romania would have left Germany "helpless."[158]

Just before the war began, Germany had some success in resolving financial and political differences, and its share of Romanian oil exports rose from 15.5 percent (874,000 tons) in 1937 to 28.9 percent (1.2 million tons) two years later.[159] The pace quickened after 1939—Romanian gasoline and lubricant exports tripled from 65.65 million RM in 1939 to 214.75 million RM two years later.[160] Yet even as Germany's share of Romanian petroleum exports rose during the first nine months of 1940 relative to 1939 (34.19 percent versus 32.61 percent), the actual

151 Reinhardt, "Bericht [...]," January 30, 1940, BA-B, R 901/112771.

152 Following the surrender of France, the Germans captured a January 1939 memorandum outlining plans to destroy the Romanian and Iraqi oilfields prior to capture. Huntziger, "Verfügung [...]," January 30, 1939, T-77/1058 (Wi/ID. 29).

153 Göring was prompted by the air force attaché in Bucharest. Military Intelligence to Defense Economy Office et al., "Rumänien," December 29, 1939, T-77/600 (Wi/IC 4.77).

154 Wi VII, "Sitzung [...]," January 29, 1940, T-77/400 (Wi/IF 5.3063). See also: Barry Leach, *German Strategy against Russia, 1939–1941* (Oxford: Oxford University Press, 1973), 43.

155 Griebel, "Rümanien," May 6, 1940, T-77/606 (Wi/IC 4.3a-b).

156 Thomas, "Vortrag [...]," April 26, 1940, Hoover Institution, Nuremberg Records, Box 996 (PS-1456).

157 When the Soviets did attack in July, the results were modest. Hayward, "Hitler's Quest," 99–100 and 123–4.

158 Bernd Wegner, "Hitlers Besuch in Finnland: Das geheime Tonprotokoll seiner Unterredung mit Mannerheim am 4. June 1942," *Vierteljahrshefte für Zeitgeschichte* 41: 1 (1993): 117–37.

159 "Mineralöl-Statistik," no date, T-77/658 (Wi/VI.109).

160 IG Farben, Report No. 4968, no date, BA-B, R 8128/950.

volume decreased (922,000 tons versus 1.03 million tons), while British imports actually increased, as did those of pro-British neutrals. Even though Romanian production was tumbling—thanks in part to the oil companies "throttling" production—the export surplus far exceeded the capacity of the transportation network (4.5 million tons versus 2.9 million tons).[161] Germany required as much oil as Romania could export, but the inadequacy of the overland transportation infrastructure kept deliveries at disappointing levels. Fewer than 20,000 tons went to Germany in January and February 1940, and exports did not crack 100,000 tons until July.[162]

During a conference in January 1940, the Economics Ministry worried that the Foreign Office's negotiations with Bucharest had been too successful: Germany was entitled to more oil—130,000 tons per month—than it was able to transport. Rather than trying to pry additional oil from Romania, Germany should concentrate on figuring out how to transport its existing share, ensure that Bucharest complied with its obligations, and frustrate any Allied sabotage. All present agreed that an emissary should go to Romania to handle these matters.[163] The Foreign Office established within its Romanian legation the position of Special Representative for Economic Questions, which Hermann Neubacher, then Mayor of Vienna, would fill.[164] In June 1940, Göring also named General Schell as Plenipotentiary for Economic Transports, with special responsibility for petroleum.[165] He concluded that Germany had a deficit of 3,500 railway cars to carry Romanian petroleum exports, which it would requisition from Belgium and France, along with barges from Holland and Germany.[166]

In May 1940, the Foreign Office calculated that the rail and riverine network was capable of delivering the amounts to which Germany was contractually entitled. Reaching the maximum throughput was impossible unless additional railway wagons became available. Luckily, the slack in deliveries from the Soviet Union, which were well below their 60,000-ton limit, meant that some wagons could be diverted to Romania.[167] Neubacher, a former IG Farben representative to Southeastern Europe, was optimistic about prospects in Romania. He believed that the primary factor holding back Romanian production was logistics.[168] Theoretically, the Danube could handle as much as 874,000 tons between April and December 1940, but "in practice" it could take around 700,000 tons—58,000 tons per month, with a high of 117,000 tons between May and November. The

[161] Neubacher, "Rümaniens [...]," November 19, 1940, BA-B, R 901/112773.

[162] Defense Economy Office, "Deutsche Mineralöl-Einfuhr [...]," December 2, 1940, IWM, FD 4809/45.

[163] Wiehl, memorandum, January 3, 1940, *DGFP* (D), viii: 598–603.

[164] *DGFP* (D), viii: 615–17 and 661. In 1941, Ribbentrop expanded Neubacher's portfolio to cover all of Southeastern Europe. Foreign Office to Neubacher, January 17, 1941, *DGFP* (D), xi: 1223–4.

[165] Göring, Decree, June 12, 1940, PAAA, 105988; Clodius to Wiehl et al., memorandum, June 14, 1940, BA-B, R 901/112772.

[166] Martius, "Tagesmeldung," July 11, 1940, BA-B, R 901/112772.

[167] Martius, note, May 15, 1940, PAAA, R 105988. Soviet deliveries only exceeded 60,000 tons per month twice between February 1940 and May 1941. Schustereit, "Mineralöllieferungen," 347.

[168] Neubacher, "Rumänische [...]," February 8, 1941, BA-B, R 901/112773.

Table 7.1. Increase in petroleum prices by 1940[169]

Type of product	% increase
Aviation fuel	347%
Gasoline	256%
Diesel fuel	258%

existing stock of railway cars contributed 25,000 tons per month, or 300,000 tons for the year, thus allowing Germany to import one million tons. Increasing imports depended upon securing additional shipping and railway wagons. Berlin also had to insist that Bucharest apply pressure on foreign-owned refineries to supply Germany. Neubacher was convinced that Romania would cooperate if Germany continued delivering weapons.[170]

Neubacher also negotiated the so-called Oil Pact in May, which formalized the exchange of Polish materiel for Romanian oil. After the war began, the price of Romanian petroleum had skyrocketed, as can be seen in Table 7.1.

The pact guaranteed that Germany would henceforth trade its weapons for Romanian oil based on prewar prices.[171] This worked out to 76.52 RM per ton of petroleum including shipping, which was more favorable than the prewar rate.[172] Bucharest also imposed various taxes to discourage the oil companies from refusing to sell to Germany.[173] Such efforts and improvements in the weather led to a quintupling of exports to Germany—21,000 tons in February 1940 to 105,000 tons by June.[174] So did a shift in Romanian policy—starting in mid-May 1940, Bucharest went out of its way to express goodwill by raising Germany's monthly quota to 190,000 tons.[175] The fact that only Germany could defend Romania from the Soviet Union certainly helped.[176]

Italy's entry into the war added new complications. Of its prewar overseas imports, 80 percent had sailed through the Suez Canal or past Gibraltar.[177] In wartime, Romania became Italy's primary oil supplier, but because of the presence of the Royal Navy in the Eastern Mediterranean, exports had to travel along the same overloaded rail lines that supplied Germany. Italy needed 60,000 tons per month

[169] Eichholtz, *Rumänisches Öl*, 33, n. 85; Reinhardt, "Bericht [...]," 30 January 1940, BA-B, R 901/112771.

[170] "Wieviel Mineralöl [...]?" May 24, 1940, attached to: Neubacher to Foreign Office, May 18, 1940 (*sic*), PAAA, R 105988.

[171] Ribbentrop to Göring, March 16, 1940, *DGFP* (D), viii: 925–6.

[172] Neubacher to Foreign Office, "Signature of the Oil Pact," May 28, 1940, *DGFP* (D), ix: 459–60. See also: Eichholtz, *Rumänisches Öl*, 38–40; Tooze, *Wages*, 381–2. In October, Göring ordered the pact to be "liquidated" and replaced by a clearing agreement. Ro I, "Besprechung [...]," October 5, 1940, T-77/606 (Wi/IC 4.40).

[173] Neubacher to Foreign Office, "Gesetz [...]," March 21, 1940, BA-B, R 901/R 901/112771.

[174] Eichholtz, *Rumänisches Öl*, 34–8.

[175] Wiehl to Ribbentrop, May 3, 1940, BA-B, R 901/112771; Fabricius to Foreign Office, May 16, 1940, *DGFP* (D), ix: 349–50.

[176] Killinger to Ribbentrop, May 29, 1940, *DGFP* (D), ix: 467–70.

[177] Kennedy, *Rise*, 294.

from Romania and transporting that quantity required 9,000 wagons, which worked out to 300 wagons in transit each day. During a June 1940 conference, the Economics Ministry suggested informing the Italians that supplying them with Romanian oil via rail might hinder exports to Germany.[178] The ministry also wanted the Italians to seize control of the Eastern Mediterranean "as quickly as possible," but the Foreign Office observed this was easier said than done.[179]

The defeat of France and the expulsion of Britain from the Continent opened up the possibility of Germany acquiring outright or shared control of those Romanian oil companies with French, Dutch, and Belgian ownership.[180] During a conference at the Geological Research Office in August 1940, Keppler suggested that, initially, Germany should only acquire control of the five largest enemy-owned companies, because of the shortage of trained personnel.[181] German diplomats were also hopeful of securing the assistance of the U.S. companies in Romania in exchange for access to Germany's postwar market.[182] Meanwhile, over the summer of 1940, Romanian exports to Britain and France ceased, while Germany's share increased to 60 percent.[183] Finally, in July 1940, Bucharest consented to the deployment of a German military mission to oversee the security of its oilfields.[184]

During a visit to Berlin in October 1940, the Romanian State Secretary for Petroleum, Basil Dimitriuc, assured Keppler and Bentz that Romania would "during wartime place at Germany's disposal all of the oil it required." The meeting highlighted the contradictory policy aims of Germany and Romania. Dimitriuc welcomed increased regulation of foreign companies in Romania, which he believed had squandered the nation's patrimony, but he was also opposed to "excessive production" from either existing or new oilfields. Keppler agreed that the foreign companies had a done a poor job, but he also complained that the Romanians were keeping German firms away from "promising territories."[185]

In February 1941, Erich Neumann, State Secretary for the Four-Year Plan, planned to visit Bucharest to lay the groundwork for a swift rise of output.[186] The following month, Göring hosted Romanian dictator Ion Antonescu in Vienna. The ostensible purpose was to convince the Romanians to raise production, but

[178] Clodius later met with the Director of Commercial Affairs at the Italian Foreign Ministry, who agreed that Italy's requirements should not cause any disruption of deliveries to Germany. Clodius, memorandum, June 22, 1940, *DGFP* (D), ix: 684.

[179] Trees, "Aufzeichnung... 14 Juni 1940," PAAA, R 105988; Junker to Clodius, July 2, 1940, BA-B, R 901/112772.

[180] Neubacher to Foreign Office, "Beteiligung [...]," September 19, 1940, PAAA, R 105989. See also: Eichholtz, *Rumänisches Öl*, 42–6. The Romanians were also keen on eliminating Western influence over their oil industry: W.F., "Neuordnung [...]," *Deutsche Wehr* (November 28, 1941).

[181] "Protokoll... 14 August 1940," PAAA, R 106289.

[182] Aschmann to Foreign Office, "Der englische Petroleumkrieg," April 2, 1940, BA-B, R 901/112771; April 18, 1940, BA-B, R 901/116641.

[183] Clodius, "Aufzeichnung [...]," August 26, 1940, PAAA, R 105989.

[184] Armed Forces High Command, Directive, September 20, 1940, reprinted in: U.S. Department of the Navy, Office of Naval Intelligence, *Fuehrer Directives and other Top-Level Directives of the German Armed Forces, 1939–1941* (Washington, DC: 1948), 115–17. See also: "Rumäniens Öl [...]," *Militär-Wochenblatt* (December 27, 1940).

[185] "Aufzeichnung [...]," October 31, 1940, reprinted in: Eichholtz, *Deutsche Ölpolitik*, 358–9.

[186] Wiehl, memorandum, February 26, 1941, PAAA, R 105991.

Göring's real aim, which he did not share with Antonescu, was to arrange for additional supplies for the invasion of the Soviet Union. Göring knew, thanks to a briefing from Thomas in February, that existing reserves would cover only two months of operational consumption.[187] He admitted to Antonescu that Germany's synthetic plants were vulnerable to air attack, and that he doubted the reliability of the Soviet Union.[188] Germany had poured resources into developing its transportation links to Romania, but this would be in vain if production continued to decline.[189] Göring beseeched Antonescu to open his country to Germany. Antonescu promised "to do everything in his power" but stressed that he would not allow any country to assume ownership of Romania's natural resources.[190]

THE CONTINENTAL OIL COMPANY, 1940–1941

The Germans had other plans. Although occupying the Romanian oilfields was not an option, Berlin was determined to bring the European oil industry under its control. Its instrument would be the same one it had used in Romania in 1918—a public–private firm to pool German resources and supplant foreign challengers.[191] Only two weeks after the Göring–Antonescu meeting, the regime would put the final touches to the latest iteration of Germany's longstanding desire to free itself from the Anglo-American oil companies. German experts understood that simply occupying the oilfields of the Soviet Union and the Middle East would not solve their energy problems—Germany would have to make massive investments of capital, labor, and equipment to maintain, expand, and distribute production from those regions. Only a vertically integrated firm with resources that rivaled those of the major U.S. and British oil companies could hope to fulfill such a task. Germany's domestic oil industry was not up to the challenge. Not only did it lack the necessary resources—financial, technical, and personnel—but Germany also controlled no significant sources of oil, nor could it acquire any through normal commercial practices. Private industry needed the Third Reich to provide it with those assets it could not procure on its own.

It is unclear when Germany's leadership decided to resurrect a challenger to Anglo-American domination of the global oil industry.[192] In June 1940, Raeder

[187] Thomas, "Aktennotiz [...]," February 27, 1941, author unclear, Hoover Institution, Nuremberg Records, Box 995 (PS-1456).

[188] The Romanians only learned of Operation Barbarossa two weeks before it began. Schmidt, "Record [...]," June 13, 1941, *DGFP* (D), xii: 996–1006.

[189] Wiehl to the Adjutant of the Foreign Minister, March 3, 1941, reprinted in: Eichholtz, *Deutsche Ölpolitik*, 359–61.

[190] Schmidt, "Record [...]," March 8, 1941, *DGFP* (D), xii: 221–7. Neumann subsequently left for Bucharest, where he met with Antonescu and Romania's Economics Minister on March 14, 1941. Neubacher to Foreign Office, March 14, 1941, PAAA, 105991. He met Deputy Prime Minister Mihai Antonescu the following day. Neubacher et al., "Notiz [...]," March 15, 1941, attached to: Neubacher to Foreign Office, "Besprechung [...]," March 17, 1941, T-120/3671.

[191] Hamlin, *Romania*, 285 and 325.

[192] The following discussion builds upon Eichholtz, *Krieg um Öl*, 45–53. See also: Hayes, *Industry*, 255–7; Anand Toprani, "Germany's Answer to Standard Oil: The Continental Oil Company and Nazi Grand Strategy, 1940–1942," *Journal of Strategic Studies* 37: 6–7 (2014): 949–73.

urged Göring to prepare to exploit British and French oil properties that would soon fall into German hands.[193] Raeder also demanded the acquisition of overseas bases to protect access to the oil.[194] Göring did not reply until November, but his response includes the first official reference to Continental Oil. A week previously, Göring had informed Thomas "that he wanted to establish a Central European fuel company." Although the company would be nominally private, in reality the regime would exercise control.[195] In his reply to Raeder, Göring announced that he had decided to establish the Continental Oil company to "take over those oil interests of the enemy powers and neutrals that fell to us now and in the future."[196]

On January 21, 1941, Neumann convened a meeting at the Economics Ministry with representatives from several banks and various officials. He revealed that a company, Continental Oil, would be formed with an initial capitalization of 50 million RM, 30 million RM of which would come from a holding corporation, Borussia, supposedly owned by three men (including Fischer) but in reality the regime's proxy.[197] The remaining 20 million RM would come from a consortium of companies, including IG Farben. Those parties that contributed the original 50 million RM would receive special "name shares" that would give them an absolute voting majority.[198] Funk would serve as chairman of the company's supervisory board, alongside Neumann (his deputy), Keppler, several bankers, and representatives from industry and the military, while the managing board would be responsible for day-to-day operations. Notably, the ubiquitous Fischer would eventually serve on both boards due to his close relationship with Göring.[199]

The banks, led by Deutsche Bank, initially rebuffed Neumann—they had no interest in participating in an unprofitable state-owned firm—and only relented after he promised that the company would operate without political interference under sound financial principles.[200] Continental Oil's immediate task was acquiring

[193] Raeder may have been prompted by the commander of Baltic naval forces, who argued that Germany should demand the "[surrender] of all rights in the Persian Gulf and the Persian-English oil facilities [...]." "Memorandum von Admiral Carls," no date (circa June 1940), reprinted in: Michael Salewski, *Die Deutsche Seekriegsleitung, 1935–1945* (Frankfurt am Main: Bernard & Graefe, 1970–1975), iii: 105–20.

[194] Raeder to Göring, "Ölversorgung," June 13, 1941 (*sic*; 1940), T-77/1399.

[195] Thomas, "Aktennotiz [...]," November 8, 1940, Nuremberg Records, Box 995 (PS-1456).

[196] Göring to Raeder, November 14, 1940, T-120/3671.

[197] "Borussia [...]," January 21, 1941, author unclear, DBA, B-63.

[198] Bearers of "name shares" would have fifty votes, whereas holders of ordinary "bearer shares" would have only one. Wangler to Teisberg, July 26, 1946, NARA, Record Group 260: Records of U.S. Occupation Headquarters, World War II, Records of the Office of Military Government for Germany (RG 260), Records of the Property Division, Records of the Office Director, General Records, Box 9; Wangler, "Research Report [...]," August 10, 1946, attached to: Murphy to Byrnes, "Transmitting Report [...]," August 29, 1946, NARA, RG 59, FW 862.6363/8–2946.

[199] "File note [...]," January 23, 1941, NARA, RG 238, T-301/89 (NI-10797). Eichholtz describes Fischer as Continental Oil's "*spiritus rector.*" Eichholtz, *Krieg um Öl*, 50. Tooze (*Wages*, 393) claims that the company was "dominated by party men," but no one doubted Fischer's competence. See Kehrl, *Krisenmanager*, 223 and 227.

[200] Abs, "Aktenvermerk," January 25, 1941, DBA, B-59.

control of Belgian and French-owned companies operating in Romania and Jersey's Hungarian holdings.[201] These purchases required the banks to raise an additional 70 million RM. The banks rejected the offer of ordinary shares in the company since they doubted its profitability. They preferred instead to provide credit for preferential non-voting shares with a fixed dividend.[202] The outstanding details were resolved during a meeting on March 18, 1941.[203] The participants largely reaffirmed the plan outlined by Neumann in January, except that the four (eventually five) participating banks would purchase 30 million RM of ordinary shares, provide a 40 million RM credit at 5 percent interest (4.5 percent after Continental Oil paid its first dividend), and receive preferential access to any future share offerings.[204] Two days later, Göring invited shareholders to commemorate the company's founding on March 27, 1941.[205]

Even before its participation in the economic exploitation of the Soviet Union, Continental Oil had demonstrated that its ambitions were truly continental in scope.[206] The French were happy to oblige to secure favorable peace terms.[207] In July 1940, furious over the Royal Navy's attack on the French fleet at Mers-el-Kébir, senior French officers even considered assisting Axis forces to disable the IPC refinery at Haifa and occupy Iraq to preserve their share of its oil production.[208]

201 Eichholtz, *Rumänisches Öl*, 42–6, and Krieg um Öl, 31–4. The task was divided between Fischer and Deutsche Bank. Wiehl to Keppler, note, October 17, 1940, BA-B, R 901/112772.

202 Abs, "Aktenvermerk," January 25, 1941, DBA, B-63; Deutsche Bank to Kadigen, "Continentale [...]," March 15, 1941, DBA, B-59.

203 For the invitation, see: "Letter of Goering [...]," March 8, 1941, NARA, RG 238, T-301/18 (NI-2020).

204 "Aktennotiz [...]," March 18, 1941, no author, NARA, RG 238, T-301/18 (NI-2016); Neumann to Abs et al., March 18, 1941, NARA, RG 238, T-301/18 (NI-2017); Deutsche Bank, "Kontinentale [...]," March 23, 1941, and Deutsche Bank to Göring, "Finanzierung [...]," March 24, 1941, both in: DBA, B-63; Deutsche Bank to Dresdner Bank, "Kontinentale [...]," March 24, 1941, DBA, P-1826; Deutsche Bank to Continental Oil, "[...] Kredit," January 8, 1942, DBA, P-1826. Deutsche Bank's official history makes only brief mention of Continental Oil: Lothar Gall et al., *Deutsche Bank* (London: Weidenfeld & Nicolson, 1995), 327 and 336. See, instead, Christopher Simpson, ed., *War Crimes of the Deutsche Bank and the Dresdner Bank: Office of the Military Government (U.S.) Reports* (New York: Holmes & Meier, 2002), 198–201.

205 Göring to Schirner et al., March 20, 1941, attached to: Neumann to Rasche, March 20, 1941, NARA, RG 238, T-301/18 (NI-2018). For records of the meeting itself, see: "Aktenvermerk," March 27, 1941, DBA, B-63, and the protocols included with DBA, B-61. Full details may be found in Schirner et al., "Foundation Report," March 27, 1941, attached to: NARA, RG 238, T-301/18 (NI-2023). For the company's financial position, see the credit reports appended to: DBA, P-1826. See also: Roswitha Czollek and Dietrich Eichholtz, "Die deutschen Monopole und der 22 Juni 1941," *Zeitschrift für Geschichtswissenschaft*, 15: 1 (1967): 64–76.

206 At the first Nuremberg Trial, Funk stressed that the Continental Oil's mission went beyond just "the Occupied Territories—this company was concerned with oil industries all over Europe." *IMT*, xiii: 188–9.

207 Osterwind to Weigelt, October 10, 1940, and Weigelt to Abs, October 18, 1940, both in: DBA, P-7068; Eichholtz, *Rumänisches Öl*, 49, n. 131; Melby, *Oil*, 177–82.

208 Huntzinger, "Note [...]," July 10, 1940, enclosed with: Armistice Commission to Armed Forces High Command, July 17, 1940, and Huntzinger to Minister of National Defense, "Naher Orient," July 18, 1940, enclosed with: Armistice Commission to Armed Forces High Command, July 17, 1940, both in: T-77/834 (OKW/1347). See also: Robert Paxton, *Vichy France: Old Guard and New Order, 1940–1944* (New York: Columbia University Press, 2001), 58–9; Eichholtz, *Krieg um Öl*, 24–5.

Two months later, the owners of Steaua and Columbia reached out to Deutsche Bank to "sell back" their shares.[209] The Belgians saw the writing on the wall—in December 1940, Deutsche Bank concluded an agreement to purchase 1,865,112 shares of the Belgian firm Concordia, responsible for 11 percent of Romanian production, for 22,439,108 RM. The following August, the bank agreed to acquire 85 percent of the shares of Colombia, which accounted for 7 percent of Romanian production, in exchange for publicly traded French debt.[210]

Continental Oil's most important acquisition in Romania was Shell's subsidiary, Astra Romana. After the board of directors of Royal Dutch fled to the West Indies in 1940, the Germans quickly took control of the company's assets in Holland to forestall the Dutch from selling their shares to a U.S. firm and preempt Bucharest from nationalizing Astra.[211] They then installed as Astra's head a pro-German director, J.H.W. Rost van Tonningen, who established a partnership with Continental Oil.[212] By the spring of 1942, Continental Oil controlled 23 percent of Romanian production. In the Soviet Union, the company put up considerable capital—250 million RM by 1942—for the reconstruction of the Caucasian oilfields.[213]

From the beginning, the foreign and domestic press claimed that Continental Oil's objectives included the acquisition of oilfields in the Middle East. Speculation became so intense that in April 1941, Iran's Prime Minister, Ahmad Matin-Daftari, confronted Minister Ettel with a wire-service report that Germany had established a company to distribute oil from Romania, Iraq, and Iran. He warned that such a development would harm German–Iranian relations. Ettel already knew about the company's formation after receiving a circular from the Foreign Office the previous December, and he requested instructions from Berlin.[214] Clodius assured him that the allegations concerning the company's intentions in Iraq and Iran were "untrue."[215] That may have been true in spring 1941, but not by February 1942, when Continental Oil was prepared to dispatch engineers to Iraq once the Afrikakorps broke through the British defenses in North Africa.[216] That same month, Fritz Grobba, the former

[209] Note to Abs and Kimmich, September 18, 1940, and Deutsche Bank to Schirner, October 31, 1940, both in: DBA, P-7068.

[210] Deutsche Bank to Continental Oil, April 16, 1941 and August 11, 1941, both in: NARA, RG 260, Property Division, Box 9; note to Badenhausen and Sopp, "Kontinentale [...]," April 21, 1941, author unclear, DBA, P-1825; Deutsche Bank to Economics Ministry, "Geschäftsreise [...]," July 24, 1941, DBA, P-1825; Continental Oil to Hemmen, "Erwerb [...]," July 29, 1941, attached to: Neumann to Hemmen, July 29, 1941, T-120/3671; Kurzmeyer to Hartleib, February 23, 1942, DBA, P-1825; Deutsche Bank to Continental Oil, March 30, 1942, DBA, P-1826.

[211] Fischer to Neubacher, April 23, 1941, enclosed with: Fischer to Keppler, April 26, 1941, BA-B, R 901/112774.

[212] Keppler to German Legation, Bucharest, December 7, 1940, BA-B, R 901/112773; Sudosteuropa-Gesellschaft, "Wirtschaftsnachrichten," July 8, 1941, DBA, B-59; Eichholtz, *Rumänisches Öl*, 45–6; Andreas Hillgruber, *Hitler, König Carol und Marschall Antonescu* (Wiesbaden: Franz Steiner Verlag, 1954), 157–8; Howarth and Jonker, *History of Shell*, 29–34 and 69.

[213] Four-Year Plan, "Ergebnisse [...]," no date (circa 1942), BA-B, R 26 I/18; "Kontinentale [...]," June 6, 1942, author unclear, DBA, B-60. For a breakdown of Continental Oil's Romanian assets, see: Halder, "Kontinentale [...]," June 18, 1946, NARA, RG 260, Property Division, Box 9.

[214] Ettel to Foreign Office, April 7, 1941, PAAA, R 106109. For the circular, see: Clodius to Ankara et al., December 4, 1940; reproduced in: Eichholtz, *Krieg um Öl*, 51.

[215] Clodius to Tehran, April 15, 1941, PAAA, R 106169.

[216] Grobba, "Vordringen [...]," February 5, 1942, PAAA, BA 61123.

German Minister to Iraq and current Plenipotentiary for the Arab World, and Rashid Ali, the exiled Prime Minister of Iraq who had fled to Germany after the Anglo-Iraqi War of May 1941, agreed to transfer the IPC's concession to Continental Oil if Germany liberated Iraq.[217]

The German business press devoted considerable attention to Continental Oil's formation. Analysts understood the company's aim was freeing Europe from the U.S. and British oil companies. Although the Four-Year Plan had done an admirable job of satisfying Germany's wartime needs, it was incapable of meeting Europe's postwar requirements. The solution lay in Romania and the Middle East, but "the international companies were in the position to create considerable difficulties," which meant that Continental Oil had to take their place.[218]

Official statements like one from the German Embassy in Spain in April 1941 boasting that Continental Oil's aim was "making Germany independent of the great Anglo-American Companies" soon brought the company to the Allies' attention. Oil executives in the United States were anxious about a rival that threatened not only their properties in Romania and the Middle East but also their postwar access to the world's largest export market.[219] News agencies beyond Germany followed the company's activities, and their reports formed the bedrock for Allied intelligence estimates.[220] Foreign journalists, diplomats, and intelligence agents were quick to grasp the significance of Continental Oil's founding. In April 1941, the U.S. chargé d'affaires in Berlin informed Washington that a "[monopoly] in Rumania and ultimate control in Iran and Iraq" was its objective.[221] Germany could only satisfy such ambitions if it won the war. The country had no ocean-going tanker fleet, and even if it did, such vessels were easy prey in wartime. Developing one in peacetime would also be difficult, since Germany would be competing against established rivals.[222] Unless Germany's could guarantee access to overseas sources of oil, Continental Oil would remain "little more than a grandiose facade"—"the greatest oil company in the world without any oil," as one historian caustically observed.[223]

Germany's escalating energy crisis reflected its broader strategic dilemma following the victory over France. Despite remarkable operational triumphs, final victory remained out of reach. Scraping by on synthetic fuel and trade with Romania and the Soviet Union was foolish—such supplies were inadequate, deepened Germany's reliance on a future adversary in the case of the latter, and left it unprepared to

[217] Grobba to Foreign Minister's Office et al., "Abschluß [...]," February 10, 1942, T-120/773.

[218] "Eine deutsche Oel-Holding," *Berliner Börsen-Zeitung* (March 30, 1941), "Die deutsche Erdölinteressen," *Wirtschafts-Beilage* (March 30, 1941), and "Kontinentale Oel-AG," *Frankfurter Zeitung* (March 30, 1941), all in: BGR 86537. See also: DBA, P-1825.

[219] "Petroleum Autarchy," April 30, 1941, attached to: Sheets to Dunn, May 14, 1941, NARA, RG 59, 862/6363/209. Emphasis in the original.

[220] See the dossier to: Mayer to Grove, September 6, 1944, NARA, RG 169, Box 2201.

[221] Morris to Hull, April 2, 1941, No. 1226, NARA, RG 59, 862.6363/205.

[222] Woods, "State Control [...]," April 7, 1941, Serial 244, attached to: Naval Attaché (Berlin) to Director of Naval Intelligence, April 15, 1941, NARA, Record Group 38: Chief of Naval Operations, Intelligence Division, Confidential Reports of Naval Attaches, Box 678.

[223] Smyser, "Formation [...]," April 2, 1941, NARA, RG 59, 862.6363/206; Karlsch, *Faktor Öl*, 208–11 (quotation from p. 211).

meet the looming Anglo-American threat. Without full control of the oilfields of the Soviet Union and the Middle East, Germany would be unable to mobilize the resources of Europe against Britain and the United States. German plans to occupy the oilfields of the Caucasus and, later, the Middle East were therefore as much a response to unexpected contingencies as the fruit of a coherent strategy.[224]

As officials grappled with the challenge of supplying Europe, one of their most important decisions was to establish Continental Oil in March 1941. The history of that forgotten company opens a window onto Germany's long-term strategy. Officials expected that Continental Oil would restore Germany's place within the international oil industry as both a leading producer and supplier of oil to Europe. In the short run, it would consolidate Germany's immediate gains in places like Romania, where it seized or compelled the sale of foreign-owned assets. Over the long run, the company also contributed to the grander project of liberating Europe from Anglo-American economic domination. In that sense, the story of Continental Oil fits within the wider history of German imperialism—a doomed effort to escape strategic irrelevance in a world dominated by the British, Soviet, and U.S. global empires by acquiring a German sphere of influence in Eastern Europe, the Soviet Union, and eventually the Middle East.[225]

[224] Helmut Mejcher, "North Africa in the Strategy and Politics of the Axis Powers, 1936–1943," *Les Cahiers de Tunisie* 117–18: 3–4 (1981): 637–42, and *Politik*, ii: 29–30 and 50–63.

[225] Tooze, *Wages*, xxii–xxvi, 9–11, 166–7, 510–11, and 658.

8

Double or Nothing
1941–1942

The decision to attack the Soviet Union in June 1941 arose from at least three factors. The first was National Socialist ideology, specifically its fixation on the acquisition of "living space." The second was the strategic quandary in which Germany found itself following its victory over France. Although Germany stood unchallenged in Europe, it the lacked the means to invade Britain. Pouring resources into peripheral theaters like the Mediterranean or the Middle East would not compel Britain's surrender and would only increase Germany's dependence on the Soviet Union. The final factor was the energy crisis that hit Axis Europe following the defeat of France. This energy shortage did not inspire Hitler to attack the Soviet Union—ideological preconceptions shaped the strategic views of most German leaders, not least Hitler.[1] Nevertheless, the energy crisis foreclosed certain options. Since his earliest days in politics, Hitler had castigated his predecessors for waging a two-front war in 1914–18. After July 1940, however, Hitler had no option but to follow their example, since he could not afford to wait until after he had subdued Britain.

Germany's supply position was already tenuous by the spring of 1941. Efforts to cajole Romania ran into the same problems that had bedeviled Berlin since 1939—an inadequate logistical network and a dearth of instruments with which to coerce Bucharest. Fuel shortages complicated but did not doom the invasion of the Soviet Union, which proved beyond Germany's capabilities once the Soviet regime had survived the initial shock. The failure of the invasion effectively cost Germany the war, since the Third Reich lacked the means to wage another theater-wide campaign to destroy the Soviet Union before the United States intervened. Having opted for war largely to acquire resources such as oil, Germany now found itself locked in a protracted struggle without the means to wage one effectively.

PRELUDE AND PLANNING, 1918–1941

Energy independence for Axis Europe was only conceivable in the short run if the Third Reich occupied the Caucasus.[2] Germany had been scheming to gain control

[1] Jörg Echternkamp et al., *German Wartime Society, 1939–1945: Politicization, Disintegration, and the Struggle for Survival*, vol. 9/1 of *Germany and the Second World War*, ed. Militärgeschichtliches Forschungsamt (Oxford: Oxford University Press, 2008): 298–301.

[2] The starting point is: Hayward, "Hitler's Quest," 94–135. See also: Karlsch, *Faktor Öl*, 213–17.

of the region since World War I. In the spring of 1918, the Germans dispatched several thousand troops to the Caucasus to forestall the adversaries, the British, and erstwhile allies, the Ottomans, who had violated the Treaty of Brest-Litovsk by invading Armenia and Azerbaijan. Besides recovering territory lost to Russia in 1878 and controlling Caucasian oil, the Ottomans wanted to incorporate the region's Turkic populations into the empire and establish friendly buffer states between the empire and Russia.[3] The British, meanwhile, wanted to deny the Central Powers access to Caucasian oil and crops and safeguard British interests in the Middle East and India.[4] German aims were a mixture of political and economic—the commander of the detachment that arrived in Baku in September 1918 recalled his purpose was to serve "as a simple travelling salesman for German war industry."[5] After the war, German officers lamented the diversion of resources, but at the time, senior officers bemoaned the lack of troops in the spring of 1918, when they might have captured Baku without a fight.[6]

Although the November Revolution halted production in the Caucasus, large quantities of crude oil and refined petroleum were still available in Baku when the city fell to the Ottomans in September, and the Nobel Oil Company, Russia's largest before the war, agreed to help the Germans restart production. Unfortunately, the political and logistical obstacles proved insurmountable. The absence of a working pipeline, insufficient rail capacity in Romania and Austria-Hungary, and a shortage of tankers prevented the Germans from using the Black Sea to move oil to Central Europe. Their only option was to ship it through Russia to the Baltic via the Volga River, which required an arrangement with the Bolsheviks. On August 27, 1918, the Germans and Bolsheviks signed the Treaty of Berlin, which granted Germany one-quarter of Baku's oil production, in exchange for coal from German-occupied Ukraine. The Germans needed the Bolsheviks' assistance to move Baku's oil, while the latter were happy to exploit the Germans' oil hunger to solidify their hold over the Caucasus and keep the Ottomans at bay. In September, the Germans and the Ottomans also concluded an agreement that granted Germany oversight over Baku's oil and control of local railways and pipelines, and negotiations between the Germans, Ottomans, and Bolsheviks were under way when Bulgaria's capitulation forced the Ottomans to evacuate the Caucasus.[7]

[3] Sean McMeekin, *Ottoman Endgame: War, Revolution, and the Making of the Modern Middle East, 1908–1923* (New York: Penguin, 2015), 365–89.

[4] Edward Lemon, "Dunsterforce or Dunsterfarce? Re-Evaluating the British Mission to Baku, 1918," *First World War Studies* 6: 2 (2015): 133–49. This policy persisted after the war but targeted the Bolshevik regime. See also: Matthew Gharzarian, "Obstruction as Power: Rethinking Britain's Caucasus Occupation, 1918–1920," *International History Review* 39: 4 (2017): 654–66.

[5] Baumgart, "Kaspi-Unternehmen," 51.

[6] Baumgart, "Kaspi-Unternehmen," 50–1, and Deutsche Ostpolitik: Von Brest-Litowsk bis zum Ende des Ersten Weltkrieges (Wien: Oldenbourg, 1966), 197, n. 84.

[7] The following discussion draws from: Baumgart, Ostpolitik, 195ff, "Kaspi-Unternehmen," 53–7 and 115–18, "Protokoll [...]," August 11, 1918, reprinted in: Winfried Baumgart, "Das 'Kaspi-Unternehmen'—Größenwahn Ludendorffs oder Routineplanung des deutschen Generalstabs: Zweiter Teil," *Jahrbücher für Geschichte Osteuropas* 18: 2 (1970): 233; Horst Boog et al., *Global War*, vol. 6 of *Germany and the Second World War*, ed. Militärgeschichtliches Forschungsamt (Oxford: Oxford University Press, 2001), 850–1; Hayward, "Hitler's Quest," 96; Suβdorf, "Feldkraftfahrwesen," 381; Zürrer, *Kaukasien*, 169–72.

As German forces withdrew from the region, civil war took their place and oil production (90 percent of Russian output) collapsed. This was a matter of indifference to Germany following the defeat of 1918. Germany's attention returned to the Caucasus, however, following the defeat of France. The impossibility of redeploying German forces to the east in 1940 in time to launch an offensive before winter forced a postponement until 1941.[8] In the interim, Germany had one of two options—it could strike at Britain or its empire. Germany lacked the means to grapple with Britain directly due to its inferiority at sea and in the air.[9] After the French campaign, Hitler did consider reorganizing armaments production to wage an air/sea campaign against Britain. This would have required a significant reduction in the strength of the army, to 120 divisions. He backtracked in August 1940 and decided that Germany needed "to prepare for every potential political situation," including a protracted war against Britain, U.S. intervention, and a "change" in relations with the Soviet Union.[10]

If an invasion was out of the question, Germany needed some other means of compelling Britain to sue for peace and escape the trap of a two-front war. One alternative was the so-called peripheral strategy of hitting Britain in the Middle East. In January 1940, Jodl first broached to Hitler the idea of using the Soviet Union to distract Britain.[11] Six months later, he repeated the idea.[12] Halder was toying with similar schemes. During a conference with Brauchitsch on July 30, 1940, he concluded that, if an invasion of Britain was impossible, Germany should collaborate with Italy and the Soviet Union to evict Britain from Europe, the Mediterranean, and the Persian Gulf.[13] The army was never convinced of the primacy of economic objectives, however, and in November Halder ruled out an offensive in the Middle East until after Operation Barbarossa was complete.[14]

The navy reached a different conclusion. Since the fall of France, Raeder had been trying to convince Hitler to concentrate on Britain by focusing on the Eastern Mediterranean.[15] In November 1940, Raeder told Hitler that an operation toward Suez and the Middle East would be of "war-decisive significance" by improving Germany's economic position through the acquisition of raw materials including oil.[16] The navy reiterated these points seven months later, following the occupation of the Balkans and Crete.[17] Many civilians also preferred the Middle East option.

[8] Gerhard Weinberg, "22 June 1941: The German View," War in History 3: 2 (1996): 229–30.

[9] Williamson Murray, *Luftwaffe: Strategy for Defeat* (Washington: Brassey's, 1996), 39–56.

[10] "[...] Entwicklung der Rüstungslage [...]," no date or author (circa August 1940), Hoover Institution, Nuremberg Records, Box 995 (PS-1456).

[11] Jodl, "Policy [...]," January 6, 1940, attached to: Keitel to Ribbentrop, January 8, 1940, *DGFP* (D), viii: 631–3.

[12] Jodl, "Weiterführung [...]," June 30, 1940, *IMT*, xxviii: 301–3.

[13] *Halder Diary*, July 30, 1940. See also: Hoffmann, "Gulf Region," 61–2.

[14] Leach, *German Strategy*, 147–50.

[15] Eichholtz, *Krieg um Öl*, 60–1. Herwig emphasizes the navy's interest in North Africa and the Atlantic Islands but only briefly mentions the Middle East. Herwig, "*Weltblitzkrieg*," 655–9.

[16] "Beurteilung Mittelmeer," attachment 3 to: Naval High Command, "Besprechung... 14.11.40.," T-1022/1727 (PG 32184).

[17] The Naval Operations Staff recommended the seizure of Alexandria and Suez, but its ambitions stretched to the Middle East and India. Aßmann, "Betrachtung [...]," no date, attachment 5 to: "Vortrag... 6. 6. 1941 [...]," T-1022/3407 (PG 32185).

In April 1941, Krosigk explained to Göring that Germany was unlikely to wrest any immediate gains from the Soviet Union—certainly less than it was receiving through trade. Taking the Suez Canal, instead, "would give us the dominant position for Africa and the Near East," plus Iraq's oil.[18] German diplomats offered similar assessments.[19] A clique in the German Foreign Office, along with Deutsche Bank, had been pushing for an anti-British strategy focused on the Middle East, particularly after the pro-Axis coup of April 1941 in Iraq.[20]

Such critiques failed to convince the only person who mattered: Hitler, to whom "peripheral" strategy was never a true alternative to Operation Barbarossa.[21] Operational successes were unlikely either to break British resistance or afford the Axis any significant short-term economic gains. German officials recognized that the oilfields of the Middle East were richer than those anywhere else, including the Soviet Union.[22] Exploiting the region's oilfields was, however, no simple matter. The logistical obstacles for German military forces operating in the region were immense.[23] Even capturing the Middle Eastern oilfields would simply create a new set of headaches. Contemporaries such as Field Marshal Erwin Rommel and subsequent generations of armchair strategists overlooked the fact that extracting, transporting, and refining Middle Eastern oil required massive expenditures of resources, not to mention peace and security in order to rebuild local infrastructure.[24] By contrast, the occupation of the Caucasus would immediately improve Germany's strategic position by eliminating one of its key rivals. With the Soviet Union gone, Germany would then be free to split the British Empire in half by overrunning the Middle East. Hitler also worried that destroying Britain would benefit the United States and Japan more than Germany. Finally, a drawn-out campaign in North Africa and the Middle East would contribute nothing toward the fulfillment of Hitler's primary objective of acquiring "living space" in the east.[25]

The risks entailed by a two-front war also appeared minimal. Germany had after all defeated Russia in 1917, and the German Army was convinced it could replicate its successes in France. Their confidence in turn fed Hitler's preference for risking a two-front war.[26] Estimates concerning occupation forces in the Soviet Union

[18] Krosigk to Göring, April 19, 1941, attached to: Krosigk to Göring, April 20, 1941, BA-B, R 2/24243.

[19] Weizsäcker, memorandum, April 28, 1941, *DGFP* (D), xii: 661–2. See also: Robert Gibbons, "Opposition gegen 'Barbarossa' im Herbst 1940: Eine Denkschrift aus der deutschen Botschaft in Moskau," *Vierteljahrshefte für Zeitgeschichte* 23: 3 (1975): 332–40.

[20] Eichholtz, *Krieg um Öl*, 62–80.

[21] Lothar Gruchmann, "Die 'Verpassten strategischen Chancen' der Achsenmächte im Mittelmeerraum 1940/41," *Vierteljahrshefte für Zeitgeschichte* 18: 4 (1970): 456–75.

[22] Mejcher, Politik, ii: 38ff.

[23] Martin van Creveld, *Supplying War: Logistics from Wallenstein to Patton* (New York: Cambridge University Press, 2004), 181–201.

[24] B.H. Liddell Hart, ed., *Rommel Papers* (New York: Harcourt Bruce, 1953), 511–15.

[25] Besides Gruchmann and Hoffmann, see: Milan Hauner, *India in Axis Strategy: Germany, Japan, and Indian Nationalists in the Second World War* (Stuttgart: Klett-Cotta, 1981), 173–93 and 259–74; Gerhard Schreiber, "The Mediterranean in Hitler's Strategy in 1940: 'Programme' and Military Planning," in: *German Military in the Age of Total War*, ed. Wilhelm Deist (Leamington Spa: Berg Publishers, 1985), 240–81, and Mediterranean, 180–301 and 557–640; Tooze, *Wages*, 395–512.

[26] Boog, *Soviet Union*, 13–51.

were also modest—besides nine security divisions, only fifty-six divisions (later reduced to fifty-four) were required, including forces earmarked for operations in Iran via the Caucasus. Of the 209 active divisions, eighty-eight could be redeployed to Axis Europe, North Africa, and the Middle East, with thirty-one in reserve and the remainder demobilized.[27]

Hitler did not issue a formal directive concerning Operation Barbarossa until December 1940.[28] When he first directed the army to begin planning, on July 31, 1940, Hitler stressed that destroying the Soviet Union would convince Britain to negotiate while distracting the United States by emboldening Japan.[29] The invasion had three immediate objectives: smashing the Red Army; capturing or destroying the Soviet Union's industrial infrastructure; and occupying Baku. Hitler did not intend to stop there. Whereas early assessments of Hitler's war aims contended that the invasion of the Soviet Union represented the apex of his ambitions, his unpublished "Second Book" proves that his objectives were grander.[30] Once it controlled the "immeasurable riches" of the Soviet Union, he explained to Ribbentrop in January 1941, Germany would have everything it needed "to wage war in future against continents."[31]

Seeking "living space" and resources in the Soviet Union did not mean, however, that the National Socialists had abandoned self-sufficiency. The Third Reich was never interested in autarky within Germany—its aim was always to establish a "Greater Economic Area." Accordingly, Hitler's idea of autarky was self-sufficiency within "Europe"—including the Soviet Union.[32] As he confessed to Fritz Todt (Speer's predecessor as Armaments Minister) on the eve of Operation Barbarossa, autarky through synthetic production was "impossible" for Germany because of labor shortages—conquest was, over the long run, the better option.[33] In Hitler's worldview, a nation's fate depended upon its access to natural resources.[34] As he explained in October 1941, since human conflict was the inevitable byproduct of the scramble for such resources, "the side that hasn't got the natural riches must end by going under."[35] It was "an absurdity," he averred, for Europeans to look overseas to solutions to their resource constraints when there was an option next door.[36]

Occupying the Soviet Union would guarantee Europe the natural resources it required indefinitely, in which case there was no need for it to compete for colonies

[27] Operations Division, "Vortragsnotiz [...]," July 15, 1941, and Menzel, "Vortragsnotiz [...]," no date, OKW, *KTB*, i: 1022–7.

[28] *Fuehrer Directives*, 127–31. [29] *Halder Diary*, July 13, 22, and 31, 1940.

[30] Contrast Hugh Redwald Trevor-Roper, "Hitlers Kriegziele," *Vierteljahrshefte für Zeitgeschichte* 8: 2 (1960): 121–33, with Meir Michaelis, "World Power," 331–60; Knox, "Conquest," 24–5.

[31] Entry for January 9, 1941, OKW, *KTB*, i: 253–9.

[32] *Table Talk*: No. 25 (September 25, 1941) and No. 35 (October 13, 1941).

[33] Thomas, "Memorandum," June 20, 1941, *NCA*, iv: 21.

[34] Hence his frustration with the Romanians: "These people for whom chance has suddenly put a petroleum well under their feet...it's contrary to the whole natural order!" *Table Talk*: No. 161 (February 26, 1942).

[35] *Table Talk*: No. 35 (October 13, 1941).

[36] A familiar refrain in Germany. Ruprecht, "Erdölbedarf und Erdölverteilung," *Militär-Wochenblatt* (April 4, 1941).

or markets after the war.[37] Finally, the occupied territories would serve as a captive market for German industry.[38] Hitler was articulating a "perverse" variant of a longstanding German ambition of reorganizing Europe as a single economic unit capable of competing with the global powers.[39] Preparing for a confrontation against the United States in particular had been an important element of his geopolitical thinking since the late 1920s, and it clearly influenced many of his key strategic decisions during the first half of World War II, including the invasion of the Soviet Union.[40]

"The decision to embark on Operation 'Barbarossa'," as one historian concludes, "was taken not on account of Whitehall's uncompromising attitude, but in spite of it."[41] If victory in the east convinced Whitehall to negotiate, or deterred Washington, so much the better—but these were not the determining factors. Hitler was committed to attacking the Soviet Union even in the summer of 1940, when he still believed that Britain might accept terms. He also recognized that any strategy that delayed Operation Barbarossa worked to his adversaries' advantage. Germany would overextend itself in secondary theaters and exhaust itself supporting Europe. Meanwhile, Germany's foes would gain either respite (Britain) or time to enhance their military capabilities (the Soviet Union and United States). Finally, while there is no evidence that the Soviet Union was planning a preemptive war—or that Hitler believed one was imminent—Soviet political demands in Eastern Europe were still troubling for the Germans because they would enhance Moscow's political and economic leverage.[42]

Hitler's decision-making had its flaws—like his generals, he was dismissive of the Soviet Union's capabilities.[43] It is nonetheless difficult to dispute the underlying logic of his assessment that only turning east offered the prospect of decisive gains—there would be many opportunities to deal with Britain later. As a testament to their confidence, the Germans began planning for the next round of campaigns in the west before Operation Barbarossa even began. Hitler laid out his intentions two weeks before in his Directive No. 32. German forces in the Caucasus would overrun Iran and Iraq from the north in concert with the Afrikakorps, which would move toward the Suez Canal, and German forces moving through Bulgaria and Turkey into the Near East.[44] On July 8, 1941, he informed Halder that he would not deploy new tanks to the east, since they were necessary for future operations in the Iberian Peninsula, North Africa, and Anatolia.[45] Three days later, in an

[37] Bene to Foreign Office, October 2, 1941, *DGFP* (D), xiii: 606–8. Hitler stressed these ideas during meetings in the autumn of 1941. See document nos. 507–11 and 519, *DGFP* (D), xiii: 849–67 and 891–4. The topic also came up frequently during his wartime monologues. *Table Talk*: No. 18 (August 19–20, 1941).

[38] *Table Talk*: No. 25 (September 25, 1941) and No. 35 (October 13, 1941).

[39] Mejcher, "North Africa," 630–1. For World War I, see: Fritz Fischer, "German War Aims 1914–1918 and German Policy before the War," in: Barry Hunt et al., ed., *War Aims and Strategic Policy in the Great War, 1914–1918* (Totowa: Rowan and Littlefield, 1977), 121–3.

[40] Tooze, *Wages*, xxii–xxvi and 656–8.

[41] Förster, "Hitler Turns East," 122.

[42] H.W. Koch, "Hitler's 'Programme' and the Genesis of Operation 'Barbarossa'," *Historical Journal* 26: 4 (1983): 891–920; Rich, *Ideology*, 204–11.

[43] Leach, German Strategy, 227ff.

[44] *Fuehrer Directives*, 186–9.

[45] *Halder Diary*, July 8, 1941.

addendum to his directive, Hitler instructed that Germany should again concentrate on the construction of naval and air platforms to fight Britain and the United States.[46]

In keeping with Hitler's instructions, in spring–summer 1941, besides preparing a blockade of the British Isles, the military began planning operations across the Mediterranean and Middle East, including occupying Gibraltar, Malta, and Tobruk (autumn 1941); closing the Mediterranean; and invading the Middle East and the Caucasus (winter 1941 to spring 1942).[47] The Germans needed to sever British access to Middle Eastern oil and procure their own supplies, since German troops could not be resupplied from Europe easily once they reached the Near East.[48] To that end, the Foreign Office would undertake negotiations to ensure that German troops received a warm welcome in Iraq. German diplomats were also considering options for reorganizing the region to secure "enduring German influence over the oil reserves there."[49] The prerequisite to all of this, however, was a successful execution of Operation Barbarossa. Only that could provide Germany with the economic and physical security it needed to wage the open-ended conflict Hitler expected between Axis Europe and the Anglo-American alliance. Hence, it was by design a war of both extermination and plunder.[50]

German assessments of the Soviet Union left no doubt as to the immensity of its oil reserves.[51] In 1941, the Geological Research Office put them at 8.6 billion tons, of which 882.4 million were proven or probable.[52] The Soviet oil industry had recovered from the Russian Revolution and civil war by 1926 thanks to massive infusions of investment (27.7 percent of the total in 1924–5), and between 1927 and 1932, the Soviet Union tripled oil exports.[53] Progress soon stalled as older fields such as Grozny gave out, and by the end of the Second Five-Year Plan (1933–7) output lagged well behind expectations—28.4 million tons instead of 47 million—which in turn provoked purges of the industry during the Great Terror.[54] Production could not keep pace with rising consumption, which increased from 6.7 million tons in 1928 to 25 million tons by 1939 due to the mechanization of agriculture, industrialization, and rearmament. Exports collapsed from 6.6 million tons in 1932 to only 931,000 tons in 1939. Whereas domestic consumption equaled 54 percent of production in 1928, by 1939 it was 76 percent.[55] The drop

[46] Hitler, Guidelines, July 14, 1941, reprinted in: Martin Moll, ed., *Führer-Erlasse, 1939–1945* (Stuttgart: F. Steiner, 1997), 183–6. See also: Weinberg, "22 June 1941," 230–2.

[47] Paulus, "Vorbereitung [...]," July 3, 1941, Brauchitsch, "Kriegsführung [...]," July 24, 1941, and Armed Forces High Command/Army High Command, "Studie [...]," August 1941, all in: T-78/647 (H 22/315).

[48] "Die Bedeutung [...]," March 5, 1941, attachment 4 to: Woermann, "Aufzeichnung [...]," March 7, 1941, PAAA, BA 61123.

[49] Woermann, "Aufzeichnung [...]," November 6, 1941, PAAA, BA 61179.

[50] Weinberg, "22 June 1941," 226–7.

[51] The following discussion draws from: Mejcher, *Politik*, ii: 38ff; Rehschuh, "Crisis," 49ff.

[52] Geological Research Office, *Erdöl in Rußland*.

[53] Mejcher, *Politik*, ii: 39–40. [54] "Erdöl [...]," *Militär-Wochenblatt* (November 25, 1934).

[55] Stahmer, "Bericht [...]," April 26, 1941, BGR 48745. See also: Mejcher, *Politik*, ii: 41; V. A. Shishkin, "Soviet Oil Exports between the Two World Wars," in: R.W. Ferrier and A. Fursenko, ed., *Oil and the World Economy* (London: Routledge, 1989), 74–82. For detailed statistics, see: "Wandlung [...]," *Ostwirtschaft* 5: 6 (1937), T-77/425 (Wi/IF 5.3444); "Mineralöl-Statistik," no date, T-77/658 (Wi/VI.109).

in exports and global oil prices also discouraged the Soviets from investing in their oil-export infrastructure.[56] Not for nothing did one German economic analyst refer to the oil industry as the "problem child of the Russian economy."[57]

German analysts also debated whether greater Soviet exports were possible. Before Operation Barbarossa, experts agreed that the most the Third Reich could expect through trade was two million tons annually.[58] Although Germany became the largest customer for Soviet oil after 1939, imports in 1940 totaled only 687,000 tons.[59] Experts also surmised that that might be significantly larger than assumed due to the sheer size of the Soviet Union—21.5 million square km. The largest short-term gains would come through better management. Although Soviet production in 1939 exceeded consumption by 7.8 million tons, due to waste and inefficiency only 1.2 million tons were available for export.[60] Increasing exports depended on "decreasing the high losses in extraction, refining, transportation and demand."[61] Higher production was "a question of organization, of industrial infrastructure and of transportation."[62]

Soviet production in 1938 was more than thirty million tons, 90 percent of which was concentrated in the Caucasus, along with more than 80 percent of the Soviet Union's refining capacity and about 65 percent of its proven or probable reserves.[63] German analysts also observed the development of the Volga-Urals. They judged that the stagnation in production and collapse of exports would end once the Soviets began large-scale production there.[64] Although proven reserves in the Volga-Urals and Kazakhstan were small—seventy-seven million tons—possible reserves—3.88 billion tons—eclipsed those of Baku. Production in both areas rose sharply in the late 1930s, and under the Third Five-Year Plan (starting in 1938), their combined output would exceed nine million tons by 1942. All told, the areas beyond the Caucasus had a planned production of 12.8 million tons in 1942, or 26.6 percent of Soviet estimated production (48.4 million tons).[65] Actual output proved disappointing, in terms not just of volume but also of quality, since oil from the Volga-Urals was sour.[66] In 1942, German and U.S. analysts poured cold water on the idea that oilfields beyond the Caucasus could supply the Soviet Union.[67]

[56] Sch., "Öl in der Sowjetunion," *Vierjahresplan*, 1942/IX; Mejcher, *Politik*, ii: 40.

[57] "Die Erdölbasis der Sowjetunion," *Vierjahresplan*, 1937/VI.

[58] Haudan, "Treibstoffversorgung [...]," attached to: Defense-Policy Institute to the General Staff, September 23, 1939, T-78/491 (H3/851).

[59] Defense Economy Office, "Die Wehrwirtschaft [...]," March 1941, LOC.

[60] Stahmer, "Bericht [...]," April 26, 1941, BGR 48745.

[61] Office for Defense-Economic Planning, "Der Wirtschaftspotential der UdSSR," January 1940, LOC.

[62] Geological Research Office, *Erdöl in Rußland*. See also: Mayer-Gürr and Stahmer, "Das Erdöl Russlands," no date (circa 1940/1), BGR 2026385.

[63] IG Farben, "Erdöl und Erdgas [...]," April 29, 1942, T-84/117 (EAP 66-c-12-44/100).

[64] Ruprecht, "Die Ergänzung [...]," *Militär-Wochenblatt* (November 10, 1939); Ruprecht, "Die Ölfelder von Baku [...]," *Militär-Wochenblatt* (May 24, 1940); Ruprecht, "Erdölbedarf und Erdölverteilung," *Militär-Wochenblatt* (April 4, 1941).

[65] Stahmer, "Russland [...]," August 27, 1942, T-84/119 (EAP 66-c-12-44/241).

[66] Mejcher, *Politik*, ii: 44.

[67] Economic Staff, East, "Rohstoff-Potential [...]," no date (circa 1941/2), T-77/1203 (Wi/ID.1375). See also: Mejcher, *Politik*, ii: 47–8.

For all of its "ruthlessness," Moscow was unlikely to achieve the goals of the Third Five-Year Plan.[68]

Thanks to the plethora of analysis by private industry and state-sponsored research institutes, the Third Reich was quite knowledgeable about the Soviet oil industry. The influence of such information on German strategy was, however, slight.[69] Most civilian analysts lacked any insight into the Third Reich's specific intentions, which meant their concerns about Soviet productivity were irrelevant. If Germany invaded the Soviet Union and exterminated or enslaved its population, it would not have to worry about allocating much oil to the Soviet economy.

Following the war, Speer told U.S. interrogators that "the need for oil certainly was a prime motive" behind Operation Barbarossa.[70] The invasion of the Soviet Union was, however, about more than just plunder. Nor was it only supposed to protect the Romanian oilfields—German planners understood that a Soviet attack against Romania would be a response to Operation Barbarossa rather than its cause.[71] Germany's energy crisis following the victory over France, the lack of strategic alternatives, and the benefits of controlling Soviet raw materials co-mingled with ideological imperatives, foremost of which were the extermination of "Judeo-Bolshevism" and the acquisition of "living space."

Planning began in July 1940 even before Hitler gave explicit instructions. Interestingly, the army's initial plans did not mention the occupation of the Caucasus—German forces would only advance to the lower Don River (Rostov).[72] This was unacceptable. As the Defense Economy Office observed in February 1941, if Germany pushed to a line stretching from Baku to Archangel, the Soviet Union would lose 90 percent of its oil production. If Germany occupied just European Russia (from the Crimea to Leningrad), it would inherit a tremendous energy deficit, as those regions contributed only 1 percent of Soviet oil production. The agricultural sector—60 percent of consumption before the war—would pose a particularly heavy burden.[73] A few days later, Thomas concluded that Germany would be unable to satisfy its military requirements by the autumn of 1941 without "recourse to Russian stocks or Russian production." Even assuming a successful end to the campaign and a return to normal levels of consumption, there was still the matter of supplying the occupied eastern territories, which was impossible without the Caucasus.[74] Göring agreed with Thomas' evaluation but was sanguine

[68] Stahmer, "Russland [...]," August 27, 1942, T-84/119 (EAP 66-c-12-44/241). See also: Bonwetsch, "Die Erdölindustrie [...]," *Ost-Express*, November 18, 1941, BGR 65936.

[69] Michael Geyer, "National Socialist Germany: The Politics of Information," in: *Knowing One's Enemies: Intelligence Assessment before the Two World Wars*, ed. Ernest May (Princeton: Princeton University Press, 1984), 314–16.

[70] Snodgrass to Davies, "German War Oil," July 16, 1945, NARA, Record Group 253: Records of the Petroleum Administration for War, Entry 10, Box 665. See also: USSBS, *Oil Industry*, 76.

[71] *Halder Diary*, July 22, 1940; Marcks, "Operationsentwurf Ost," August 5, 1940, T-84/217 (EAP 21-g-16/4q). For Hitler's perspective, see: *Table Talk*: No. 223 (May 20, 1942).

[72] Boog, *Soviet Union*, 257ff.

[73] Defense Economy Office, "Vortragsnotiz," February 7, 1941, T-77/1066 (Wi/ID. 73).

[74] Thomas, "Die wehrwirtschaftlichen Auswirkungen [...]," February 13, 1941, reprinted in: Thomas, *Geschichte*, 514–32. Emphasis in the original. Thomas doubted, however, that the loss of the Caucasus would cause the collapse of the Soviet Union. That would only occur "following the loss of

about capturing the oilfields intact—he was convinced the Soviet regime would collapse too quickly to sabotage the oilfields.[75]

Planning for the capture of the Caucasian oilfields began well before Operation Barbarossa.[76] In May 1941, the Armed Forces High Command noted that Axis consumption was 1.15 million tons per month, against a total supply (excluding the Soviet Union) of only 850,000 tons. At this rate, stocks would run out by August. This meant the case for occupying at least Maikop and Grozny was "irrefutable."[77] Once the invasion began, however, the Defense Economy Office determined there was little chance of preventing the destruction of Baku's oilfields. Accordingly, it made sense to focus on capturing the North Caucasian oilfields and refineries. The latter, which boasted an annual capacity of 19.1 million tons, were especially important, as they were easier to destroy and harder to rebuild than oilfields.[78]

German plans for the Soviet oil industry were only a small part of a wider program for the ruthless exploitation of the Soviet Union.[79] Göring had been disengaged from economic policy since the outbreak of the war, but he took a special interest in this task.[80] Hitler empowered Göring to take all necessary measures to exploit or repair Soviet economic assets a week after the invasion began, but Göring had begun preparations months before.[81] The Economic Command Staff, East, established in March 1941, would oversee the plunder of the Soviet Union and coordinate between relevant agencies in Berlin, while the Economic Staff, East (codenamed "Oldenburg") was responsible for the execution.[82] Göring chaired the former, but his deputy State Secretary Paul Körner would run it, while Thomas was responsible for all "military matters."[83]

the industrial areas of the Urals." Thomas, "Voraussichtliche Entwicklung [...]," October 1, 1941, attached to: Thomas to Halder, October 22, 1941, T-77/1066 (Wi/ID. 73). See also: Leach, *German Strategy*, 144–5.

75 "Aktennotiz [...]," February 27, 1941, author unclear, Hoover Institution, Nuremberg Records, Box 995 (PS-1456).

76 Besides Eichholtz, see: Titus Kockel, "Geologie und Krieg: Die Erdölpolitik des Reichsamt für Bodenforschung und seiner Vorläufer 1933–1945," Magisterarbeit (Technische Universität Berlin, 1995), 43–53.

77 To preempt the oilfields' destruction, the high command considered an air landing but was skeptical of the prospects. "Unterlagen [...]," no date, attached to: Armed Forces High Command, "Das kaukasische Ölgebiet," May 4, 1941, T-77/1066 (Wi/ID. 73). See also: Defense Economy Office to Joint-Operations Staff, May 19, 1941, T-77/1066 (Wi/ID. 73). The military also considered creating an autonomous Caucasian nation. Wi VI, "Sicherung [...]," July 9, 1941, T-77/1066 (Wi/ID. 73). Leach concludes that the Germans could have secured the oilfields, but he underestimates the logistical difficulties. Leach, German Strategy, 146ff.

78 Chef Wi, "Inbesitznahme [...]," July 16, 1941, T-77/1066 (Wi/ID. 73).

79 The following discussion draws from: Boog, *Soviet Union*, 118–224; Eichholtz, *Krieg um Öl*, 40–6 and 84–90; Alex Kay, *Exploitation, Resettlement, Mass Murder: Political and Economic Planning for German Occupation Policy in the Soviet Union, 1940–1941* (New York: Berghahn, 2006), 26–46; Norman Rich, *Hitler's War Aims: The Establishment of the New Order* (New York: W.W. Norton, 1974), 326–93.

80 Kehrl, *Krisenmanager*, 223.

81 "Erlaß [...]," June 29, 1941, reprinted in: Moll, *Führer-Erlasse*, 179–80.

82 Kay, "Mass Starvation," 690–1.

83 Hamann, "Conference [...]," March 1, 1941, Thomas, "Report [...]," March 20, 1941, and "Conference... 29th April 1941," all in: *NCA*, iii: 911–13 and 808–16.

In his "Green Folder" instructions of June 1941, Göring announced that the "primary economic aim" of Operation Barbarossa was "to win as much as much foodstuffs and petroleum for Germany as is possible." Since petroleum was of "decisive" importance, Göring decreed that maintaining its production and shipment from the Soviet Union was to take "precedence in all circumstances" but one—supplying the agricultural sector.[84] This had nothing to do with feeding the captured population—they would starve by the millions to depopulate and deindustrialize the Soviet Union. The Germans planned to sever food "deficit" areas in European Russia from "surplus" regions such as the Ukraine and Caucasus, which would generate an agricultural-export surplus for Europe.[85] In this instance, the regime's murderous hatred for Slavs and Jews served a practical political purpose, for the National Socialists did not want to risk the loyalty of the civilian population by imposing additional rationing.[86] In practice, the Germans only succeeded in destroying the occupied territories' economic productivity.[87]

In his decree of July 27, 1941, Göring stipulated that Germany would initially focus on repairing only vital industries, among them oil.[88] Hitler had assigned oversight of the occupied territories to Rosenberg, but this was a meaningless assignment due to overlapping authorities.[89] Rosenberg's remit did not extend to economic affairs, where Göring was supreme.[90] Göring also decreed that four "monopoly" firms would operate in the occupied territories: Continental Oil, the Central Trading Company for Agricultural Marketing and Requirements, Eastern Mining and Metallurgy, and Eastern Fibers.[91] On July 22, 1941, the Economics Ministry provided Continental Oil with instructions delineating its responsibilities, as well as those of the "Petroleum Commandos." The company enjoyed the "exclusive right of mining, processing and trading in mineral oil products," along with the authority to expropriate any physical infrastructure serving the Soviet oil industry. The "Petroleum Commandos" were units composed of military and oil industry personnel.[92] They would secure captured Soviet oilfields before turning them over to Continental Oil.[93]

[84] Göring, "Richtlinien [...]," June 1941, *IMT*, xxviii: 3–15.

[85] Kay, "Mass Starvation," 685–700.

[86] Lizzie Collingham, *Taste of War: World War Two and the Battle for Food* (London: Allen Lane, 2011), 32–8.

[87] Compare Aly, *Architects*, 234–52, with Jonathan Steinberg, "Third Reich Reflected: German Civil Administration in the Occupied Soviet Union, 1941–4," *English Historical Review* 110: 437 (1995): 637–43; Timothy Snyder, *Bloodlands: Europe between Hitler and Stalin* (New York: Basic Books, 2010), 166–9.

[88] Decree, July 27, 1941, BA-B, R 26 I/13.

[89] Steinberg, "Civil Administration," 624–7 and 629–31.

[90] "Erlaβ [...]," July 17, 1941, OKW, *KTB*, i: 1027–8. Rosenberg supported Göring's economic mission, but he believed the best way to accomplish it was by cultivating disgruntled Soviet minorities. "Rede...20 Juni 1941," *IMT*, xxvi: 610–27. Hitler was indifferent to the wishes of the local population. "Memorandum [...]," July 17, 1941, *NCA*, vii: 1086–93.

[91] See also: Hirschfeld to Bremen Chamber of Commerce, "Wirtschaftsentwicklung [...]," October 11, 1941, DBA, R 8119/1825. Kehrl recalled that the inspiration for the creation of Eastern Fiber came from Fischer. Kehrl, *Krisenmanager*, 227.

[92] Karlsch, *Faktor Öl*, 214–15.

[93] Economics Minister to Continental Oil, "Russian Mineral Oil Economy," July 22, 1941, NARA, RG 238, T-301/18 (NI-2021).

As Göring explained during an interdepartmental conference in November 1941, all captured Soviet economic assets would become the property of the Third Reich except oil installations, which Continental Oil would administer. Although the company was acting on behalf of the regime, it would operate autonomously.[94] Later that month, Günther Schlicht, an engineer with Germany's largest independent oil company, was appointed Continental Oil's "Chief Representative" in the occupied territories.[95] Neubacher, meanwhile, would oversee Germany's economic interests in the Caucasus.[96]

On July 24, 1941, following Hitler's supplement to Directive No. 33 ordering an immediate advance across the Don River, the General Staff completed their initial planning for an invasion of the Caucasus.[97] Assuming forces were in place by November, the Germans did not expect to reach Batum and Baku until the end of June 1942. They would then invade northern Iran, occupying Tabriz and Kermanshah, before crossing the Iraqi border and seizing Rawanduz and Khanaqin by September 1942.[98] Preparations continued until the end of October 1941, and on November 3, 1941, Brauchitsch informed Army Group South that Hitler still expected them to capture Stalingrad and Maikop.[99] All of these plans were rendered moot four days later when Hitler called off the invasion of the Caucasus in 1941, in large part because it was impossible to keep the German spearheads supplied with fuel.[100]

DISAPPOINTMENT IN THE EAST, 1941–1942

Germany's oil position was poor even before Operation Barbarossa began, due to difficulties in Romania and the unexpected Balkan campaign of April–May 1941.[101] Thomas warned Keitel in February 1941 that existing supplies would be inadequate by the autumn.[102] The following month, General Eduard Wagner, the Quartermaster-General, gave an equally gloomy prognosis to Halder. Once Germany lost Soviet imports, existing stocks would last for only two months of "large-scale offensive" consumption.[103] German leaders and planners accepted that their reach exceeded

[94] Göring to Keitel et al., November 20, 1941, NARA, RG 238, T-301/5 (NI-440).

[95] "Vollmachterteilung für Herrn Bergassessor a.D. Schlicht," November 26, 1941, BA-B, R 176/1.

[96] Thomas, "Ergebnis [...]," July 18, 1941, Hoover Institution, Nuremberg Records, Box 996 (PS-1456); Karlsch, *Faktor Öl*, 215.

[97] OKW, *KTB*, i: 1220; Leach, *German Strategy*, 198–200.

[98] "Operation [...]," no date, OKW, *KTB*, i: 1038–40.

[99] Operations Division, "Besprechung [...]," no date, OKW, *KTB*, i: 1072–3.

[100] Chief of Staff, "Ergebnis [...]," October 24, 1941, T-77/175 (Wi/IF 5.694); *Halder Diary*, November 7, 1941. See also: Hayward, "Hitler's Quest," 107ff.

[101] Eichholtz claimed that the shortages prior to Operation Barbarossa arose because of German operations against Yugoslavia and Greece in April–May 1941. Eichholtz, *Krieg um Öl*, 40–6 and 80–100. Thomas was in fact worried well before this. See Hedler, "Die Mineralöle [...]," August 31, 1941, T-77/438 (Wi/IF 5.2726); Leach, German Strategy, 140ff.

[102] Entry for February 11, 1941, OKW, *KTB*, i: 316–17.

[103] *Halder Diary*, March 13, 1941. The situation was even worse for Italy, whose reserves would "touch bottom" by June 1941 without an immediate transfer of 260,000 tons. *Halder Diary*, May 27, 1941.

their grasp because they expected that the Soviet Union would collapse swiftly. This was a plausible expectation considering the state of the Red Army on June 22, 1941, but it soon proved unfounded. In the midst of the German high tide in July 1941, Halder noted that fuel consumption was unexpectedly high: 11,500 cubic meters per day instead of the estimated 9,000. The Germans had the good fortune of seizing large quantities of petroleum during Operation Barbarossa's initial stages.[104] These proved to be a mixed blessing. Unlike German tanks, Soviet armored vehicles ran on diesel, and Soviet aviation fuel was useless because of its low octane rating.[105] In a conspicuous example of foresight, before the invasion the army created mobile facilities to raise the octane rating of captured gasoline with benzol.[106] A week into the invasion, German forces were already dependent on captured Soviet supplies for one-third of their consumption.[107]

They would not be so fortunate in the future. No additional oil was available in Europe. Demand there between October 1941 and April 1942 would be 1.01 million tons per month, and Germany needed 1.17 million tons per month if it wished to stockpile for future offensives.[108] Production within Germany and Romania was only 770,000 tons, leaving a six-month deficit of 1.44 million tons. Germany could meet its minimum requirements for gasoline and aviation fuel during the six months to come (1.4 million tons and 726,000 tons, respectively), but diesel and fuel oil production was inadequate (68 percent and 61 percent of consumption, respectively).[109] Germany's only hope was imports from the Soviet Union—it needed 400,000 tons per month for six months to cover the existing shortfall and accumulate new stockpiles, plus another 100,000 tons to expand aviation fuel production.[110] The Defense Economy Office revised its figures a month later. It slashed expected six-month production in Romania by 400,000 tons (from 4.6 million tons to 4.2 million tons), although estimated demand did not change—6.37 million tons including the Eastern Front.[111]

On July 17, 1941, Göring asked Thomas to undertake a "quick assessment" of Germany's ability to raise imports from Romania and ship oil from the Caucasus.[112] Thanks to unexpected Soviet resilience, the Armed Forces High Command now conceded that Soviet sabotage would prevent any oil exports for a significant period. Germany had to extract "considerable savings" from the civilian economy

[104] Leach, *German Strategy*, 205–6. [105] Hayward, "Hitler's Quest," 101.

[106] *Halder Diary*, May 20, 1941.

[107] *Halder Diary*, July 1, 1941. One cubic meter of oil equals 6.3 bbl of oil. Over the course of the entire campaign, the average daily consumption was 5,400 tons. "Aktenvermerk," no date or author, T-77/704 (Wi/IE 4.40).

[108] "Laufender Bedarf [. . .]," no date, attachment 1 to: Defense Economy Office, "Beurteilung [. . .]," July 7, 1941, T-77/500 (Wi/I. 36). These figures did not include the 45,000 tons per month required by the forces on the Eastern Front, who had to live off the land.

[109] "Deutsche und rumänische Treibstoff-Erzeugung [. . .]," no date, attachment 2 to: Defense Economy Office, "Beurteilung [. . .]," July 7, 1941, T-77/500 (Wi/I. 36).

[110] Defense Economy Office, "Beurteilung [. . .]," July 7, 1941, T-77/500 (Wi/I. 36).

[111] Defense Economy Office, "Deutsche und rumänische Treibstoff-Erzeugung [. . .]," August 5, 1941, T-77/704 (Wi/IE 4.40).

[112] Thomas, "Ergebnis [. . .]," July 18, 1941, Hoover Institution, Nuremberg Records, Box 996 (PS-1456).

and the occupied territories plus troops not deployed on the Eastern Front.[113] By August, the situation was so serious that even Halder had to pay attention after learning that Germany would have a monthly deficit of 120,000 tons starting on October 1, 1941.[114] On August 14, the Defense Economy Office, Foreign Office, Economics Ministry, and the Office of the Four-Year Plan agreed on a strategy to deal with the crisis. First, production in Germany, Romania, and Hungary had to rise "to the highest degree that was technically feasible."[115] Second, because it was impossible to reduce either military or agricultural consumption, all forces stationed within Germany or on garrison duty would suffer a 20 percent reduction in their fuel allocation, while the Economics Ministry imposed further cutbacks on both the civilian and the transportation sectors. This was the end of the line for reducing non-essential consumption—any further cuts "would weaken the defense-economic potential of the Reich to a considerable extent."[116]

German planners expected that operational demand in the Soviet Union would peak in August 1941 at 216,000 tons and shrink to 45,000 tons by December, plus 17,000 tons for special mechanized divisions starting in October. Meanwhile, the allocation for forces stationed in Germany or on occupation duty would drop from 89,900 tons in July to 41,500 tons by September.[117] Despite such cuts, there was still insufficient fuel. Assuming the war would be protracted, Thomas concluded that accumulating "large quantities for army operations starting at the beginning of 1942 [is] impossible... without Russian imports." Without drastic reductions in consumption—20 percent for everyone besides forces conducting operations in the Soviet Union or North Africa—Germany would end the year with a deficit of 1.27 million tons. Such cutbacks would be "so substantial that negative political, military, and economic repercussions are to be expected." Even if Germany survived 1941, no reserves were available for divisions in formation or undertaking refit. Addressing the 1942 shortfall—127,000 tons during the first quarter alone, or 230,000 including the residual 1941 deficit—was also only possible if Germany "at least" secured Maikop.[118]

On the supply side, in late August, the Defense Economy Office estimated that Romania would produce 2.7 million tons of refined petroleum during the second half of 1941. Meanwhile, German monthly production would increase by 33,000 tons between July and December, for a six-month total of 2.18 million tons. Even assuming the suggested reductions in consumption, demand would still be at least

[113] Keitel, "Mineralölversorgung [...]," July 30, 1941, T-77/429 (Wi/IF 5.3563).

[114] *Halder Diary*, August 7, 1941.

[115] To that end, in December, Neubacher received a new commission to promote "the greatest possible increases of crude oil production" in Romania, Hungary, and Croatia. Göring to Neubacher, December 16, 1941, T-77/606 (Wi/IC 4.3a-b).

[116] Löbbecke to Thomas, "Treibstofflage [...]," August 19, 1941, T-77/594 (Wi/I. 273).

[117] Griebel, "Vortragsnotiz [...]," August 14, 1941, T-77/704 (Wi/IE 4.40).

[118] Thomas, "Mineralölversorgungslage Europas [...]," August 26, 1941, T-77/704 (Wi/IE 4.40). Emphasis in the original. See also: Defense Economy Office, "Mineralölversorgungslage in Europa [...]," August 26, 1941, T-77/704 (Wi/IE 4.40). The office reiterated these conclusions in October using the latest figures. Defense Economy Office, "Stand [...]," October 7, 1941, T-77/704 (Wi/IE 4.40). See also: Hayward, "Hitler's Quest," 104.

3.68 million tons.[119] Keitel ordered that all measures be taken to reduce the expected shortfall during the first quarter of 1942.[120] Göring gave his approval two days later, although he was still hopeful about capturing Maikop, which he estimated would take six months to repair.[121]

By this point, however, Operation Barbarossa had already stalled short of any of its key geographical objectives. The resiliency of the Soviet regime and the Red Army were certainly the most important factors, but logistical constraints also played a role. The German Army that went to war in 1939 was not a mechanized or even motorized one.[122] The bulk of its striking power came from unmounted infantry divisions that, like their Wilhelmine predecessors, relied on horses to carry their supplies.[123] Building an army around the internal combustion engine was a non-starter because Germany simply lacked the necessary petroleum or rubber.

The army could compensate for these problems within the confined spaces of Central and Eastern Europe, where its small armored spearheads could not significantly outrun the supporting infantry divisions that would annihilate encircled enemy units or their supply columns. The Soviet Union was another story. Advancing units quickly outran their infantry and supplies. Meanwhile, the long distances and harsh conditions of the Eastern Front caused higher rates of mortality for horses and malfunction for trucks. Repair of the latter was complicated by the shortage of spare parts, since much of the German motor pool was composed of captured vehicles of various makes and models. Even if the German Army was thoroughly motorized, the dearth of good roads would have hindered resupply. Getting fuel and ammunition to the front depended making Germany optimal use of captured Soviet railways and rolling stock to get supplies to the railheads closest to the front. It would then use its small supply of trucks to carry supplies the rest of the way. This system had already failed repeatedly during World War I.[124] It was probably the best Germany could do under the circumstances, but the results were no different than before—operational and strategic failure.[125]

During preparations for their assault on Moscow, Operation Typhoon, the Army High Command requested twenty-two to twenty-seven fuel trains per day in September 1941 and twenty-nine in October after the offensive began. The Armed Forces High Command responded that it could only provide twenty-one trains daily in October unless Germany made significant cuts to allocations to other theaters, the air force, and construction.[126] Thanks to lower demand in

[119] Defense Economy Office, "Rumänische Mineralöversorgung [...]," August 27, 1941, Defense Economy Office, "Deutsche Erzeugung [...]," August 27, 1941, and Defense Economy Office, "Deutscher Bedarf [...]," August 30, 1941, all in: T-77/704 (Wi/IE 4.40).

[120] Ro, "Aktenvermerk [...]," September 2, 1941, IWM, FD 4809/45.

[121] Thomas, "Aktennotiz [...]," September 5, 1941, IWM, FD 4809/45.

[122] Only 6 percent of the army's divisions before the invasion of France and the Low Countries in 1940 were armored. Murray, *Balance*, 33–4; DiNardo, *Panzer Arm*, 5–26.

[123] R.L. DiNardo and Austin Bay, "Horse-Drawn Transport in the German Army," *Journal of Contemporary History* 23: 1 (1988): 129–42.

[124] Van Creveld, *Supplying War*, 142–80.

[125] H.G.W. Davie, "The Influence of Railways on Military Operations in the Russo-German War 1941–1945," *Journal of Slavic Military Studies* 30: 2 (2017): 321–46.

[126] *Halder Diary*, September 11, 1941; Division of National Defense, "Betriebstoff-Forderung [...]," September 15, 1941, T-77/175 (Wi/IF 5.694). See also: Hayward, "Hitler's Quest," 101–4.

other sectors of the Eastern Front and ruthless cuts elsewhere, the Germans had accumulated enough fuel for Army Group Center when Operation Typhoon commenced in October.[127] Between June and December 1941, in fact, deliveries to the Eastern Front exceeded consumption, such that the army technically had a surplus of 100,000 tons by year's end. Meanwhile, civilian allocations of gasoline and diesel in December 1941 were one-quarter and 42 percent, respectively, of those of September 1939.[128] Getting it to the spearheads was another matter. Besides the misery they inflicted on both sides' troops, the harsh winter conditions also caused increased fuel consumption—quantities once adequate for 100km now sufficed for only 35–40km. The railway network got so overloaded that by the end of October there was a backlog of 142 trains.[129] These supplies proved invaluable in repelling the Soviet counteroffensive the following month, but that was doubtless of little comfort to the Germans.[130]

Operation Barbarossa's failure left Germany on the hook for supplying both Axis Europe and the occupied eastern territories. There was also the matter of supplying German forces on the Eastern Front—they required at least 100,000 tons per month for defensive purposes, but the most the Defense Economy Office could initially spare was 48,000 tons. Once the existing surplus disappeared in March 1942, the army's ability to conduct even "defensive military operations would be endangered."[131] The German and Italian navies were even worse off.[132] Germany needed the Caucasus more than ever, but as the New Year dawned, German forces would not enjoy "the same operational freedom," since there were no longer any reserves left to supplement production.[133] Another offensive along the entire Eastern Front was out of the question, as Axis Europe would have a 595,000-ton shortfall during the first quarter of 1942 alone.[134] The great hope in 1939, Romania, would never live up to expectations—although that did not stop Keitel sending a groveling letter to Antonescu in May 1942 begging him to set aside 320,000 tons for the impending invasion of the Caucasus.[135]

[127] Monthly military consumption in 1941 averaged 102,000 tons of aviation fuel, 135,000 tons of gasoline, 68,000 tons of diesel, 183,000 tons of fuel oil (including the requirements of the Italian Navy), and 10,500 tons of lubricants. Defense Economy Office, "Anlage [...]," April 29, 1942, T-77/606 (Wi/IC 4.3a-b).

[128] Defense Economy Office, "Die deutsche Treibstoffversorgung [...]," February 16, 1942, T-77/668 (Wi/VI. 216).

[129] Chief of Staff, "Ergebnis [...]," October 24, 1941, T-77/175 (Wi/IF 5.694).

[130] Hayward, "Hitler's Quest," 105–6.

[131] Defense Economy Office, "Die deutsche Treibstoffversorgung [...]," February 16, 1942, T-77/668 (Wi/VI. 216). Emphasis in the original.

[132] "Bedeutung des Maikop-Ölgebietes," no date or author, T-1022/1770 (PG 32198b).

[133] Armed Forces High Command, "Entwicklung [...]," November 3, 1941, Thomas, "Gegenüberstellung [...]," January 12, 1942, and "Aktenvermerk," no date or author, all in: T-77/704 (Wi/IE 4.40). Quotation from the second source. See also: Joint-Operations Staff, "Wehrkraft 1942," June 6, 1942, quoted in: OKW, *KTB*, ii: 170–3.

[134] Griebel, "Voraussichtliche Mineralöversorgungslage [...]," October 24, 1941, T-77/704 (Wi/IE 4.40).

[135] Griebel to Keitel, April 27, 1942, Defense Economy Office, "Beitrag [...]," April 27, 1942, and Keitel to Antonescu, May 6, 1942, all in: T-77/606 (Wi/IC 4.3a-b).

This was only a temporary expedient. As the Defense Economy Office recognized, Germany could only satisfy the needs of Axis Europe if it secured Maikop and Grozny.[136] Getting there would require further petroleum, but less was available than was the case the year before. Thomas informed Halder in November 1941 that the military's monthly fuel quota for the first quarter of 1942 would be reduced by 20,000 tons from the 95,000 tons it would receive in December. "This is for us," Halder moaned, "the end of operational freedom."[137] The Third Reich would do whatever it could to supply the military for the upcoming campaign season—hence Göring's decree of January 1942, backed by Hitler, that the "raising of petroleum production" in Axis Europe would "have precedence over all other undertakings."[138] There was, however, a limit to what Germany could accomplish on its own—increasing synthetic fuel output could only come at the expense of munitions production. In March, Göring rebuked the General Staff—the military had to "adjust" its operational requirements to the supply situation. It would yet again have to do more with less.[139]

Setbacks in Romania compounded Operation Barbarossa's failure.[140] Even before the invasion of the Soviet Union, Thomas had concluded that the Axis war effort depended upon increasing Romanian exports.[141] Germany boosted imports between September 1940 and August 1941 by one million tons from the previous twelve months, even as Romanian production continued to decline.[142] In October 1941, the Germans consented to a hefty price increase of 15 percent for imported crude oil.[143] For their part, the Romanians insisted they had made every effort to satisfy Germany's needs during Operation Barbarossa—deliveries between July and December 1941 averaged 267,000 tons per month and over 500,000 tons in July and August, largely by depleting Romanian stockpiles.[144] Exports to Germany in 1941 were also double those of 1940—3.06 million tons versus 1.39 million tons—and 576,000 tons went to Germany, the Eastern Front, and Italy in August alone. None of this would have been possible without massive investment to increase transportation capacity. The number of tankers plying the Danube increased from

136 Raw Materials Division to Thomas, "Mineralöllage in Rumänien," October 27, 1941, T-77/606 (Wi/IC 4.3a-b); Armed Forces High Command, "Entwicklung [...]," November 3, 1941, T-77/704 (Wi/IE 4.40).

137 *Halder Diary*, November 19, 1941.

138 Göring, Decree, January 7, 1942, T-77/1057 (Wi/ID. 26); Hitler, "Rüstung 1942," January 10, 1942, Document 131 in: Moll, *Führer-Erlasse*, 219–21. For discussion of material and labor requirements, see: "Auszug aus dem Vortrag Dr. Ritter [...]," January 21, 1942, T-77/649 (Wi/VI. 34).

139 Thomas, "Aktennotiz [...]," March 9, 1942, IWM, FD 1434/46. See also: Thomas to Keitel, "Betriebstoffversorgung im Sommer 1942," March 16, 1942, T-77/704 (Wi/IE 4.40).

140 A.E. Gunther, "German War for Crude Oil in Europe, 1934–1945," *Petroleum Times* (January 17, 1948), 65–8 and 84.

141 Thomas to Gercke, May 21, 1941, T-77/377 (Wi/IF 5.2694).

142 Defense Economy Office, "Mineralöl-Einfuhr [...]," December 22, 1941, IWM 4809/45. For a comparison of Romanian production vs. exports to Germany in 1940–1, see: Göring, "Vergleich [...]," November 30, 1941, T-77/594 (Wi/I. 273).

143 Neubacher to Marinescu and Marinescu to Neubacher, October 31, 1941, BA-B, R 901/112776.

144 Killinger to Foreign Office, January 5, 1942, BA-B, R 901.112777; Eichholtz, *Rumänisches Öl*, 58–9. Of course, the Economics Ministry had been expecting monthly exports to reach 288,000 tons since May 1940. Ro V, "Voraussichtliche Mineralöl-Einfuhr [...]," January 11, 1941, IWM, FD 4809/45.

138 in September 1939 to 505 by December 1941, while 23,000 railway cars were in service by the winter of 1941/2, compared to 1,800 in that of 1939/40.[145]

During the entire war, Romania sent between 55 and 70 percent of its production to the Axis.[146] Yet the energy crisis of late 1941 was attributable in part to the Romanians' inability to deliver required supplies because of their failure to restrict domestic consumption, especially of fuel oil.[147] Romanian consumption in 1941 was 200,000 tons higher than in 1938: 1.99 million tons versus 1.78 million tons–134,000 tons of this increase was for additional fuel oil.[148] Meanwhile, in November 1941, the Germans were receiving only 41,000 of the 92,000 tons of fuel oil they expected. As far as the Defense Economy Office was concerned, the fault lay with the Romanians.[149] Fischer disagreed—the Krauch Plan was diverting resources needed in Romania.[150] Whatever the exact reason, Bentz forecast that Romania would not meet its target for the foreseeable future. He estimated output at 5.2 million tons for 1941. On the bright side, German production in 1942 would probably rise above five million tons.[151]

Ribbentrop, Göring, and Hitler all brought up the need to raise oil production with Deputy Prime Minister Mihai Antonescu in November 1941. Ribbentrop and Göring still hoped that Germany would occupy the North Caucasus before end of campaign, but the latter reckoned it would take at least a year to restart production. Until then, Germany needed Romania to step up, even at the risk of exhausting its oilfields. In exchange, Göring promised the Romanians compensation in the Soviet Union and Iran.[152] Ion Antonescu replied that he had made every effort to live up to his promises, as evidenced by the fact that daily production had increased by 2,500 tons. While Romania's total production had fallen, Germany and Italy's share of its exports had risen from 25 percent in 1936 to 83 percent in 1941.[153] Antonescu actually understated Romania's contribution—Germany and Italy's share that year was 90 percent: 3.66 million tons of 4.05 million tons.[154] As Operation Barbarossa stalled, however, Romanian exports

[145] Foreign Office, "Die rumänische Mineralölwirtschaft," no date (circa 1942), T-120/2618.

[146] Karlsch, *Faktor Öl*, 206–7.

[147] Foreign Office, "Aufzeichnung [...]," February 17, 1942, T-120/2618.

[148] Foreign Office, "Die rumänische Mineralölwirtschaft," no date (circa 1942), T-120/2618.

[149] Defense Economy Office, "[...] Mineralölbezüge aus Rumänien," November 3, 1941, T-77/606 (Wi/IC 4.3a-b).

[150] "Grundsätzliche Stellungnahme [...]," no date, attached to: Spalke to Thomas, "Mineralöllage in Rumänien," October 9, 1941, T-77/606 (Wi/IC 4.3a-b). The Defense Economy Office agreed that efforts in Romania were suffering due to "competition with a great number of proposed plans," but it disagreed with the suggestion to take over production and distribution of Romanian oil. Raw Materials Division to Thomas, "Mineralöllage in Rumänien," October 27, 1941, T-77/606 (Wi/IC 4.3a-b).

[151] Ro Vf, "Mineralölerzeugung," November 18, 1941, and Krauch, "Treibstoff-Produktionsplan 1942," December 2, 1941, both in: T-77/385 (Wi/IF 5.2816).

[152] *DGFP* (D), xiii: 844–8, 870–6, and 891–4.

[153] Antonescu to Hitler, December 5, 1941, *DGFP* (D), xiii: 963–4.

[154] IG Farben, "Das rumänische Erdöl," April 11, 1939, T-77/611 (Wi/IC 4.16); Foreign Office, "Die rumänische Mineralölwirtschaft," no date (circa 1942), T-120/2618; "Zahlen [...]," September 16, 1942, attached to: Continental Oil to Bentz, September 29, 1942, T-401/5 (RBF 142); Eichholtz, *Rumänisches Öl*, 67.

declined—from 361,600 tons in August to 104,000 tons in December.[155] Thereafter, neither the German military nor the Economics Ministry trusted the Romanians. In preparation for a meeting between Hitler and Antonescu in February 1942, the Economics Ministry stressed that Romanian consumption had actually increased.[156]

No one in Berlin was pleased with the situation, but German leaders shrank from taking a hard line. Göring considered Antonescu to be "the only one in Romania who still defends the German line."[157] The experience of 1916 ruled out the "extreme measures" Germany had "so frequently adopted elsewhere, often with far less provocation."[158] The farthest Berlin was prepared to go was requesting that Bucharest reduce its consumption of key products.[159] This was essential if Germany was to fuel the tractors it required to exploit the agricultural lands of occupied Ukraine—every 1,000 tons of fuel could yield 50,000 tons of grain.[160] Bentz also urged the Foreign Office to expedite the acquisition of the two largest French-owned companies in Romania: Colombia and Concordia.[161]

Thanks to mediation by Fischer, the Germans secured passage of a new mining law in 1942, but their problems did not end.[162] The previous year, the Romanians had begun to drag their heels with regard to authorizing the transfer of shares of the largest French company in Romania to Continental Oil.[163] Nevertheless, the failure of Operation Barbarossa had one unexpected benefit—it allowed Germany to divert to Romania drilling equipment intended for the Caucasus.[164] This infusion of resources temporarily reversed the waning of Romanian production, which increased by 3 percent in 1942.[165] Such success was short-lived—by June 1942, Continental Oil realized that it could not stabilize its production in Romania.[166] Output in 1943 was the lowest on record: 5.27 million tons—a fall

155 Hayward, "Hitler's Quest," 102.

156 "Entwurf [...]," no date, attached to: Economics Minister to Defense Economy Office, "Besuch [...]," February 3, 1942, and talking points for Keitel, "Mineralöl [...]," February 4, 1942, both in: T-77/606 (Wi/IC 4.3a-b). Given its expectation of occupying the Caucasus, the Armed Forces High Command also wanted Romanian assistance in rebuilding the destroyed oil-loading facilities in Odessa. "Mineralöl-Umschlagseinrichtungen [...]," February 7, 1942, T-77/606 (Wi/IC 4.3a-b).

157 Antonescu was complaining that the Germans were not delivering sufficient quantities of coal or drilling equipment. Thomas, "Aktennotiz [...]," March 9, 1942, IWM, FD 1434/46.

158 Rich, *Establishment*, 255–6.

159 German Legation, Bucharest, to Antonescu, "Note," February 27, 1942, T-120/2618. See also: Economics Ministry to Neubacher, no date, attached to: Rosencrantz to Foreign Office, March 10, 1942, T-120/2618.

160 Four-Year Plan, "Treibstoffversorgung [...]," October 23, 1942, BA-B, enclosed with: Gramsch to Wiehl, "Rumänische Mineralöllieferungen," November 12, 1942, BA-B, R 901/68737.

161 Bentz to Foreign Office, July 2, 1942, NARA, RG 238, T-301/18 (NI-1934).

162 Neubacher to Foreign Office, "Stand [...]," February 23, 1942, BA-B, R 901/112777; Pearton, *Oil*, 226–44. Whereas the Romanians had complained in 1941 that they could not fulfill German expectations because of the lack of equipment, after this they found new excuses. Bentz worried that the Romanians would block implementation of German drilling plans in 1942 through the restrictive mining law of 1937. Bentz to Wiehl, "Aufzeichnung [...]," February 17, 1942, T-120/2618.

163 Stelzer to Foreign Office, November 28, 1941, BA-B, R 901/112776.

164 Memorandum for Wiehl, February 17, 1942, author unclear, BA-B, R 901/112777.

165 Dr. vdW/Hof., "Memorandum [...]," September 17, 1943, and "Schaubilder [...]," no date, both in: DBA, B-60.

166 "Kontinentale [...]," June 6, 1942, author unclear; reports to the Supervisory Board; all in: DBA, B-60.

of 500,000 tons since 1940. Despite its investments (60,000 tons of steel annually between 1942 and 1944), Germany's imports and its share of production from Romania (53 percent) both peaked in 1941.[167] Finally, German–Romanian relations took a turn for the worse following the Stalingrad debacle, which cost Romania 150,000 casualties. Even before Allied bombing slashed output in 1944, German ambitions in Romania were already in ruins. A year before Romania defected from the Axis, the German Minister in Bucharest was complaining of pro-Allied sympathies among Romanian businessmen and describing Romania's demands for compensation as "blackmail." In combination with synthetic fuel, Romanian oil would preserve Germany's ability to wage a defensive war until 1944. The dream of victory, however, had evaporated years before.[168]

Following Operation Barbarossa's failure, Germany returned to the offensive in June 1942, with oil now its explicit objective.[169] Even as Army Group Center desperately parried Soviet counterattacks outside of Moscow, Continental Oil was already preparing for the next campaigning season. The second meeting of the company's supervisory board, on January 13, 1942, was devoted to the company's current and future operations in the Soviet Union. Karl Blessing of the managing board (later President of the Bundesbank) announced that the firm had established a subsidiary (Eastern Oil) to operate in the Caucasus.[170] Another subsidiary, Continental Transport, would purchase, lease, or construct tankers, loading and offloading facilities, and pipelines, both in the Black Sea and the Mediterranean. Fischer outlined the company's long-term plans. Rehabilitating antiquated Soviet infrastructure was impossible without the support of German industry. Soviet pipeline capacity was insufficient (more than 40 percent of Soviet oil shipments moved by rail before the war) and the refining required modernization.[171] Fischer fretted, however, over the shortage of equipment due to German industry's focus on munitions production.[172]

During the invasion of the Caucasus in the summer of 1942 (Case Blue), Continental Oil would have the assistance of the Technical Petroleum Brigade, a lavishly equipped successor of the "Petroleum Commandos" of 1941.[173] The fact that the firm contributed the bulk of the brigade's technical staff makes it difficult

[167] Eichholtz, *Rumänisches Öl*, 63–4 and 67; Pearton, *Oil*, 259–63. The development of Hungary, where production rose from 231,000 tons in 1940 to 809,000 tons four years later, softened the blow somewhat. Estonia was also contributing 107,000 tons annually by 1943. Krammer, "Fueling," 408–9.

[168] Eichholtz, *Deutsche Ölpolitik*, 332–50, 487–93, and 520–3.

[169] The following discussion draws from: Anand Toprani, "The First War for Oil: The Caucasus, German Strategy, and the Turning Point of the War on the Eastern Front, 1942," *Journal of Military History* 80: 3 (July 2016): 815–54.

[170] Eastern Oil's capitalization was later raised to four million RM. Continental Oil to the Supervisory Board, "Berichterstattung [...]," February 3, 1943, T-401/2 (RBF 35).

[171] Mejcher, *Politik*, ii: 44–7.

[172] "Minutes...13 January 1942," NARA, RG 238, T-301/83 (NI-10162). Planning for the 1942 campaign is summarized in Gunther, "Caucasus Expedition, 1942–1943," March 1946, BNA, WO 252/1151; Kockel, "Geologie und Krieg," 54–64.

[173] Göring to Thomas, "Mineralölwirtschaft [...]," July 16, 1942, T-77/1058 (Wi/ID. 29).

to distinguish between the two organizations.[174] Finally, in July 1942, the Third Reich concluded a formal concession agreement with Continental Oil—the firm's "Magna Carta."[175] The company would enjoy "the exclusive right to the prospecting, production, refining and marketing of crude oil, natural gas and all other bituminous materials" in the occupied Soviet Union. In exchange, it would pay a quarterly royalty equivalent to 7.5 percent of its oil and gas production.[176]

Chastened by the failure of Operation Barbarossa, during Case Blue the Third Reich curbed some its murderous tendencies out of economic necessity. In October 1942, Hitler ordered that captured Soviet oil workers, along with their families, should receive adequate rations, but it was too late—the Soviet counteroffensive to relieve Stalingrad began a month later.[177] Moreover, due to Soviet sabotage, the Germans produced only a trickle of oil during their occupation of Maikop between October 1942 and January 1943.[178] Despite its preparation and ruthlessness, the Third Reich's campaign to exploit the Soviet Union was as much an economic failure as a military one. The value of the resources Germany extracted from Belgium (9.3 billion RM) was double that from the Soviet Union (4.5 billion RM), while that from France (35 billion RM) was eight times greater.[179] In terms of oil, Germany probably received more from the Soviet Union through trade before June 22, 1941 than it did through coercion during the occupation.

Operation Barbarossa was irresistible because it offered a resolution to Germany's overarching strategic dilemma after June 1940 while pandering to the ideological preconceptions of its leaders. Hitler and his advisors were convinced that Germany could annihilate the Soviet Union within weeks. Besides eliminating Britain's only potential ally of consequence in Eurasia, the conquest of the Soviet Union would provide Germany with a plethora of resources, including oil, coal, minerals, arable land, and slave labor. With these in hand, the Third Reich could look forward to the looming struggle against the Anglo-American coalition with every hope of success.

The resilience of the Soviet regime wrecked the German war plan. If the Soviet Union failed to disintegrate, Germany could not compel its surrender. Operation Barbarossa's failure did not signify that defeat was inevitable, but rather (as Hitler recognized in December 1941) that Germany lacked the means of achieving victory on its own terms. The Germans could still, however, smash Soviet war-making capacity and gain some breathing room—hence the decision to strike toward the

[174] Continental Oil to Funk et al., November 6, 1942, DBA, B-60. For the brigade's instructions, see: Keitel to Technical Petroleum Brigade, "Mineralölwirtschaft [...]," July 28, 1942, T-77/1058 (Wi/ID. 29).

[175] Schlicht, "Notiz [...]," August 4, 1942, BGR 65936.

[176] Göring to Continental Oil, "Einsatz [...]," July 10, 1942, BA-B, R 176/111.

[177] Hitler, "Wiederaufbau [...]," October 12, 1942, reprinted in: Moll, *Führer-Erlasse*, 286–7.

[178] 1,000 tons of oil between December 1942 and January 1943 at the cost of 10,000 tons of equipment abandoned when German forces withdrew. Gunther, "Caucasus Expedition, 1942–1943," March 1946, BNA, WO 252/1151. See also: Wolfgang Birkenfeld, "Illusionen am Kaukasus 1942/43." In *Wissenschaft, Wirtschaft und Technik: Studien zur Geschichte*, ed. Karl-Heinz Manegold (München: Bruckmann, 1969), 85–91.

[179] Christoph Buchheim, "Die Besetzten Länder im Dienste der deutschen Kriegswirtschaft während des Zweiten Weltkrieges," *Vierteljahrshefte für Zeitgeschichte* 34: 1 (January 1986): 123ff and 142–3.

Caucasus in 1942. Case Blue was not a rerun of Operation Barbarossa. Even as the Third Reich accelerated the mass murder of Europe's Jews and expanded the war's scope by including the United States, it fought for increasingly modest geopolitical objectives. The acquisition of oil was indisputably at the center of German strategy, but the illusions of 1940–1 had vanished. As Hitler explained to Raeder in August 1942, Germany was not seeking an unlimited victory, but rather to keep the Soviet Union at bay and acquire the means of waging a protracted war against Britain and the United States until they were "ready to make peace."[180] Germany's subsequent failure to capture the Caucasian oilfields, followed shortly afterwards by the destruction of the Sixth Army at Stalingrad, did not turn the tide of World War II so much as hasten its end.[181]

[180] Boog, *Global War*, 1206–15; Hillgruber, *Hitlers Strategie*, 549–56.

[181] See also: Joel Hayward, "Too Little, Too Late: An Analysis of Hitler's Failure in August 1942 to Damage Soviet Oil Production," *Journal of Military History* 64: 3 (2000), 769–94.

Conclusion

Oil and the End of European Primacy, 1914–1945

The shift from coal to oil as the developed world's dominant source of energy was the most significant change in energy-consumption patterns of the twentieth century. It had profound consequences for the global balance of power by limiting the autonomy of Britain, Germany, and other powers lacking domestic sources of oil. Conversely, the shift enhanced the power of the United States and the Soviet Union, both of which possessed extensive supplies. Because of its geopolitical significance, the history of oil is an important chapter in the story of Europe's decline in the twentieth century, much as the history of coal is central to explaining Europe's rise a century before.

During the nineteenth century, the coal and oil industries coexisted, since the two commodities served different purposes—the former as a source of power generation, the latter for illumination or lubrication. The two enjoyed a symbiotic relationship—just as coal brought people together in urban and industrial environments, oil allowed them to function in the absence of natural light. Coal also enabled firms to mass-produce the iron and steel needed to drill for and transport oil.[1] When the transition from coal to oil came, it was not due to efficiency alone—interstate and corporate competition, as well as internal class conflict, all contributed.[2]

Furthermore, while the coal industry grew complacent, the oil industry moved into the transportation sector once it recognized the threat posed to kerosene-based illumination by electrification, since coal has historically been the primary source of energy for electricity generation. Coal companies focused their attention on national or regional markets—in 1912, 90 percent of British exports (two-thirds of global exports) went no farther than Europe or the Mediterranean.[3] By contrast, the oil industry had extensive experience developing global supply, marketing, distribution, and marketing chains, which enhanced its political power and reduced its vulnerability to labor unrest relative to the coal

[1] Christopher Jones, *Routes of Power: Energy and Modern America* (Cambridge: Harvard University Press, 2014), 89–121.

[2] Podobnik, *Shifts*, 1–17; Timothy Mitchell, "Carbon Democracy," *Economy and Society* 38: 3 (2009): 399–432, and *Carbon Democracy*, 12–42. Consumer preferences were less significant. Jones, *Routes*, 6–10.

[3] Mitchell, *Carbon Democracy*, 37.

industry. By the time oil's superiority as a propulsion fuel became clear, the coal industry was too weak to strangle its rival.[4]

The eclipse of the coal industry by its oil rival had immediate consequences for the balance of power. Oil's rise ensured that some nations that had vaulted to economic great-power status during the nineteenth century thanks to their abundant supplies of coal—Britain and Germany—now found themselves at an immense disadvantage relative to superpower rivals—the United States and the Soviet Union—that commanded access to much of the world's oil production. The effects of this disparity became apparent during World War II—a conflict whose origins, conduct, and outcome revolved around access to energy (oil, coal, and food) in both the European and Pacific theaters. Although raw materials such as coal and iron ore (for steel) played vital roles, pride of place went to petroleum.[5] The war was not simply "fought on oil," one historian observed, but "indeed sometimes seemed fought *for* it."[6]

What was clear to contemporaries has been lost in hindsight, since "resources shaped the conduct and outcome of World War II in ways that political and military historians often neglect."[7] Two of the most significant strategic decisions of World War II were motivated by the need to acquire access to secure supplies of oil: Japan's attack on Pearl Harbor in 1941 to screen its advance into the East Indies, and Germany's concentration of its entire offensive effort on the Eastern Front in 1942 upon the Caucasus. We should not overlook the long-term consequences, either. Oil was an essential ingredient in the process of transforming Britain into a client of the United States. Ample supplies also helped the United States to crush Germany, thereby preventing Germany from mimicking the strategy of landed expansion and self-sufficiency that the United States had used to become a great power in the nineteenth century.

This study's findings are not limited to Europe—Japan, too, possessed the requisite blend of capabilities and intentions to pursue a strategy of energy independence. Japan's experience was an amalgam of Britain and Germany's. Like Britain, it was a maritime power that was undone by its inability to secure its overseas access to oil. Despite being one of the victors of World War I, Japan (as well as Italy) ended up on Germany's side during World War II after forsaking participation in an Anglo-American world order in favor of continental expansion.[8] Like both Britain and Germany, Japan came to grief because of overextension.

[4] Nowell, *Cartel*, 224–36. Podobnik, *Shifts*, 38–91 ascribes the shift to the internal combustion engine and does not refer to electrification. See also: Martin Melosi, *Coping with Abundance: Energy and Environment in Industrial America* (Philadelphia: Temple University Press, 1985), 50 and 138–59.

[5] David Edgerton, "Controlling Resources: Coal, Iron Ore, and Oil in the Second World War," in: Total War: Economy, Society, and Culture, ed. Michael Geyer and Adam Tooze, vol. 3 of Cambridge History of the Second World War (Cambridge: Cambridge University Press, 2015), 122–48.

[6] Carl Solberg, *Oil Power: The Rise and Imminent Fall of an American Empire* (New York: Mason/Charter, 1976), 131–2. Emphasis in the original.

[7] Eckes, *Minerals*, 89.

[8] Michael Barnhart, "Japan's Economic Security and the Origins of the Pacific War," *Journal of Strategic Studies* 4: 2 (1981): 105–24, and Japan Prepares for Total War: The Search for Economic Security, 1919–1941 (Ithaca: Cornell University Press, 1987).

To sustain its continental empire in China, it provoked a war with the United States to conquer the oilfields of the East Indies. Although the Japanese were spectacularly successful at occupying the latter, U.S. submarines and aircraft severed Japan's access to those oilfields in 1943–4.[9]

Scrutinizing the role of oil in the great-power struggles of the first of the twentieth century affords us another way of understanding the eclipse of the multipolar world order after 1945 by a bipolar international system dominated by the United States and Soviet Union, in which the European powers and Japan served as auxiliaries. Such a prospect had terrified the European political and military leaders of 1914, but it became inevitable thanks to the war they unleashed. None of these developments is comprehensible without grasping the interrelationship between oil and power—economic or military. Yet the role of oil remains absent or underexplored in many histories of the twentieth century, and few historians have examined how oil influenced the strategic imagination of civilian and military leaders.[10] One of the reasons why may be that policymakers were careful to cover their tracks, especially in liberal-democratic societies such as Britain, where there was a possibility of public scrutiny. This fact alone enjoins us to remember that the absence of oil in the documentary record is not necessarily evidence of its absence, for as one historian of British strategy observes, "Oil was not mentioned because it was always in mind."[11]

DECLINE OR CATASTROPHE

Britain's decline as a great power still provokes spirited debate.[12] By "decline," scholars are referring to Britain's diminishing geopolitical standing or the steady deindustrialization of the British economy during the twentieth century. Economic decline is often conflated with geopolitical decline, since they were contemporaneous. Most ordinary people cared little about the latter and, based on general standards for quality of life, they were better off after Britain ceased to be an empire. For British elites, however, decline was traumatic because of the loss of status and purpose it entailed. Because of their disproportionate influence in British society, they constructed a narrative of decline to rebuke contemporaries for failing to preserve the accomplishments of earlier generations.[13]

[9] USSBS, *Oil in Japan's War* (Washington, DC: GPO, 1946), 33–88; Daniel Yergin, *The Prize: The Epic Quest for Oil, Money, and Power* (New York: Simon & Schuster, 1991), 305–27.

[10] Paul Kennedy references oil-related statistics and suggests how a lack of oil posed a strategic vulnerability or precipitated conflict, but he does not reflect on the consequences of the disparity in oil supplies for the balance of power. Kennedy, *Rise*, 151, 233–4, 243–4, 259, 269, 289, 294, 302–3, 306–8, 314, 340–1, and 343.

[11] Helmut Mejcher, *Imperial Quest for Oil: Iraq, 1910–1928* (London: Ithaca Press, 1976), 165–6.

[12] David Edgerton, "The Decline of Declinism," *Business History Review* 71 (1997): 201–6.

[13] Barry Supple, "Fear of Failing: Economic History and the Decline of Britain," *Economic History Review* 47: 3 (1994): 444–7 and 454ff.

Decline had both exogenous and endogenous causes. External factors included the rise of the United States and Germany, as well as certain technological developments, particularly in the field of transportation. As Halford Mackinder argued in 1904, the proliferation of railroads marked the end of the "Columbian" era by allowing Britain's continental adversaries in the Eurasian "heartland" to mass superior manpower against British positions along the "rim land"—especially the Middle East.[14] Overstretch arising from the British dispersal of resources in marginal regions only exacerbated this growing imbalance of power. The cumulative effects of this variegated process were disastrous. The commitment of British assets to the defense of the empire dissuaded Whitehall from imposing additional costs on the British public to maintain the balance of power in Europe, which reduced the credibility of British deterrence in both 1914 and 1939 to the point that the Germans felt they could risk a war.[15]

In terms of economics, scholars have attributed Britain's decline to its relative backwardness in terms of innovation and productivity compared to the United States and Germany after the 1880s. There were also internal factors such as free trade, the cost of imperial defense, and the influence of the financial-services sector, whose ascent came at the expense of heavy industry, with second-order effects for Britain's war-making capacity.[16] World War I did immense damage, although the tenfold increase in national debt was less significant than the erosion of Britain's position in international trade and finance. As a result, there would be no repeat of the sustained growth following the Napoleonic Wars.[17] Britain's economic track record after the Great War was unimpressive, and the economic depressions of 1920–1 and 1929 did not help matters. Britain had barely recovered from the first before the second hit, and in both cases, its recovery was anemic. After 1929, Britain took longer than any other great power to return to its 1913 rate of manufacturing productivity, even though its fall was not as sharp as that of Germany or the United States.[18]

While there is agreement that Britain's decline was relative, the debate over its timing and causes persists. For many decades, the "declinist" school contended that Britain's fall was due to choices made by British elites from the late nineteenth century onward.[19] Others focused on developments beyond Britain, such as the rise of the United States, Germany, Russia, and Japan, which made it impossible for Britain to sustain its naval predominance, even as its dependence on overseas trade grew. Also, the balance between naval and land power was shifting back to the latter. Britain could neither defend many of its imperial possessions against numerically superior foes, nor fall back upon the traditional "British way of war"

[14] Halford Mackinder, "Geographical Pivot of History," *Geographical Journal* 23: 4 (1904): 421–7.

[15] Patrick O'Brien, "Imperialism and the Rise and Decline of the British Economy, 1688–1989," *New Left Review* 1: 238 (1999): 76ff.

[16] Patrick O'Brien, "The Costs and Benefits of British Imperialism, 1846–1914," *Past & Present* 120 (1988): 163–200.

[17] Kennedy, "Finance," 52–3.

[18] Murray, "Collapse," 403.

[19] Besides Barnett, see also: Murray, *Balance*, 50–92, and "Collapse," 393–427.

of providing naval and financial assistance to European proxies to check an aspiring hegemon.[20] Finally, Britain was a victim of its own success. The pre-1914 empire was an expensive, unwieldy contraption of varying strategic value, and the post-World War I realm in the Middle East was "a classic, and gigantic, example of strategic over-extension," whose value was incommensurate with the costs incurred to acquire and maintain it.[21]

In recent years, however, critics have challenged the narrative of Britain as a broken power after World War I. Britain's experience since 1900 is proof that "relative decline and absolute growth can and do co-exist"—growth rates in the period 1951–73 were on average higher than those in 1873–1913.[22] Others argue that the Great War heralded not the end of the Pax Britannica, but rather a new chapter. Britain, they contend, was in a superior geopolitical, economic, and even financial position relative to its potential adversaries following World War I than after the Napoleonic Wars, and its position of global leadership was unmatched after the United States declined to exert its power beyond the financial realm.[23]

We can enrich this debate by establishing how changes in global energy consumption contributed to the rise and fall of British power. Strangely, the role of natural resources in Britain's decline remains underexplored.[24] Over the nineteenth century, coal became Britain's "Black Indies"—a replacement for the Caribbean sugar that had swelled British coffers in the eighteenth century—accounting for one-third of British exports by 1913, and arguably, the greatest single economic advantage it enjoyed relative to its competitors.[25] As the Sea Lords explained in 1931, "Fuel was a large factor in England's greatness"—from wind in the eighteenth century to coal in the nineteenth century and oil potentially thereafter, if Britain could acquire secure supplies.[26]

Britain's embrace of oil after 1912 turned out to have profound consequences for its national and economic security. The rise of oil undermined the basis of the Pax Britannica, which rested upon Britain's navy, overseas bases, and industry.[27] Contemporaries understood that one of the foundations of British power in the nineteenth century was Britain's "unlimited supply of marine coal" and stranglehold over its international distribution.[28] Even as the United States supplanted Britain as the world's dominant coal producer, the latter's control over naval fuel

20 Kennedy, "Mahan," 39–66.

21 Barnett, *Collapse*, 71–83 (quotation from p. 80); Murray, "Collapse," 395.

22 Supple, "Fear," 442.

23 Gordon Martel, "The Meaning of Power: Rethinking the Decline and Fall of Great Britain," *International History Review* 13: 4 (1991): 686–93.

24 One exception is Peter Hugill, "The American Challenge to British Hegemony, 1861–1947," *Geographical Review* 99: 3 (2009): 403–5 and 418–23.

25 John Darwin, *Empire Project: The Rise and Fall of the British World-System* (Cambridge: Cambridge University Press, 2009), 9 and 275, and Unfinished Empire: The Global Expansion of Britain (New York: Bloomsbury Press, 2013), 188.

26 "Oil [...]," no date or author, attached to: Alexander, "Oil [...]," February 17, 1931, C.P. 51 (31), CAB 24/220.

27 Paul Kennedy, *Rise and Fall of British Naval Mastery* (London: A. Lane, 1976), 157.

28 Miles, "Supply [...]," April 14, 1920, BNA, ADM 203/48.

supplies remained unchallenged.[29] Coal was a critical factor behind Britain's maintenance of naval predominance after Trafalgar. British ships may not have been superior to those of France or the United States, but British supplies of (and prices for) coal certainly were, while the empire provided ample coaling stations.[30] By the turn of the twentieth century, Britain exercised the power of life or death over the world's navies. Without British coal, for example, it would have been impossible for the U.S. Asiatic Fleet to strike Manila in April 1898, while the lack of access to British coal, coaling stations, and colliers hamstrung the Russian Baltic Fleet as it sailed around the world to relieve Port Arthur in 1904–5 during the Russo-Japanese War.[31]

While the shift to oil extended the range of Royal Navy ships, it also meant that Britain's coaling stations and colliers no longer determined the movements of foreign navies. The main beneficiary proved to be the United States. Although blessed with natural resources, the United States lacked coal reserves on its West Coast suitable for naval propulsion, which hindered the projection of U.S. naval power into the Pacific. In the Age of Coal, the U.S. ships either consumed coal shipped from Norfolk around Cape Horn or purchased supplies from British coaling stations and colliers. During its 1907–9 cruise, for instance, the "Great White Fleet" consumed almost ten times as much coal supplied from British colliers as it did from U.S. colliers.[32] The shift to oil, combined with discovery of oilfields in California during the 1890s, did more than end U.S. dependence on Britain's coal-supply network—it fundamentally upset the balance of power in the Pacific by allowing the United States to exert its naval power independently of Britain.[33]

Meanwhile, the shift to oil left Britain saddled with an "obsolete" coal production and distribution infrastructure.[34] Bases that guarded access to overseas oil supplies, served as staging and refurbishment points for imperial forces, or possessed dry docks would retain their value.[35] Most of the coaling stations, however, would have to be discarded and the remainder refurbished as oil-storage facilities.[36] Refurbishing or liquidating obsolescent assets was not a straightforward matter. Domestic coal mines and overseas stations represented enormous sunk costs that Britain could not easily liquidate, since the fate of the coal industry was as much a domestic policy issue as a strategic one. In the aftermath of World War I, both the coal and textile industries—the foundations of the first Industrial Revolution—became sources of weakness for the British economy.[37] In particular, the rise of oil,

[29] Ediger, "Geopolitics," 6–8. [30] Brodie, *Sea Power*, 17–45 and 115ff.

[31] Lamar J.R. Cecil, "Coal for the Fleet That Had to Die," *American Historical Review* 69: 4 (1964): 990–1005.

[32] Brown, "Fuel Supplies," 259 (n. 3).

[33] The preceding discussion of naval strategy also draws from: Maurer, "Battle Fleet," 60–77.

[34] Babij, "Japan," 88.

[35] Andrew Lambert, "The Royal Navy and the Defence of Empire, 1856–1918," in: *Imperial Defence: The Old World Order 1856–1956*, ed. Greg Kennedy (London: Routledge, 2008), 127.

[36] Geoffrey Kemp and John Maurer, "The Logistics of Pax Britannica: Lessons for America," in: *Projection of Power: Perspectives, Perceptions, and Problems*, ed. Uri Ra'anan, et al. (Hamden: Archon Books, 1982), 28–49.

[37] Darwin, *Empire Project*, 371 and 431.

combined with labor militancy and competition from other coal suppliers in Europe, Africa, and Asia, proved the death knell for Britain's once vibrant coal-export industry.[38]

The substitution of oil for coal at sea also created a host of difficulties. It reduced sales of coal at British coaling stations and eliminated a valuable source of export earnings. Because Britain's finished exports took up less space in cargo ships than bulky imports, shippers had compensated by using coal as ballast on their outbound voyages, which they would then sell abroad before returning. This allowed them to charge full freight to and from Britain, thereby reducing the shipping costs for British imports.[39] Oil imports also strained Britain's balance of payments. After the United States and Germany began eroding its industrial position in the 1880s, Britain only balanced its current account through exports to captive markets such as India; invisible income from the sale of services rather than goods, especially shipping and insurance fees; and earnings on foreign investment.[40] Imported oil—by 1933, 13 percent of Britain's daily imports—added yet another financial burden while degrading Britain's ability to pay for these imports by depressing demand for coal.[41]

Although its empire was at its greatest extent following World War I, Britain's pretensions to world-power status were hollow without oil. Whitehall recognized it had to develop a new strategy to compensate for its unfavorable energy position in a world where geopolitical power presumed access to oil. Britain had two options. It could subordinate itself to the United States to guarantee reliable supplies. Besides curtailing Britain's freedom of action, depending on the United States was inadvisable if rising U.S. consumption eliminated any exportable surplus, officials on both sides of the Atlantic feared. There was also the matter of Britain's war debt to the United States and shortage of foreign exchange—the former so poisoned political relations that it become impossible for financiers to resolve the latter. Alternatively, Britain could produce oil in territories it controlled directly or indirectly, which is the option Whitehall chose. Unfortunately, the most important of these—in the Middle East—was geographically difficult to access, strategically vulnerable, and politically unstable, and its defense further taxed Britain's overextended military forces. The end result of either strategy would have been the same—the end of British energy independence—but it seems fair to surmise that the consequences of dependence on the United States for Britain were less dire than they might have been if trusting its fate to an empire beset by internal and external challengers.

In an era of continental superpowers, the Royal Navy and the empire were the economic and security foundations of Britain's claims to great-power status. On

[38] Farnie, *Suez*, 567, 569, 581.

[39] J. Ellis Barker, "Coal and Shipping: The American Danger," *Fortnightly Review* 115 (February 1, 1921): 255–66; Brodie, *Sea Power*, 115–16; Wilkinson, "Influence of Oil [...]," *RUSI* 68: 469 (1923): 109–14.

[40] M.W. Kirby, *Decline of British Economic Power since 1870* (London: Routledge, 1981), 1–23 and 138; Kennedy, "Finance," 48–9.

[41] "Ölstrategie," *Militär-Wochenblatt* (January 15, 1937).

the eve of World War II, the latter was indeed a valuable supplier of raw materials, minerals, and foodstuffs, as well as a peacetime market for British goods and investment.[42] Its contributions to British security in the interwar period were, however, negligible.[43] Mobilizing the full potential of the empire after 1939 also depended upon the support of the United States, since Britain could not provide the necessary funds, raw materials, and munitions.[44] Imperial resources, even when denominated in Britain's own currency, were also not free. During World War II, in exchange for goods and services supplied to Britain, members of the sterling area accumulated "sterling balances" (debts) in London that were seven times greater than British hard currency reserves by war's end. The recipients of these funds naturally wished to spend these balances but could not due to currency controls imposed by Britain on all members of the sterling area. The existence of such liabilities threatened the stability of sterling for several decades until the British were able to negotiate the gradual liquidation of these balances.[45] Finally, the empire still produced little oil. On the eve of World War II, even including Bahrain, Brunei, and Sarawak, the empire accounted for only 2 percent of global production and only one-quarter of Britain's annual wartime requirements.[46] This meant that Britain's war-making capacity, even if it had exchanged economic liberalism for Imperial Preference earlier than 1932, was not in fact "limitless," as some scholars claim.[47]

The notion that Britain had nothing to worry about because "it could import oil rather than waste resources on synthetic oil" overlooks the empire's dearth of oil and understates the difficulty of moving or paying for oil from elsewhere.[48] For these reasons, the notion that only the military disasters of 1942 in North Africa and Singapore "made the Empire dependent on the United States" is untenable. So, too, is the contention that at the war's outset Britain was trying "to avoid the neutral USA as a source" of oil.[49] While Britain did not depend on U.S. oil before 1938, much of the empire did. British planning for a "dual contingency" after 1938 was predicated on a "friendly" United States, since the tanker shortage ruled out importing Iranian oil into Britain. Moreover, well before the attack on Pearl Harbor, Britain required U.S. financial assistance to pay for its oil imports. Finally, while Britain's wartime mobilization compared favorably to Germany's efforts until 1942, it was only possible with U.S. aid.[50]

[42] Steiner, Triumph, 798ff.

[43] G.C. Peden, "The Burden of Imperial Defense and the Continental Commitment Reconsidered," *Historical Journal* 27: 2 (1984): 415ff.

[44] Kennedy, "Mahan," 63–4.

[45] P.J. Cain and A.G. Hopkins, *British Imperialism: Crisis and Deconstruction, 1914–1990* (London: Longman, 1993), 269–85. For a contrary view, see: Price, *Failure*, 184–6.

[46] Kent, *Moguls*, 148.

[47] Contra Price, *Failure*, xiii.

[48] David Edgerton, *Warfare State: Britain, 1920–1970* (Cambridge: Cambridge University Press, 2006), 57.

[49] David Edgerton, *Britain's War Machine: Weapons, Resources, and Experts in the Second World War* (New York: Oxford University Press, 2011), 4–5, 76–8, and 84–5.

[50] Murray, "Collapse," 404–6. For British munitions production, see: Mark Harrison, "The Economics of World War II: An Overview," in: *The Economics of World War II: Six Great Powers in International Comparison*, ed. Mark Harrison (Cambridge: Cambridge University Press, 1998), 5–18.

Critics of the "declinist" school have underestimated the handicaps Britain suffered due to its shortage of oil. Granted, ordinary Britons (and Germans) benefited from the coal-to-oil transition through higher standards of living. For them, decline was a positive development. Such people, however, did not make policy in Britain, and beyond questions of employment, British elites did not give much weight to domestic policy considerations when developing their oil strategy. Furthermore, the strategies they adopted were supposed to enhance the viability of another elite project, the empire. This turned out to be a mistake—rather than alleviating its dependence on imported oil, the empire turned out to be "a sprawl which entangled Britain in great-power rivalries both in the Mediterranean-Middle East region and in the Far East."[51]

Consequently, at no point during the era of the world wars could Britain have contemplated a conflict against any of its major rivals without U.S. oil supplies. Synthetic fuel was not a viable alternative considering the sunk costs of building Britain's capacity to import oil, not to mention the fact that even an energy independent Britain would still require imports of other crucial commodities. Moreover, there is a difference between titular ownership of foreign oilfields and the ability to exploit such resources, particularly in wartime. Ultimately, the British proved better at acquiring titular control of foreign oilfields than at tailoring their national and military strategy to compensate for the empire's perilous oil position. How else to explain the fact that Britain considered the ramifications of a "closed" Mediterranean only after the break with Italy and not before?[52]

Britain's relative decline correlates with the rise of oil, and this study demonstrates that there was also a causal link between the two phenomena. Oil was the not the cause of Britain's decline. Rather, its emergence aggravated difficulties that were already in existence and created others. Recognizing the challenges posed by oil validates the broader historiographical consensus concerning Britain's strategic dilemma during the first half of the twentieth century. As an island nation that depended upon exports while defending a global empire, "Britain could either possess adequate defences and be bankrupt, or remain solvent yet strategically vulnerable, *but she could not achieve her aims in both areas*."[53] In fact, actions to promote economic security came at the expense of national security, and vice versa.[54] A similar dynamic was at work with oil. British officials could seek energy independence and risk strategic vulnerability, or they could embrace security and accept dependence on a rival great power, but they could not have their cake and eat it too.

After 1918, Britain still produced many first-rate weapons, as well as possessing a shipbuilding industry that churned out both naval and merchant vessels. This war-making infrastructure, combined with its advanced financial-services sectors and the resources of the empire, meant that Britain was more deserving of the title

[51] Barnett, *Strategy*, 6–7. See also: Murray, *Balance*, 91–2. [52] Steiner, *Triumph*, 793.

[53] Paul Kennedy, "'Appeasement' and British Defence Policy in the Inter-War Years," British Journal of International Studies 4: 2 (1978): 168. Emphasis in the original.

[54] Kennedy, "Finance," 58.

of "great power" than Germany before 1939.[55] During World War II, Britain produced more combat aircraft, artillery, mortars, and small arms than the Third Reich, and more tanks between 1940 and 1942.[56] Unlike during the Age of Coal, however, Britain no longer possessed the means to fuel its hydrocarbon-dependent platforms. Britain's descent to the status of U.S. client was due at least in part to its lack of energy security, which exacerbated the other economic, financial, and military challenges besetting Whitehall.[57]

Britain did, thanks largely to U.S. assistance, supply itself and its empire with petroleum during World War II. The scale of this accomplishment was enormous. British oil imports in 1944 were roughly double those of 1938—20.34 million tons, versus 11.61 million tons.[58] This success cannot, however, be judged in isolation—it must be compared with British expectations following World War I. Britain's survival between 1939 and 1945 does not mitigate the fact that Britain's strategy between the wars had failed to make Britain energy independent. Quite the contrary—by 1941, due to Britain's dependence on the Western Hemisphere and U.S. tankers, "[relations] with America became the key factor in determining how much oil Britain could import."[59] Luckily, Halifax had been correct in 1939—the United States could not allow Britain to fall, for that would allow the Axis to dominate Eurasia. Washington therefore supplied the means—oil not the least of them—that enabled the Allies to bludgeon the Axis into submission. Britain survived World War II, but not on terms of its own choosing, and now incapable of escaping the fate that awaited it: subordination to the United States. Germany would reach a similar destination but only after traveling a much harder road.

The notion that oil was Germany's Achilles' heel is a popular leitmotif among historians.[60] This was also the verdict of many contemporaries.[61] Energy supplies are a critical vulnerability of any modern society and war machine. No industrialized nation besides the United States and the Soviet Union could meet its entire wartime oil requirements from internal sources. Europe was doubly disadvantaged by having both minuscule local production and relatively high consumption. Before World War II, the deficit was as high as twenty-six million tons. At the beginning of 1936, excluding Romania, Poland, and the Soviet Union, the Continent's reserves totaled only ten million tons, against 24.4 million tons of annual demand annually among its great powers.[62]

[55] Edgerton, *War Machine*. [56] Harrison, "Economics," 15–16.

[57] Snyder, "Petroleum," 255–74, makes this point in passing.

[58] Edgerton, *War Machine*, 181–94; Payton-Smith, *Oil*, 483.

[59] Payton-Smith, *Oil*, 473–4.

[60] Collingham, *Taste*, 518, n. 96; Hayward, "Hitler's Quest," 96; Mazower, *Hitler's Europe*, 290; Steiner, *Triumph*, 674.

[61] Fritz Sternberg, *From Nazi Sources: Why Hitler Can't Win* (New York: Stratford Press, 1939), 103–10.

[62] Germany and Romania accounted for almost 90 percent of the 11.63 million tons produced in Europe in 1938. Statistical Office, "Rohstoffversorgung [...]," July 1940, T-84/70 (EAP 66-c-12/10); Statistical Office, "Rohstoffversorgung [...]," October 1940, T-84/69 (EAP 66-c-12/8a); Institute for Economic Research, "Rohstoffbilanz [...]," 1941, T-84/69 (EAP 66-c-12/4); Statistical Office, "Versorgung [...]," January 1942, T-84/69 (EAP 66-c-12/11/9).

Once Germany had adopted an aggressive foreign policy supported by a strategy of energy independence, its options were limited to synthetic fuel and imports from Romania. Choice is a luxury for great powers whose ambitions are commensurate with their means. Moreover, synthetics could only make Germany self-sufficient in lighter fuels such as gasoline and aviation fuel.[63] Fortunately, these were the primary fuels required by the army and air force. The navy ended up with the short end of the stick. It had never put much stock in the Third Reich's plans for self-sufficiency, preferring to put its faith in stockpiling.[64] These reserves quickly dwindled, however, once it had to share fuel with the army and Italy after 1940.[65]

Another drawback was expense. Besides assorted currency controls, taxes, and tariffs to keep synthetic fuel competitive with natural petroleum, limit imports denominated in foreign exchange, and slow the growth in consumption, the financial and material costs were enormous.[66] One German military journal estimated in 1936 that a nation with a force structure of 300 divisions (largely infantry with a mechanized core), 9,000 aircraft, and a navy composed of smaller craft and submarines would consume 12.65 million tons of petroleum annually in wartime. Substituting synthetic fuels for natural petroleum would require an additional forty million tons of coal production, a capital expenditure of four billion RM, and the labor of 250,000 men.[67] These figures turned out to be eerily prescient—the petroleum industry claimed 41 percent of the 13.25 billion RM Germany invested in boosting raw materials production between 1936 and 1942. This worked out to 50 percent of total industrial investment during that time span even though output by 1942 was only 45 percent of the prewar plan.[68] The only way to justify such an enormous investment was if it guaranteed appropriate long-term dividends.

After 1933, Germany's petroleum strategy mirrored its overall one. Both were essentially sequential strategies—a "series of visible, discrete steps, each dependent on the one that preceded it," aimed at securing incremental gains. That is, each foreign policy or military success would lay the foundation for accomplishing the next objective.[69] Germany's economic position on the eve of World War II was not ideal. In certain respects, it was better prepared than in 1914—but it was also still dependent on imports it could neither afford nor import safely.[70] The prospects for petroleum supplies appeared, at first glance, especially grim. Despite massive investments of labor, raw materials, and intermediate goods required by other

63 "Produktion [...]," *Wehrtechnische Monatshefte* 2 (1936), T-77/425 (Wi/IF 5.3444).

64 Wulf, "Die Ölversorgung [...]," no date, T-1022/4076 (PG 33957).

65 Grossman to Naval Operations Staff, December 6, 1939, T-1022/1735 (PG 32198b); Naval Defense Economy Office, "Bedeutung [...]," August 5, 1940, T-1022/1772 (PG 32209); Toprani, "German Navy," 107.

66 Between 1930 and 1936, Germany quadrupled the duty on imported fuel. Stranges, "Fuel Industry," 159 and 162. By 1936, gasoline taxes provided one-third of Germany's customs revenue (421 million RM). Tooze, *Wages*, 151.

67 Steinberger, "Der Treibstoffverbrauch [...]," *Die Deutsche Volkskraft* (January 16, 1936).

68 Petzina, "Vierjahresplan," 79.

69 Gerhard Weinberg, "22 June 1941," 227; Wylie, *Strategy*, 119.

70 Compare Hedler, "Deutsche Rohstofflage [...]," *Militär-Wochenblatt* (February 25, 1936), with Tomberg, "Deutschlands Wehrwirtschaftspotential [...]," February 26, 1940, T-77/425 (Wi/IF 5.3442).

branches of the war economy, the synthetic fuel industry could only satisfy a fraction of Germany's requirements, which were already disproportionately low. In 1938, Germany ranked twenty-seventh in global per-capita consumption, at 0.67 bbl—one-thirteenth that of the United States.[71] Meanwhile, Germany's petroleum production that year was only one-fifteenth that of the United States.[72]

Nevertheless, there was a cautious optimism that the Third Reich had enough petroleum to fight a major war if it had access to Romania. German bluster in 1939 and 1940 that the country's petroleum position was stronger than that of Britain and France had some merit.[73] Oil consumption on the eve of the war was small, comprising only 3 percent of the country's energy requirements, compared to 90 percent for coal.[74] Although certain kinds of petroleum products—mainly naval fuel oil and diesel fuel—were in short supply, thanks to the prewar decision to eschew the creation of a large mechanized and motorized force, the army was not handicapped by a lack of fuel. As the Defense Economy Office observed in 1942, "serious difficulties" did not emerge until Operation Barbarossa.[75] Two years later, it concluded that "the supply situation until the end of 1940 proved on average to become progressively better."[76]

Such assessments downplayed the role of contingency. Before 1941, the Third Reich had been the beneficiary of some remarkably good fortune. For example, it could not have tripled crude oil production between 1938 and 1943 (from 885,000 tons to 1.88 million tons) without the timely annexation of Austria.[77] The most significant factor was the expansion of synthetic production, which was the product of considerable prewar planning and preparation. In a truly awesome feat of modern engineering and industrial mobilization, Germany expanded its domestic petroleum supply almost out of thin air. Synthetic production in 1939—1.28 million tons from fourteen synthetic plants—was greater than Germany's imports from Romania that year.[78] The magnitude of Germany's accomplishment should not be underestimated. The technical difficulties surrounding synthetic fuel were considerable—Japan's attempt between 1937 and 1944 to mass-produce synthetic fuel was a conspicuous failure.[79] One should also not overlook the Germans' success in finding alternative foreign suppliers. In 1938, Germany imported 5.1 million tons of petroleum, half of which came from the United States and

[71] USSBS, *Oil Industry*, 7.

[72] 7.5 million tons versus 164 million tons. USSBS, *Over-All Report (European War)*, (Washington, DC: GPO, 1945), 39.

[73] Mende, "Zur Treibstofflage," *Deutsche Wehr* (February 9, 1940).

[74] Stokes, "Oil Industry," 254.

[75] Defense Economy Office, "Die deutsche Treibstoffversorgung [...]," February 16, 1942, T-77/668 (Wi/VI. 216). Emphasis in the original. See also: Hedler, "Die Mineralöle [...]," August 31, 1941, T-77/438 (Wi/IF 5.2726).

[76] Armed Forces High Command, "Die Arbeiten [...]," December 1944, T-77/183 (Wi/IF 5.751).

[77] German Institute for Economic Research, "Erdölgewinnung [...]," October 1944, T-77/341 (Wi/IF 5.2164, 2687).

[78] Stranges, "Fuel Industry," 162.

[79] Anthony Stranges, "Synthetic Fuel Production in Prewar and World War II Japan: A Case Study in Technological Failure," *Annals of Science* 50: 3 (1993): 229–65.

Venezuela, to satisfy 7.12 million tons of consumption.[80] After World War II began, Germany transitioned to relying exclusively on internal production and overland imports from Romania and (until June 1941) the Soviet Union.

This was not much compared to Allied supplies, but it was enough for Germany to exploit its relative military advantage in the period 1939–41, even though the results never lived up to expectations.[81] Under the best-case scenario, Germany would produce eleven million tons of petroleum by 1944.[82] In 1943, the United States produced 228 million tons of liquid hydrocarbons, and its *daily* average (625,000 tons) was two-thirds of Germany's projected *monthly* average.[83] Thomas later complained that the Third Reich never took the expansion of Germany's raw materials production seriously enough, preferring to focus on weapons production. "[The] day will come," he warned Göring in July 1942, "when responsibility for the inadequate expansion of petroleum production will be sought."[84] At war's end, Thomas was still railing at those who failed to grasp "that the provision of necessary fuels can be many times more important as the production of arms and munitions."[85]

The results were nonetheless remarkable. In 1936, Germany produced 1.55 million tons of petroleum—two years later, this figure had increased by less than 50 percent, but after 1938, production tripled, from 2.28 million tons to 7.6 million tons by 1943. The regime had "transformed" Germany's petroleum position.[86] By 1941, Germany was producing 75 percent of its requirements, as opposed to 31 percent in 1936.[87] During the first half of the war, Germany was self-sufficient in aviation fuel and certain lubricants.[88] By 1944, twelve hydrogenation plants were producing at a rate of three million tons per annum, two-thirds of which was aviation fuel—95 percent of the air force's requirements.[89] Germany achieved these results despite its resource scarcity. Except for the first three-quarters of 1939, the synthetic fuel industry never received its full steel allocation for plant construction and expansion.[90] Although synthetic output and imports from Romania were consistently below target, higher domestic crude oil production, booty, new sources such as Austria and Hungary, rationing, prewar stockpiles, and lower consumption

[80] "Mineralöl-Statistik," no date, T-77/658 (Wi/VI.109).

[81] Although the Germans never gave up on the Krauch Plan—they still planned to reach their production targets until the Allied strategic bombing campaign of 1944. "Probleme [...]," February 14, 1944, no author, T-77/90 (Wi/IF 5.399).

[82] Office for Economic Development, "Neuer Mineralölplan," July 12, 1938, T-77/423 (Wi/IF 5.3398). Production in 1943 was 7.6 million tons. After adding imports and booty, the total supply was ten million tons. Jonas Scherner, "Bericht zur deutschen Wirtschaftslage 1943/44," *Vierteljahrshefte für Zeitgeschichte* 55: 3 (2007): 522.

[83] DeGolyer, *Historical Data*, 20.

[84] Thomas, memorandum, July 6, 1942, IWM, FD 4809/45.

[85] Thomas, *Geschichte*, 111–15 (quotation from p. 114).

[86] Office for Economic Development, "Die deutsche Rohstoffversorgung [...]," June 20, 1940, T-84/216 (EAP 66-c-12-62/29).

[87] Four-Year Plan, "Chemischer Erzeugungsplan," June 10, 1942, T-77/649 (Wi/VI. 34).

[88] Krauch, "Die Kriegsleistungen [...]," *Vierjahresplan*, 1942/I.

[89] Stranges, "Fuel Industry," 207.

[90] Office for Economic Development, "Treibstoffversorgung," January 15, 1940, T-84/216 (EAP 66-c-12-62/29).

as the tide of war turned against the Axis meant that German supplies sufficed until the spring of 1944 (see Map 8).

During the war, German officials never wavered from their belief that the Third Reich's most important source of petroleum was synthetic fuel.[91] The second Four-Year Plan of September 1940 sought nothing less than "the systematic rationalization and mechanization of the entirety of German economic life."[92] Besides continuing to stockpile enough for six months of consumption—4.9 million tons—Germany would produce more than 500,000 tons per month, against a future mobilization demand of 845,000 tons.[93] The major point of departure between the two Four-Year Plans was their long-term objective. The first was only supposed to satisfy Germany's needs, whereas the second aimed to meet Axis Europe's requirements "in the struggle against the Anglo-Saxon world and Soviet Russia [...]."[94]

Germany did not put all of its eggs in the synthetic basket—to believe so would be to confuse the propaganda of the Four-Year Plan with reality. The system in place by 1939—synthetics, stockpiling, and overland imports—was the product of technological fortitude and improvisation, with a dollop of optimism. Its defining characteristic was its instrumentalism—it was simply a means to an end.[95] The objectives of National Socialism—most importantly the acquisition of "living space"—were fixed, but in terms of execution, the Third Reich displayed considerable flexibility and created a unique economic system to achieve its aims.[96] Opportunism—the capacity to seize unexpected opportunities—was, after all, at the heart of National Socialist economic thinking.[97]

The fact that German petroleum supplies were "inadequate" to fight a protracted war is irrelevant, because that is not the sort of conflict the Germans expected to fight.[98] The more pertinent question is, were Germany's supplies sufficient to achieve its short-term objectives? The answer is, yes. The fact that Germany's prewar strategy could not deliver self-sufficiency does not mean it failed. Rather, German leaders adopted strategies that enabled their military to compensate for Germany's economic inferiority. While it fell short of making Germany energy independent, synthetic fuel contributed tremendously to German war-making capacity by limiting oil imports, which conserved foreign exchange—in short supply due to rearmament. It also stimulated demand for steel and coal, while complementing the production of ammonia, methanol, and synthetic rubber. Investments in synthetic fuel, combined with expanded trade with Eastern Europe, also invalidated British and French assumptions concerning Germany's vulnerability to economic warfare.[99]

[91] Ritter, "Mineralöl," July 21, 1942, BA-B, R 3112/175.

[92] Körner, "Der zweite Vierjahresplan," *Vierjahresplan*, 1941/I, III. Hitler renewed Göring's mandate in October 1940. Hitler, "Decree... 18 October 1940," *NMT*, xii: 535.

[93] Office for Economic Development, "Gesichtspunkte [...]," September 5, 1940, NARA, RG 238, T-301/73 (NI-8845).

[94] Krauch, "Die Kriegsleistungen [...]," *Vierjahresplan*, 1942/I.

[95] Karlsch, *Faktor Öl*, 243.

[96] Barkai, *Nazi Economics*, 1–20.

[97] Barkai, "Grundanschauungen," 384.

[98] Contra Krammer, "Fueling," 404.

[99] Kennedy, "Mahan," 59–61.

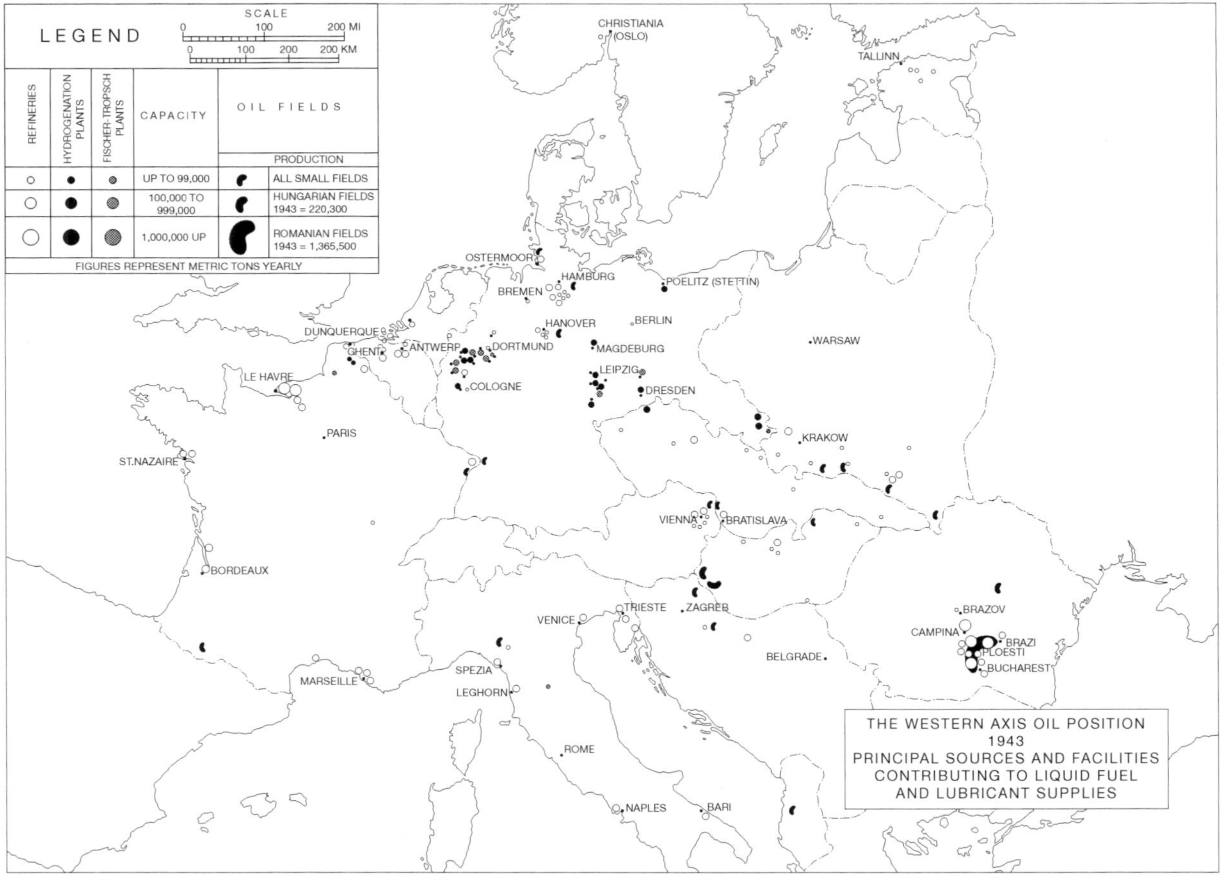

Map 8. "The Western Axis Oil Position," 1943

Source: NARA, RG 59, Lot File 77D141, PED, Box 27.

The German petroleum strategy was clearly successful at fueling a limited war of short duration. What it could not do was sustain an empire after June 1940, when Germany found itself "responsible for the oil supplies of practically the whole European Continent."[100] At the moment of its greatest triumph, Germany came down with a case of "imperial overstretch," having become responsible for waging a protracted air/sea campaign against Britain (backed by the United States) while supporting resource-poor allies and occupied Europe. In the final judgment, synthetics, along with Romania, enabled Germany to wage war but not to win it.

Following the war, German generals scoffed at analysts who urged the conquest of the Caucasus in 1942, claiming that Germany "managed to carry on the war until 1945 without ever securing the Caucasus oil."[101] In fact, the German war effort was running on fumes even before the final collapse. Starting in the spring of 1944, the Allies stripped Germany of its remaining sources of petroleum. The synthetic facilities were tempting targets now that the Allies had a capability to wage sustained aerial campaigns to obliterate individual sites. Bombing could disrupt production even indirectly, through attacks on railways and distribution hubs that supplied the synthetic fuel facilities.[102] Bombing did not destroy German refining—capacity was halved between March and December 1944—but it proved devastating to the synthetic fuel plants, where output fell from a high of 348,000 tons in April 1944 to 26,000 tons five months later. Combined with efforts against Romania, German supplies dropped from 968,000 tons in March 1944 to 281,000 tons in October. Raids against hydrogenation facilities also reduced supplies of ammonia, essential for both munitions and fertilizer production, by 75 percent between March and December 1944. Finally, bombing created a negative feedback loop by reducing training for German pilots, which in turn affected the air force's ability to defend the synthetic plants.[103]

Historians still debate strategic bombing's value, but neither German nor U.S. military leaders had any doubts as far as petroleum was concerned. During his first postwar debriefing, Göring claimed that Allied attacks against Germany's synthetic fuel industry and the transportation sectors had been crippling.[104] General Carl Spaatz, commander of the U.S. bombing effort in Europe, conceded the benefits of targeting the German transportation infrastructure but warned in March 1944 that it would neither contribute to the success of Operation Overlord (the Allied invasion of France) nor enable the destruction of the German Air Force.

[100] Levy, "Oil and War," May 9 (or September 5), 1941, NARA, Record Group 169: Records of the Foreign Economic Administration, Entry 360, Box 2191, and "Oil Drives Hitler's Army," March 1942, Levy Papers, Box 1.

[101] B.H. Liddell Hart, *Other Side of the Hill: Germany's Generals, Their Rise and Fall, with Their Own Account of Military Events, 1939–1945* (London: Cassell, 1973), 305–6.

[102] Alfred Mierzejewski, *Collapse of the German War Economy, 1944–1945* (Chapel Hill: University of North Carolina Press, 1988), xii–xiv and 177–87.

[103] United States Air Forces in Europe, "Contribution of Air Power to the Defeat of Germany," August 7, 1945, LOC, Papers of Carl Spaatz, Boxes 272–3; USSBS, *Defeat of the German Air Force* (Washington, DC: GPO, 1945), *Effects*, 73–83, *Oil Division*, 1–3, 14ff and 74ff, *Oil Industry*, 51–68, and *Over-All Report*, 39–45; Yergin, *Prize*, 346–8.

[104] Warburg, "Rise and Fall [...]," June 1, 1945, CARL.

Accomplishing the latter required destroying Germany's fighter strength, but it was unlikely that the Germans would oblige the Americans simply for the sake of "marshalling yards or rail installations." The Americans needed a target the Germans would "defend at all costs." The best option was the synthetic fuel industry. "We believe," Spaatz concluded, that the Germans "will defend oil to their last fighter plane."[105] The apostles of strategic bombing had been wrong about many things, but not this.

OIL AND WORLD POWER SINCE WORLD WAR II

Whereas the Great War was largely fought and won with the weapons of the nineteenth century, the following world war was a contest between twentieth-century technologies. Much like World War I, however, the outcome of World War II owed more to the Allies' economic and financial superiority than to their military performance. By 1942, the United States, Soviet Union, and Britain boasted a combined GDP (adjusted in 1990 dollars) double that of Germany, Japan, and Italy—$1.9 trillion versus $903 billion. Meanwhile, the 1938 GDP of pro-Allied neutrals was 1.7 times greater than that of pro-Axis ones—$259.4 billion versus $151 billion. The Allies' numerical advantage in soldiers in 1942—1.5 times on the Eastern Front and 1.1 times on the Western and Pacific fronts—was not significant. The real disparity was in materiel output. Leaving aside qualitative differences, between 1942 and 1944, the United States, Britain, and Soviet Union produced three times as many rifles, machine guns, artillery pieces, and combat aircraft as the Axis, and five times as many tanks and naval vessels.[106]

The historical consensus today is that "[the] Allies won the Second World War because they turned their economic strength into effective fighting power."[107] Whether one credits the Allies' victory primarily to the contributions of the Red Army, which inflicted the bulk of German casualties, or to those of Anglo-American naval and air forces, which destroyed the greatest share of German war production, neither could have accomplished their missions without copious supplies of petroleum.[108]

"It would be wrong to argue that oil determined the outcome of the war on its own," one historian concludes, "though there could scarcely have been a resource more vital to waging modern combat."[109] Oil played this special role because it enabled the Allies to "bridge" their economic strength and fighting power.[110] Once again, the United States played a decisive role as both the arsenal and gas station of

[105] Spaatz, "Employment [...]," March 24, 1944, Spaatz Papers, Box 144.

[106] Harrison, "Economics," 1–42.

[107] Richard Overy, *Why the Allies Won* (New York: W.W. Norton, 1995), 325.

[108] A significant oversight in Phillips Payson O'Brien, *How the War Was Won: Air-Sea Power and Allied Victory in World War II* (Cambridge: Cambridge University Press, 2015), 1–94.

[109] Overy, *Why*, 233.

[110] Henry Eccles, *Logistics in the National Defense* (Harrisburg: Stackpole Company, 1959), 41, 46–7, and 52–6.

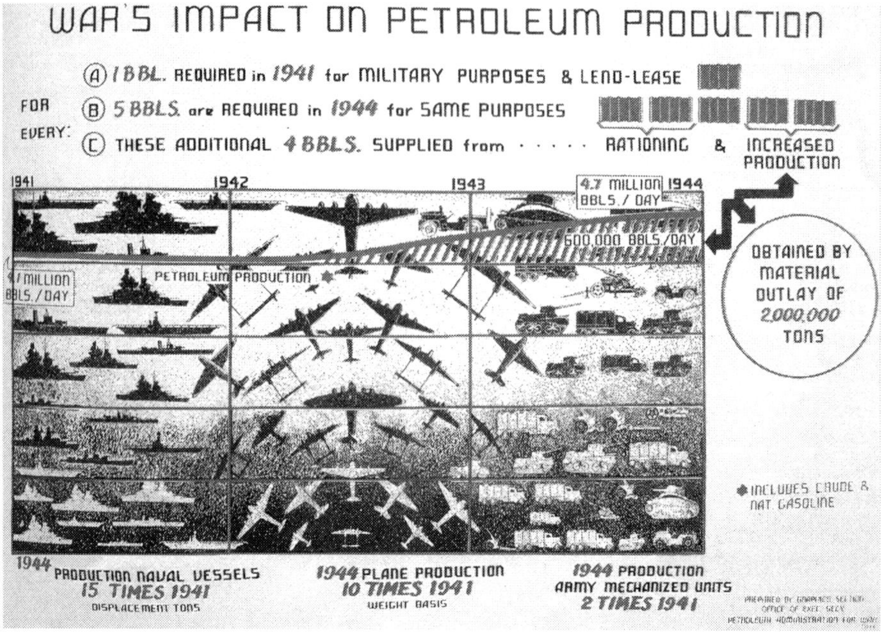

Illustration 12. "War's Impact on Petroleum Production," no date (circa 1945)
Source: NARA, RG 59, 800.6363/2-145.

democracy—quintupling oil supplies for military and Lend-Lease consumption between 1941 and 1944 even as the production of ships, planes, and vehicles soared (see Illustration 12).

The Allies' quantitative edge did not count for much during the conflict's early stages due to the Axis' qualitative advantages, but it did give the Allies the ability to survive their early defeats and learn from their mistakes.[111] Germany and Japan both gambled that operational/tactical superiority, plus a head start at rearmament, would guarantee a quick victory before their adversaries could mobilize and unite.[112] Once these plans failed, the Axis countries were in dire straits. Not only did they lack the economic means to confront their opponents on anything approaching equal terms after 1942, but the Allies also gradually narrowed the qualitative disparity. As Churchill remarked, after Pearl Harbor, the victory of the Grand Alliance depended "merely [upon] the proper application of overwhelming force."[113] Economic and financial power was not distributed equally among the

[111] Raymond Goldsmith, "The Power of Victory: Munitions Output in World War II," *Military Affairs* 10: 1 (1946): 69–80.

[112] Geyer, "German Strategy," 575.

[113] Kennedy, *Rise*, 347–57 (quotation from p. 347). For a detailed study of how the Allies used their materiel superiority to bludgeon the Axis into submission, see: John Ellis, *Brute Force: Allied Strategy and Tactics in the Second World War* (London: Deutsch, 1990).

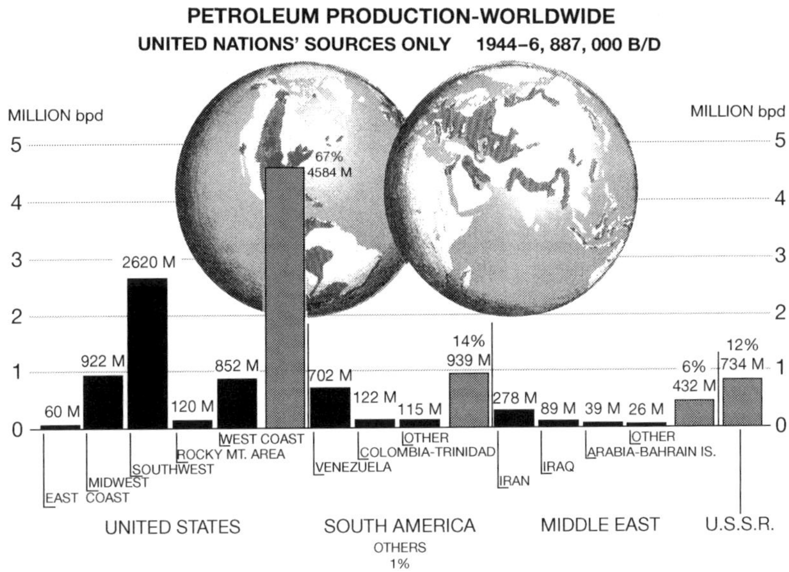

Illustration 13. "Petroleum Production—Worldwide," 22 February 1945
Source: NARA, RG 59, 800.6363/5-2645.

combatants, and while the Allies enjoyed superiority across the spectrum, nowhere was the difference as wide as in "energy supplies, the most basic driver of modern urban and industrial society" (see Illustration 13).[114] German analysts ruefully concluded in 1944 that Allied petroleum supplies were so plentiful that the only meaningful challenge they faced was a logistical one of getting enough fuel to the various battlefields of the European theater.[115]

They had no one but themselves to blame for that predicament. In hindsight, the idea that a resource-starved and relatively poor nation like Germany could defeat a global coalition comprising the British Empire, the Soviet Union, and the United States appears preposterous.[116] Short of a miracle, there was no way for Germany to reconcile its "vast ambition and the modest military and economic means at its disposal."[117] Even assuming Germany made more efficient use of the economic resources of Axis Europe, there was an insuperable gulf between Germany's capacity to produce more weapons and its ability to utilize them due to

[114] Tooze, *Wages*, 411–12 and 666 (quotation from p. 411).
[115] Air Force High Command, "Die Mineralölversorgung [...]," December 25, 1944, NARA, RG 242, National Archives Microfilm Publication T-321, Reel 236 (OKL 854).
[116] For Germany's relative economic backwardness, see: Tooze, *Wages*, 135–45. For Britain's advantages, see: Edgerton, *War Machine*, 11–85. For appraisals of Germany's prospects, see: Boog, *Global War*, 110–44 and 843–1215; Tooze, *Wages*, 368–512; Bernd Wegner, "The Road to Defeat: The German Campaigns in Russia, 1941–1943," *Journal of Strategic Studies* 13: 1 (1990): 105–27.
[117] Adam Tooze, "No Room for Miracles: German Industrial Output in World War II Reassessed," *Geschichte und Gesellschaft* 31 (2005): 439–64 (quotation from p. 464).

its shortage of petroleum. Notwithstanding assertions that "[by] the summer of 1942 the resources of the area under the military control of the Axis Powers were in no way substantially inferior to those controlled by the Allies," Germany commanded only a fraction of the petroleum at its adversaries' disposal.[118] In 1943, Germany's total supply peaked at eighty million bbl.[119] Production in the U.S. that year was 1.51 billion bbl—133 times or more than two orders of magnitude greater.[120] The picture did not change when both alliances were included: in 1941, the Anglo-American coalition produced twenty-two times more petroleum than Axis Europe—78.9 percent versus 3.5 percent of global production (295 million tons versus 10.4 million tons). Even if the Axis occupied the Caucasus, the deficit would still have been 5.4:1.[121]

The findings of this study demonstrate that the quest for energy independence can be counterproductive and have deleterious strategic consequences. Accordingly, they are relevant even for a superpower such as the United States, which has never confronted the sorts of material or financial handicaps that Britain and Germany endured between 1914 and 1945.

At a time when economic nationalism is on the rise, it is worth recalling that, during the Cold War, the United States often gained more from economic interdependence than independence.[122] Encouraging the consumption of internal sources of oil by imposing import quotas between 1959 and 1973 not only accelerated the exhaustion of U.S. reserves but also harmed the United States' national interests. For example, limiting U.S. imports depressed international prices and kept domestic prices artificially high, thereby putting the U.S. oil industry at a disadvantage. Lower global prices also encouraged the formation of the Organization of Petroleum Exporting Countries in 1960, which wrested control over prices from private firms in the 1970s.[123]

Finally, the depletion of U.S. reserves undermined the United States' ability to play the role of "swing" producer—a country able to boost output quickly to compensate for supply disruptions elsewhere. On several occasions after World War II, most notably during the Iranian crisis of 1951–4, the Suez crisis of 1956–7, and the Arab Oil Embargo of 1967, the United States used its slack capacity to guarantee supplies for its friends and stabilize global prices. It could no longer play this role after 1970, when U.S. oil production peaked. This, in turn, reduced U.S. leverage over the global oil industry, leaving it at the mercy of price shocks in 1973–4 and 1979–80 that contributed to a decade of low economic growth and high inflation (stagflation).

Increasing dependence on oil imports may have bruised the United States' ego, but the effects were not entirely negative for U.S. foreign policy. Many oil-exporting

[118] Contra Gerhard Weinberg, "Some Myths of World War II," *Journal of Military History* 75: 3 (2011): 702.

[119] USSBS, *Effects*, 75. [120] DeGolyer, *Historical Data*, 5.

[121] Statistical Office, "Produktionskapazitäten [...]," February 1942, T-84/46 (EAP 66-c-2-34/7).

[122] The following discussion draws from: David Painter, "Oil and the American Century," *Journal of American History* 99: 1 (2012): 24–39.

[123] Painter, *Oil*, 206–7.

nations had pro-U.S. governments, which used their oil revenues to buy popular support and blunt the appeal of communism or radical nationalism. They also recycled their petrodollars by purchasing U.S. goods and services (especially weapons), depositing them with Western banks, and supporting U.S. strategic initiatives, as Saudi Arabia did by underwriting the U.S. covert campaign against the Soviet Union in Afghanistan after 1979.[124] These governments could also supply forward-deployed U.S. military forces with petroleum, which came in handy when a U.S.-led coalition evicted Iraq from Kuwait in 1991.

Conversely, if any nation had the capacity to thrive through autarky, it was the Soviet Union—boasting vast natural resources, ruled by a ruthless regime willing to accept lower living standards in exchange for greater industrial and military production, and animated by an ideology that valued self-sufficiency over trade with capitalist states. The outcome of the Cold War, however, proved otherwise. Every U.S. President since Richard Nixon has proclaimed energy independence to be a national-security priority. Nevertheless, despite its failure to restore its self-sufficiency, the United States recovered from the Vietnam War and the economic challenges of the 1970s to win the Cold War; meanwhile, the energy-rich Soviet Union failed to modernize its economy, bankrupted itself through defense spending and by propping up satellite regimes, and imploded after oil prices collapsed in the 1980s.[125]

The outcome of the U.S.–Soviet superpower rivalry of 1945–90 clearly supports the proposition that economic interdependence through trade is a better guarantor of national power than independence through autarky—a fact worth recalling in the context of the emerging Sino-American rivalry in East Asia. Since the early 2000s, China has adopted a strategy of "energy mercantilism" based on many of the same principles that Churchill identified in 1914. Like Britain's, China's approach stresses geographic dispersal of supplies, but it also leverages political economic relationships with oil and gas producers and companies across Eurasia, Africa, and South America to blunt the effects of any future U.S. sanctions or blockade.[126] Will China's strategy prove more successful than Britain's? Time will tell. What is not in doubt is that China cannot become a geopolitical peer of the United States while leaving its oil supplies at the mercy of the United States, even at the cost of dissipating resources in peripheral theaters such as Central Asia or putting China's energy future in the hands of unreliable suppliers such as Russia, Venezuela, and Iran.

Scholars often ascribe the decline of great powers to self-destructive decisions, or policies contributing to "imperial overstretch."[127] Europe's dependence on imported

[124] See also: David Painter, "From the Nixon Doctrine to the Carter Doctrine: Iran and the Geopolitics of Oil in the 1970s," in: *American Energy Policy in the 1970s*, ed. Robert Lifset (Norman: University of Oklahoma Press, 2014), 61–92.

[125] David Painter, "Oil and Geopolitics: The Oil Crises of the 1970s and the Cold War," *Historical Social Research* 39: 4 (2014): 186–208, and "From Linkage to Economic Warfare: Energy, Soviet-American Relations, and the End of the Cold War," in: Perović, Energy, 283–318.

[126] Jennifer Lind and Daryl Press, "Markets for Mercantilism? How China Secures Its Energy Supplies," *International Security* 42: 4 (2018): 170–204.

[127] Kennedy, *Rise*, xv–xvi.

oil between 1914 and 1945, like that of China today, however, "was a matter of geography and had nothing to do with politics."[128] Although at opposite ends of the political spectrum, Britain and Germany had at least one thing in common—if they wished to remain great powers, they had to have oil, no matter the cost. Recognizing the gravity of their predicament, both nations spent a generation trying to evade the handicaps imposed by their unfavorable geographic endowments. This random factor—the unequal distribution of global oil reserves—proved to be one of the key factors determining the outcome of the great-power rivalries that convulsed the world during the twentieth century. Resource availability is never a sufficient explanation for complex geopolitical outcomes, but it is a necessary one. The accelerating mechanization of society and warfare after World War I ensured that only great powers with assured access to oil—either their own or from reliable trading partners—could hope to impose their will upon their rivals with any expectation of success.

This history remains relevant to contemporary strategists. For the past forty years, politicians and pundits have trumpeted the value of energy independence without considering exactly how it will promote either U.S. energy security or national interests. After President George W. Bush pledged in 2006 to end the United States' "addiction" to oil imported from the Middle East, the CEO of Exxon—and later U.S. Secretary of State—Rex Tillerson sought in vain to convince the administration that energy security served U.S. interests better than independence.[129] Such arguments have usually fallen on deaf ears, but the stakes have grown since the "shale boom" of 2014 transformed the global oil industry.

Thanks to new production, U.S. oil production has rocketed back to the heights of the 1970s. Imports have meanwhile plummeted, even though annual U.S. petroleum consumption today is equal to total Allied consumption during all of World War II (roughly seven billion bbl). These developments have ignited spirited debates over whether to lift the 1975 ban on crude oil exports (repealed in 2015), reduce—if not close—the Strategic Petroleum Reserve, and withdraw from the Middle East. It seems fair to surmise that energy independence will not serve U.S. interests if it contributes to political and economic dysfunction in the Middle East, which continues to supply oil-poor Europe, and South and East Asia, or hinders efforts to slow anthropogenic climate change. Whatever the exact consequences, Britain and Germany's ill-fated bids for self-sufficiency should serve as reminders to the United States that energy independence is not synonymous with its security—energy or otherwise.

[128] Levy, *Oil*, 9.

[129] Steve Coll, *Private Empire: ExxonMobil and American Power* (Penguin: New York, 2012): 435–41.

Select Bibliography

ARCHIVES

Federal Republic of Germany
Bundesanstalt für Geowissenshaften und Rohstoffe, Hannover
Bundesarchiv, Berlin-Lichterfelde
Deutsche Bank, Frankfurt am Main
Politisches Archiv des Auswärtigen Amtes, Berlin

United Kingdom
Imperial War Museum, Duxford
National Archives, Kew
British Petroleum Archive, University of Warwick

United States of America
Franklin D. Roosevelt Library, Hyde Park, NY
Library of Congress, Washington, DC
National Archives and Records Administration, College Park, MD
Hoover Institution Archives, Stanford University, Palo Alto, CA
American Heritage Center, University of Wyoming, Laramie, WY
Sterling Memorial Library, Yale University, New Haven, CT

Digital archives
Combined Arms Research Library, U.S. Army Command and General Staff College: https://cgsc.contentdm.oclc.org/
Fischer-Tropsch Archive, Emerging Fuels Technology: http://www.fischer-tropsch.org/
Digital Churchill Archive, Churchill College, Cambridge: http://www.churchillarchive.com/

Microfilm collections
U.S. Military Intelligence Reports: France, 1919–1941. Frederick: University Publications of America, 1985.
U.S. Military Intelligence Reports: Germany, 1919–1941. Frederick: University Publications of America, 1983.

Governmental reports, official histories, and documentary collections
Auswärtiges Amt. *Weißbuch Nr. 6: Die Geheimakten des französischen Generalstabs.* Berlin: 1941.
Auswärtiges Amt. *Documents on German Foreign Policy, 1918–1945.* Washington, DC: GPO, 1949–1983.
Berndt, Alfred-Ingemar. *Gebt mir vier Jahre Zeit! Dokumente zum ersten Vierjahresplan des Führers.* München: Zentralverlag der NSDAP. Franz Eher Nachf., G.m.b.H., 1938.
DeGolyer and MacNaughton. *Twentieth Century Petroleum Statistics: Historical Data.* Dallas: DeGolyer and MacNaughton, 2004.

DeGolyer and MacNaughton. *Twentieth Century Petroleum Statistics*. Dallas: DeGolyer and MacNaughton, 2012.

Domarus, Max. *The Complete Hitler: A Digital Desktop Reference to His Speeches and Proclamations, 1932–1945*. Wauconda: Bolchazy-Carducci, 2007.

Eichholtz, Dietrich and Wolfgang Schumann. *Anatomie des Krieges: Neue Dokumente über die Rolle des deutschen Monopolkapitals bei der Vorbereitung und Durchführung des zweiten Weltkrieges*. Berlin: VEB Deutscher Verlag der Wissenschaften, 1969.

Frey, John W. and H. Chandler Ide. *History of the Petroleum Administration for War*. Washington, DC: GPO, 1946.

Greiner, Helmuth and Percy Ernst Schramm, ed. *Kriegstagebuch des Oberkommandos der Wehrmacht*. Frankfurt am Main: Bernard & Graefe, 1961–5.

International Military Tribunal. *Trial of the Major War Criminals*. Nuremberg, 1947–9.

Kimball, Warren F., ed. *Churchill and Roosevelt: The Complete Correspondence*. Princeton: Princeton University Press, 1984.

Lissance, Arnold, ed. *War Journal of Franz Halder*. 1950, <http://cgsc.contentdm.oclc.org/cdm/search/collection/p4013coll8/searchterm/War%20journal%20of%20Franz%20Halder/field/title/mode/all/conn/and>.

Militärgeschichtliches Forschungsamt. *Germany and the Second World War*. New York: Oxford University Press, 1990–2008.

Minuth, Karl-Heinz (Volume I) and Friedrich Hartmannsgruber (Volumes II–V), ed. *Akten der Reichskanzlei: Regierung Hitler, 1933–1938*. Boppard am Rhein: Boldt, 1983–2008.

Moll, Martin, ed. *Führer-Erlasse, 1939–1945*. Stuttgart: F. Steiner, 1997.

Nuernberg Military Tribunal. *Trials of the War Criminals*. Washington, DC: GPO, 1953.

Office of United States Chief of Counsel for Prosecution of Axis Criminality. *Nazi Conspiracy and Aggression*. Washington, DC: GPO, 1946.

Payton-Smith, D.J. *Oil: A Study of War-time Policy and Administration*. London: HMSO, 1971.

Reichsamt für Bodenforschung. *Erdöl in Vorderasien*. Berlin: 1942.

Reichstelle für Bodenforschung. *Erdöl in Europa und im Nahen Orient*. Berlin: 1940.

Reichstelle für Bodenforschung. *Erdöl in Rußland*. Berlin: 1941.

United Kingdom of Great Britain and Northern Ireland. *Correspondence between His Majesty's Government and the United States Ambassador Respecting Economic Rights in Mandated Territories*. Cmd. 1226. London: HMSO, 1921.

United Kingdom of Great Britain and Northern Ireland. *Correspondence with the Mexican government regarding the expropriation of oil properties in Mexico. Mexico, April 8 to May 20, 1938*. Cmd. 5758. London: HMSO, 1938.

United Kingdom of Great Britain and Northern Ireland. *Official History of the Ministry of Munitions*. East Sussex: Naval & Military Press, 2009.

U.S. Department of State. *Papers Relating to the Foreign Relations of the United States*. Washington, DC: GPO, 1870–1946.

U.S. Department of State. *Foreign Relations of the United States*. Washington, DC: GPO, 1948–.

U.S. Department of the Interior. *Oil Shale: An Historical, Technical, and Economic Study*. Washington, DC: GPO, 1924.

U.S. Department of the Navy. Office of Naval Intelligence. *Fuehrer Directives and other Top-Level Directives of the German Armed Forces*. Washington, DC: 1948.

U.S. Senate. *Bolshevist Movement in Russia*. 66th Congress, 2nd Session, Document No. 172. Washington, DC: GPO, 1920.

United States Strategic Bombing Survey. *Effects of Strategic Bombing on the German War Economy*. Washington, DC: GPO, 1945.

United States Strategic Bombing Survey. *German Oil Industry, Ministerial Report Team 78.* Washington, DC: GPO, 1945.
United States Strategic Bombing Survey. *Oil Division: Final Report.* Washington, DC: GPO, 1945.
United States Strategic Bombing Survey. *Over-All Report (European War).* Washington, DC: GPO, 1945.
United States Strategic Bombing Survey. *Oil in Japan's War.* Washington, DC: GPO, 1946.
United States Strategic Bombing Survey. *Defeat of the German Air Force.* Washington, DC: GPO, 1947.
Wrangel, Georg Freiherr von. *Deutschlands Kampf um Rohstoff-Freiheit.* Berlin: Propaganda-Verlag Paul Hochmuth, 1942.

SECONDARY SOURCES

Adams, Walter, James W. Brock, and John M. Blair. "Retarding the Development of Iraq's Oil Resources: An Episode in Oleaginous Diplomacy, 1927–1939." *Journal of Economic Issues* 27: 1 (1993): 69–93.
Alpert, Michael. "The Spanish Civil War and the Mediterranean." *Mediterranean Historical Review* 13: 1–2 (1998): 150–67.
Aly, Götz and Susanne Heim. *Architects of Annihilation: Auschwitz and the Logic of Destruction.* Princeton: Princeton University Press, 2002.
Anderson, Irvine H. *ARAMCO, the United States, and Saudi Arabia: A Study of the Dynamics of Foreign Oil Policy.* Princeton: Princeton University Press, 1981.
Avery, Peter, Gavin Hambly, and Charles Melville, ed. *From Nadir Shah to the Islamic Republic.* Vol. 7 of *The Cambridge History of Iran.* Cambridge: Cambridge University Press, 2008.
Baer, George. *Test Case: Italy, Ethiopia, and the League of Nations.* Stanford: Hoover Institution Press, 1976.
Bailey, Gavin. "The Narrow Margin of Criticality: The Question of the Supply of 100-Octane Fuel in the Battle of Britain." *English Historical Review* 123: 501 (2008): 394–411.
Bamberg, James H. *The Anglo-Iranian Years, 1928–1954.* Vol. 2 of *The History of the British Petroleum Company.* Cambridge: Cambridge University Press, 1994.
Barkai, Avraham. "Wirtschaftliche Grundanschauungen und Ziele der N.S.D.A.P.: Ein unveröffentlichtes Dokument aus dem Jahre 1931." *Jahrbuch des Instituts für Deutsche Geschichte* 7: 2 (1978): 355–85.
Barkai, Avraham. *Nazi Economics: Ideology, Theory, and Policy.* New Haven: Yale University Press, 1990.
Barnett, Correlli. *The Collapse of British Power.* London: Eyre Methuen, 1972.
Barnett, Correlli. *Strategy and Society: The Spenser Wilkinson Memorial Lecture, 1974.* Manchester: Manchester University Press, 1975.
Barnhart, Michael. "Japan's Economic Security and the Origins of the Pacific War." *Journal of Strategic Studies* 4: 2 (1981): 105–24.
Barnhart, Michael. *Japan Prepares for Total War: The Search for Economic Security, 1919–1941.* Ithaca: Cornell University Press, 1987.
Baumgart, Inge and Horst Benneckenstein. "Die Interessen der Deutschen Bank am mesopotamischen Erdöl." *Jahrbuch für Wirtschaftsgeschichte* 1 (1988): 49–65.
Baumgart, Winfried. *Deutsche Ostpolitik: Von Brest-Litowsk bis zum Ende des Ersten Weltkrieges.* Wien: Oldenbourg, 1966.

Baumgart, Winfried. "Das 'Kaspi-Unternehmen'—Größenwahn Ludendorffs oder Routineplanung des deutschen Generalstabs: Erster Teil." *Jahrbücher für Geschichte Osteuropas* 18: 1 (1970): 47–126.

Baumgart, Winfried. "Das 'Kaspi-Unternehmen'—Größenwahn Ludendorffs oder Routineplanung des deutschen Generalstabs: Zweiter Teil." *Jahrbücher für Geschichte Osteuropas* 18: 2 (1970): 231–78.

Beck, Peter J. "The Anglo-Persian Oil Dispute, 1932–1933." *Journal of Contemporary History* 9: 4 (1974): 123–51.

Beck, Peter J. " 'A Tedious and Perilous Controversy': Britain and the Settlement of the Mosul Dispute, 1918–1926." *Middle East Studies* 17: 2 (1981): 259.

Berridge, Geoff. *British Diplomacy in Turkey, 1583 to the Present: A Study in the Evolution of the Resident Embassy*. Leiden: Martinus Nijhoff Publishers, 2009.

Birkenfeld, Wolfgang. "Leuna 1933." *Tradition: Zeitschrift für Firmengeschichte und Unternehmerbiographie* 8 (1963): 97–112.

Birkenfeld, Wolfgang. *Der Synthetische Treibstoff, 1933–1945*. Göttingen: Musterschmidt-Verlag, 1964.

Birkenfeld, Wolfgang. "Stalin als Wirtschaftspartner Hitlers, 1939–1941." *Vierteljahrschrift für Sozial- und Wirtschaftsgeschichte* 53: 4 (1966): 477–510.

Blair, John M. *The Control of Oil*. New York: Pantheon Books, 1976.

Boyce, Robert W.D. *British Capitalism at the Crossroads, 1919–1932: A Study in Politics, Economics and International Relations*. Cambridge: Cambridge University Press, 1987.

Brew, Gregory. "In Search of 'Equitability': Sir John Cadman, Rezā Shah and the Cancellation of the D'Arcy Concession." *Iranian Studies* 50: 1 (2017): 125–48.

Brodie, Bernard. "American Security and Foreign Oil." *Foreign Policy Reports* 23: 24 (March 1, 1948): 298–311.

Brodie, Bernard. *Sea Power in the Machine Age*. New York: Greenwood, 1969.

Brown, Jonathan C. "Why Foreign Oil Companies Shifted Their Production from Mexico to Venezuela during the 1920s." *American Historical Review* 90: 2 (1985): 362–85.

Brown, Jonathan C. "Domestic Politics and Foreign Investment: British Development of Mexican Petroleum, 1889–1911." *Business History Review* 61 (1987): 387–416.

Brown, Jonathan C. *Oil and Revolution in Mexico*. Berkeley: University of California Press, 1993.

Brown, Jonathan C. and Alan Knight, ed. *The Mexican Petroleum Industry in the Twentieth Century*. Austin: University of Texas Press, 1992.

Brown, Warwick Michael. "The Royal Navy's Fuel Supplies, 1898–1939: The Transition from Coal to Oil." Ph.D. Dissertation, KCL, 2003.

Brown, Warwick Michael. "When Dreams Confront Reality: Replenishment at Sea in the Era of Coal." *International Journal of Naval History* 9: 1–3 (2010). Last modified December1, 2010.http://www.ijnhonline.org/2010/12/01/when-dreams-confront-reality-replenishment-at-sea-in-the-era-of-coal/.

Buchheim, Christoph. "Die Besetzten Länder im Dienste der deutschen Kriegswirtschaft während des Zweiten Weltkrieges: Ein Bericht der Forschungstelle für Wehrwirtschaft." *Vierteljahrshefte für Zeitgeschichte* 34: 1 (January 1986): 117–45.

Bud-Frierman, Lisa, Andrew Godley, and Judith Wale. "Weetman Pearson in Mexico and the Emergence of a British Oil Major, 1901–1919." *Business History Review* 84: 2 (2010): 275–300.

Burk, Kathleen. "Great Britain in the United States, 1917–1918: The Turning Point." *International History Review* 1: 2 (1979): 228–45.

Cain, P.J. and A.G. Hopkins. *British Imperialism: Innovation and Expansion, 1688–1914*. London: Longman, 1993.

Cain, P.J. and A.G. Hopkins. *British Imperialism: Crisis and Deconstruction, 1914–1990.* London: Longman, 1993.

Cameron, Norman and R.H. Stevens, trans. *Hitler's Table Talk: His Private Conversations.* New York: Enigma, 2000.

Carroll, Berenice A. *Design for Total War: Arms and Economics in the Third Reich.* The Hague: Mouton, 1968.

Cecil, Lamar J.R. "Coal for the Fleet That Had to Die." *American Historical Review* 69: 4 (1964): 990–1005.

Chandler, Alfred D. *Scale and Scope: The Dynamics of International Capitalism.* Cambridge: Harvard University Press, 1990.

Churchill, Winston S. *The World Crisis, 1911–1914.* Toronto: Macmillan, 1923.

Coll, Steve. *Private Empire: ExxonMobil and American Power.* New York: Penguin, 2012.

Collingham, Lizzie. *The Taste of War: World War Two and the Battle for Food.* London: Allen Lane, 2011.

Cooke, Ronald and Roy Nesbit. *Target, Hitler's Oil: Allied Attacks on German Oil Supplies, 1939–1945.* London: W. Kimber, 1985.

Corley, T.A.B. *History of the Burmah Oil Company, 1886–1924.* New York: Heinemann, 1983.

Corley, T.A.B. "Oil Corporations and Public Policy: A US–UK Comparison, 1900–1975." *Business and Economic History* 21 (1992): 138–46.

Czollek, Roswitha and Dietrich Eichholtz. "Die deutschen Monopole und der 22 Juni 1941: Dokumente zu Kriegszielen und Kriegsplanung führender Konzerne beim Überfall auf die Sowjetunion." *Zeitschrift für Geschichtswissenschaft*, 15: 1 (1967): 64–76.

Darwin, John. *The Empire Project: The Rise and Fall of the British World-System, 1830–1970.* Cambridge: Cambridge University Press, 2009.

Darwin, John. *Unfinished Empire: The Global Expansion of Britain.* New York: Bloomsbury Press, 2013.

Davenport, E.H. and Sidney Russell Cooke. *The Oil Trusts & Anglo-American Relations.* London: Macmillan and Co., Ltd, 1923.

Davie, H.G.W. "The Influence of Railways on Military Operations in the Russo-German War 1941–1945." *Journal of Slavic Military Studies* 30: 2 (2017): 321–46.

Deese, David A. "Oil, War, and Grand Strategy." *Orbis* 25: 3 (1981): 525–55.

Deist, Wilhelm, ed. *The German Military in the Age of Total War.* Leamington Spa: Berg Publishers, 1985.

Dennis, Michael Aaron. "Drilling for Dollars: The Making of U.S. Petroleum Reserve Estimates, 1921–1925." *Social Studies of Science* 15: 2 (1985): 241–65.

DeNovo, John. "Petroleum and the United States Navy before World War I." *Mississippi Valley Historical Review* 41: 4 (1955): 641–56.

DeNovo, John. "The Movement for an Aggressive American Oil Policy Abroad, 1918–1920." *American Historical Review* 61: 4 (1956): 854–76.

DeNovo, John. *American Interests and Politics in the Middle East, 1900–1939.* Minneapolis: University of Minnesota Press, 1963.

DiNardo, R.L., *Germany's Panzer Arm.* Westport: Greenwood Press, 1997.

DiNardo, R.L. and Austin Bay. "Horse-Drawn Transport in the German Army." *Journal of Contemporary History* 23: 1 (1988): 129–42.

Divine, Robert A. *The Illusion of Neutrality.* Chicago: University of Chicago Press, 1962.

Earle, Edward Mead. "The Turkish Petroleum Company—A Study in Oleaginous Diplomacy." *Political Science Quarterly* 39: 2 (1924): 265–77.

Eccles, Henry E. *Logistics in the National Defense.* Harrisburg: Stackpole Company, 1959.

Eckes, Alfred E. *The United States and the Global Struggle for Minerals.* Austin: University of Texas Press, 1979.

Edgerton, David. "The Decline of Declinism." *Business History Review* 71 (1997): 201–6.

Edgerton, David. *Warfare State: Britain, 1920–1970*. Cambridge: Cambridge University Press, 2006.

Edgerton, David. *Britain's War Machine: Weapons, Resources, and Experts in the Second World War*. New York: Oxford University Press, 2011.

Ediger, Volkan and John Bowlus. "A Farewell to King Coal: Geopolitics, Energy Security, and the Transition to Oil." *Historical Journal* (2018), <https://doi.org/10.1017/S0018246X18000109>.

Eichholtz, Dietrich. "Zum Anteil des IG-Farben-Konzerns an der Vorbereitung des zweiten Weltkriegs: Ein Dokument zur staatsmonopolistischen Kriegsplanung des faschistischen deutschen Imperialismus." *Jahrbuch für Wirtschaftsgeschichte* 2 (1969): 83–105.

Eichholtz, Dietrich. "Der Raubzug des faschistischen deutschen Imperialismus zu den Erdölquellen des Kaukasus, 1941–1943." *Jahrbuch für Geschichte* 14 (1976): 445–502.

Eichholtz, Dietrich. "Die 'Großraumwehrwirtschaft' für den großen Krieg: Zwei geheime Memoranden der Reichstelle für Wirschaftsausbau vom Früjahr/Sommer 1939." *Bulletin des Arbeitskreises "Zweiter Weltkrieg"* 1–4 (1986): 70–160.

Eichholtz, Dietrich. *Deutsche Politik und rumänisches Öl, 1938–1941: Eine Studie über Erdölimperialismus*. Leipzig: Leipziger Universitätsverlag, 2005.

Eichholtz, Dietrich. *Krieg um Öl: Ein Erdölimperium als deutsches Kriegsziel, 1938–1943*. Leipzig: Leipziger Universitätsverlag, 2006.

Eichholtz, Dietrich. "Öl, Krieg, Politik: Deutscher Ölimperialismus, 1933–1942/43." *Zeitschrift für Geschichtswissenschaft* 51: 6 (2006): 493–510.

Eichholtz, Dietrich. *Deutsche Ölpolitik im Zeitalter der Weltkriege*. Leipzig: Leipziger Universitätsverlag, 2010.

Eichholtz, Dietrich and Titus Kockel. *Von Krieg zu Krieg: Zwei Studien zur deutschen Erdölpolitik in der Zwischenkriegszeit*. Leipzig: Leipziger Universitätsverlag, 2008.

Ellis, John. *Brute Force: Allied Strategy and Tactics in the Second World War*. London: Deutsch, 1990.

Elm, Mostafa. *Oil, Power, and Principle: Iran's Oil Nationalization and Its Aftermath*. Syracuse: Syracuse University Press, 1992.

Ericson, Edward E. *Feeding the German Eagle: Soviet Economic Aid to Nazi Germany, 1933–1941*. Westport: Praeger, 1999.

Farnie, D.A. *East and West of Suez, 1854–1956*. Oxford: Clarendon Press, 1967.

Fensch, Dorothea and Olaf Groehler. "Imperialistische Ökonomie und militärische Strategie: Eine Denkschrift Wilhelm Groeners." *Zeitschrift für Geschichtswissenschaft* 19: 3 (1971): 1167–77.

Ferrier, Ronald W. *The Developing Years, 1901–1932*. Vol. 1 of *The History of the British Petroleum Company*. Cambridge: Cambridge University Press, 1982.

Ferrier, R.W. and A. Fursenko, ed. *Oil and the World Economy*. London: Routledge, 1989.

Fetzer, Fritz. *Ölpolitik der Großmächte unter kriegswirtschaftliche Gesichtspunkten: Das japanische Beispiel*. Hamburg: Hanseatische Verlagsanstalt, 1935.

Fink, Carole, Frank Hadler, and Tomasz Schramm, ed. *1956: European and Global Perspectives*. Leipzig: Leipziger Universitätsverlag, 2006.

Fischer, Fritz. *Germany's Aims in the First World War*. New York: W.W. Norton, 1967.

Fisher, John, Effie G.H. Pedaliui, and Richard Smith. *Foreign Office, Commerce and British Foreign Policy in the Twentieth Century*. London: Palgrave, 2016.

Fitzgerald, Edward Peter. "The Iraq Petroleum Company, Standard Oil of California, and the Contest for Eastern Arabia." *International History Review* 13: 3 (1991): 441–60.

Fitzgerald, Edward Peter. "Business Diplomacy: Walter Teagle, Jersey Standard, and the Anglo-French Pipeline Conflict in the Middle East, 1930–1931." *Business History Review* 67: 2 (1993): 207–45.

Fitzgerald, Edward Peter. "France's Middle Eastern Ambitions, the Sykes-Picot Negotiations, and the Oil Fields of Mosul, 1915–1918." *Journal of Modern History* 66: 4 (1994): 697–725.

Forbes, Neil. *Doing Business with the Nazis: Britain's Economic and Financial Relations with Germany, 1931–1939.* London: Cass, 2000.

Foreign Policy Association. *Great Decisions 2017.* New York: 2016.

Forstmeier, Friedrich and Hans-Erich Volkmann, ed. *Wirtschaft und Rüstung am Vorabend des Zweiten Weltkriegs.* Düsseldorf: Droste, 1975.

Frank, Alison. *Oil Empire: Visions of Prosperity in Austrian Galicia.* Cambridge: Harvard University Press, 2005.

Friedensburg, Ferdinand. *Das Erdöl im Weltkrieg.* Stuttgart: Ferdinand Enke Verlag, 1939.

Friedensburg, Ferdinand. *Rohstoffe und Energiequellen im neuen Europa.* Berlin: G. Stalling, 1943.

Fritz, Stephen G. *Ostkrieg: Hitler's War of Extermination in the East.* Lexington: University Press of Kentucky, 2011.

Fursenko, A.A. *The Battle for Oil: The Economics and Politics of International Petroleum Conflict over Petroleum, 1860–1930.* Greenwich: JAI Press, 1990.

Gadea, James. "'Ulei Română'" during World War II and Beyond: Development of the Romanian Oil Industry." *Yale Review of International Studies* 5: 1 (2014): 75–85.

Gall, Lothar, Gerald D. Feldman, Harold James, Carl-Ludwig Holtfrerich, and Hans E. Büschgen. *The Deutsche Bank, 1870–1995.* London: Weidenfeld & Nicolson, 1995.

Galpern, Steven G. *Money, Oil, and Empire in the Middle East: Sterling and Postwar Imperialism, 1944–1971.* Cambridge: Cambridge University Press, 2009.

Gantzel, Klaus Jürgen and Helmut Mejcher, ed. *Oil, the Middle East, North Africa and the Industrial States: Development and International Dimensions.* Paderborn: F. Schöningh, 1984.

Garner, Paul. *British Lions and Mexican Eagles: Business, Politics, and Empire in the Career of Weetman Pearson in Mexico, 1889–1919.* Stanford: Stanford University Press, 2011.

Geyer, Michael. *Aufrüstung oder Sicherheit: Die Reichswehr in der Krise der Machtpolitik.* Wiesbaden: Steiner, 1980.

Geyer, Michael. "The Crisis of Military Leadership in the 1930s." *Journal of Strategic Studies* 14: 4 (Dec. 1991): 448–62.

Geyer, Michael and Adam Tooze, ed. *Total War: Economy, Society, and Culture.* Vol. 3 of *Cambridge History of the Second World War.* Cambridge: Cambridge University Press, 2015.

Gharzarian, Matthew. "Obstruction as Power: Rethinking Britain's Caucasus Occupation, 1918–1920." *International History Review* 39: 4 (2017): 654–66.

Gibbons, Robert. "Opposition gegen 'Barbarossa' im Herbst 1940: Eine Denkschrift aus der deutschen Botschaft in Moskau." *Vierteljahrshefte für Zeitgeschichte* 23: 3 (1975): 332–40.

Gibson, William. *Britain's Quest for Oil: The First World War and the Peace Conferences.* Solihull: Helion, 2017.

Goldsmith, Raymond W. "The Power of Victory: Munitions Output in World War II." *Military Affairs* 10: 1 (1946): 69–80.

Goralski, Robert and Russell W. Freeburg. *Oil & War: How the Deadly Struggle for Fuel in WWII Meant Victory or Defeat.* New York: Morrow, 1987.

Grobba, Fritz. *Männer und Mächte im Orient: 25 Jahre diplomatischer Tätigkeit im Orient.* Göttingen: Musterschmidt, 1967.

Gruchmann, Lothar. "Die 'Verpassten strategischen Chancen' der Achsenmächte im Mittelmeerraum 1940/41." *Vierteljahrshefte für Zeitgeschichte* 18: 4 (1970): 456–75.

Gunther, Albert E. "The German War for Crude Oil in Europe, 1934–1945." *Petroleum Times* (1947–8).

Haber, Stephen, Noel Maurer, and Armando Razo. *The Politics of Property Rights: Political Instability, Credible Commitments, and Economic Growth in Mexico.* Cambridge: Cambridge University Press, 2003.

Haber, Stephen, Noel Maurer, and Armando Razo. "When the Law Does Not Matter: The Rise and Decline of the Mexican Oil Industry." *Journal of Economic History* 63: 1 (2003): 1–32.

Hall, Linda B. *Oil, Banks, and Politics: The United States and Postrevolutionary Mexico, 1917–1924.* Austin: University of Texas Press, 1995.

Hamlin, Daniel. *Germany's Empire in the East: Germans and Romania in an Era of Globalization and Total War.* Cambridge: Cambridge University Press, 2017.

Hammond, Richard. "An Enduring Influence on Imperial Defence and Grand Strategy: British Perceptions of the Italian Navy, 1935–1943." *International History Review* 39: 5 (2017): 810–35.

Harrison, Mark, ed. *The Economics of World War II: Six Great Powers in International Comparison.* Cambridge: Cambridge University Press, 1998.

Hattendorf, John B. *Naval Policy and Strategy in the Mediterranean: Past, Present and Future,* ed. John Hattendorf. London: Frank Cass, 2000.

Hauner, Milan. "Did Hitler Want a World Dominion?" *Journal of Contemporary History* 13: 1 (1978): 15–32.

Hauner, Milan. *India in Axis Strategy: Germany, Japan, and Indian Nationalists in the Second World War.* Stuttgart: Klett-Cotta, 1981.

Hayes, Peter. "Carl Bosch and Carl Krauch: Chemistry and the Political Economy of Germany, 1925–1945." *Journal of Economic History* 47: 2 (1987): 353–63.

Hayes, Peter. "Industrie und Ideologie: Die IG Farben in der Zeit des Nationalsozialismus." *Zeitschrift für Unternehmensgeschichte* 32: 2 (1987): 124–36.

Hayes, Peter. *Industry and Ideology: IG Farben in the Nazi Era.* Cambridge: Cambridge University Press, 1987.

Hayward, Joel. "Hitler's Quest for Oil: The Impact of Economic Considerations on Military Strategy." *Journal of Strategic Studies* 18: 4 (1995): 94–135.

Hayward, Joel. "Too Little, Too Late: An Analysis of Hitler's Failure in August 1942 to Damage Soviet Oil Production." *Journal of Military History* 64: 3 (2000): 769–94.

Herbst, Ludolf. *Der Totale Krieg und die Ordnung der Wirtschaft: Die Kriegswirtschaft im Spannungsfeld von Politik, Ideologie und Propaganda.* Stuttgart: Deutsche Verlags-Anstalt, 1982.

Herring, George C. *From Colony to Superpower: U.S. Foreign Relations since 1776.* Oxford: Oxford University Press, 2008.

Herwig, Holger H. "Prelude to *Weltblitzkrieg*: Germany's Naval Policy toward the United States of America, 1939–41." *Journal of Modern History* 43: 4 (1971): 649–68.

Herwig, Holger H. *"Luxury" Fleet: The Imperial German Navy, 1888–1918.* London: Ashfield, 1980.

Herwig, Holger H. "The German Reaction to the *Dreadnought* Revolution." *International History Review* 13: 2 (1991): 273–82.

Herwig, Holger H. *The First World War: Germany and Austria-Hungary, 1914–1918.* London: Bloomsbury, 2014.

Hillgruber, Andreas. *Hitler, König Carol und Marschall Antonescu: Die Deutsch-Rumänischen Beziehungen.* Wiesbaden: Franz Steiner Verlag, 1954.

Hillgruber, Andreas. *Hitlers Strategie: Politik und Kriegsführung, 1940–1941.* Frankfurt am Main: Bernard & Graefe Verlag für Wehrwesen, 1965.

Hillgruber, Andreas. *Deutsche Großmacht- und Weltpolitik im 19. und 20. Jahrhundert.* Düsseldorf: Droste, 1977.

Hoffmann, Peter. "The Gulf Region in German Strategic Projections, 1940–1942." *Militärgeschichtliche Mitteilungen* 44: 2 (1988): 61–73.

Hogan, Michael J. *Informal Entente: The Private Structure of Cooperation in Anglo-American Economic Diplomacy, 1918–1928.* Columbia: University of Missouri Press, 1977.

Howarth, Stephen and Joost Jonker. *Powering the Hydrocarbon Revolution, 1939–1973.* Vol. 2 of *History of Royal Dutch Shell.* Oxford: Oxford University Press, 2007.

Hughes, Thomas Parke. "Technological Momentum in History: Hydrogenation in Germany, 1898–1933." *Past & Present* 44: 1 (1969): 106–32.

Hugill, Peter J. "The American Challenge to British Hegemony, 1861–1947." *Geographical Review* 99: 3 (2009): 403–25.

Hunt, Barry and Adrian Preston, ed. *War Aims and Strategic Policy in the Great War, 1914–1918.* Totowa: Rowan and Littlefield, 1977.

Jack, Marian. "The Purchase of the British Government's Shares in the British Petroleum Company, 1912–1914." *Past & Present* 39 (1968): 139–68.

Jensen, W.G. "The Importance of Energy in the First and Second World Wars." *Historical Journal* 11: 3 (1968): 538–54.

Jersak, Tobias. "A Matter of Foreign Policy: 'Final Solution' and 'Final Victory' in Nazi Germany." *German Studies Review* 21: 3 (2003): 369–91.

Jones, Christopher. "A Landscape of Energy Abundance: Anthracite Coal Canals and the Roots of American Fossil Fuel Dependence, 1820–1860." *Environmental History* 15 (July 2010): 449–84.

Jones, Christopher. *Routes of Power: Energy and Modern America.* Cambridge: Harvard University Press, 2014.

Jones, Geoffrey G. "The British Government and the Oil Companies, 1912–1924: The Search for an Oil Policy." *Historical Journal* 20: 3 (1977): 647–72.

Jones, Geoffrey G. *The State and the Emergence of the British Oil Industry.* London: Macmillan, 1981.

Jones, Geoffrey and Maurice Kirby, ed. *Competitiveness and the State: Government and Business in Twentieth-Century Britain.* Manchester: Manchester University Press, 1991.

Jonker, Joost and Jan Luiten van Zanden. *From Challenger to Joint Industry Leader, 1890–1939.* Vol. 1 of *History of Royal Dutch Shell.* Oxford: Oxford University Press, 2007.

Kaiser, David E. *Economic Diplomacy and the Origins of the Second World War: Germany, Britain, France, and Eastern Europe, 1930–1939.* Princeton: Princeton University Press, 1980.

Kalicki, Jan H. and Goldwyn, David L., ed. *Energy and Security: Strategies for a World in Transition.* Washington, DC: Woodrow Wilson Press, 2013.

Karlsch, Rainer and Raymond G. Stokes. *Faktor Öl: Die Mineralölwirtschaft in Deutschland, 1859–1974.* München: C.H. Beck, 2003.

Kasper, Hanns-Heinz. "Das Erdöl in den Raubplänen des deutschen Faschismus in Vorbereitung und bei der Durchführung des Zweiten Weltkrieges." *Jahrbuch für Wirtschaftsgeschichte* 3 (1976): 55–77.

Kay, Alex J. *Exploitation, Resettlement, Mass Murder: Political and Economic Planning for German Occupation Policy in the Soviet Union, 1940–1941*. New York: Berghahn Books, 2006.

Kay, Alex J. "Germany's Staatssekretäre, Mass Starvation and the Meeting of 2 May 1941." *Journal of Contemporary History* 41: 4 (2006): 685–700.

Keating, Aileen. *Mirage: Power, Politics, and the Hidden History of Arabian Oil*. Amherst: Prometheus Books, 2005.

Kehrl, Hans. *Krisenmanager im Dritten Reich: 6 Jahre Frieden—6 Jahre Krieg, Erinnerungen*. Düsseldorf: Droste Verlag, 1973.

Kelanic, Rosemary A. "The Petroleum Paradox: Oil, Coercive Vulnerability, and Great Power Behavior." *Security Studies* 25: 2 (2016): 181–213.

Kennedy, Greg. *The Merchant Marine in International Affairs, 1850–1950*. London: Frank Cass, 2000.

Kennedy, Greg, ed. *Imperial Defence: The Old World Order, 1856–1956*. London: Routledge, 2008.

Kennedy, Paul. "Mahan versus Mackinder: Two Interpretations of British Sea Power." *Militärgeschichtliche Mitteilungen* 16 (1974): 39–66.

Kennedy, Paul. *The Rise and Fall of British Naval Mastery*. London: A. Lane, 1976.

Kennedy, Paul. " 'Appeasement' and British Defence Policy in the Inter-War Years." *British Journal of International Studies* 4: 2 (1978): 161–77.

Kennedy, Paul. "Strategy versus Finance in Twentieth-Century Great Britain." *International History Review* 3: 1 (1981): 44–61.

Kennedy, Paul. *The Rise and Fall of the Great Powers: Economic Change and Military Conflict from 1500 to 2000*. New York: Random House, 1987.

Kent, Marian. *Oil and Empire: British Policy and Mesopotamian Oil, 1900–1920*. London: Macmillan, 1976.

Kent, Marian. *Moguls and Mandarins: Oil, Imperialism, and the Middle East in British Foreign Policy, 1900–1940*. London: Frank Cass, 1993.

Kimball, Warren F. *The Juggler: Franklin Roosevelt as Wartime Statesman*. Princeton: Princeton University Press, 1991.

King, Robert and Robin Kilson, ed. *The Statecraft of British Imperialism: Essays in Honor of Wm. Roger Louis*. London: Frank Cass, 1999.

Kirby, M.W. *The Decline of British Economic Power since 1870*. London: Routledge, 1981.

Knowlton, Evelyn H. and George S. Gibb. *History of Standard Oil Company: Resurgent Years 1911–1927*. New York: Arno Press, 1956.

Knox, MacGregor. *Mussolini Unleashed, 1939–1941: Politics and Strategy in Fascist Italy's Last War*. Cambridge: Cambridge University Press, 1982.

Knox, MacGregor. "Conquest, Foreign and Domestic, in Fascist Italy and Nazi Germany." *Journal of Modern History* 56: 1 (1984): 1–57.

Knox, MacGregor. *Hitler's Italian Allies: Royal Armed Forces, Fascist Regime, and the War of 1940–43*. Cambridge: Cambridge University Press, 2000.

Knox, MacGregor. *To the Threshold of Power, 1922/33: Origins and Dynamics of the Fascist and National Socialist Dictatorships*. Cambridge: Cambridge University Press, 2007.

Knox, MacGregor and Williamson Murray, ed. *The Dynamics of Military Revolution, 1350–2050*. Cambridge: Cambridge University Press, 2001.

Koch, H.W. "Hitler's 'Programme' and the Genesis of Operation 'Barbarossa'." *Historical Journal* 26: 4 (1983): 891–920.

Kockel, Titus. "Geologie und Krieg: Die Erdölpolitik des Reichsamt für Bodenforschung und seiner Vorläufer 1933–1945." Magisterarbeit, Technische Universität Berlin, 1995.

Kockel, Titus. "Eine Quelle zur Vor- und Gründungsgeschichte der Kontinentale Öl AG aus dem Jahr 1940: E. R. Fischer (Reichswirtschaftsministerium, II Min. Öl), 'Die Versorgung Europas mit Mineraloel vor dem Kriege, Ermittlung des Nachkriegsverbrauchs und Sicherung der Belieferung, 1940,' September 1940." *Jahrbuch für Wirtschaftsgeschichte* 1 (2003): 175–208.

Kockel, Titus. "Geologie und deutsche Ölpolitik, 1928 bis 1938: Die frühe Karriere des Erdölgeologen Alfred Theodor Bentz." Ph.D. Dissertation, Technische Universität Berlin, 2003.

Koehl, Robert Lewis. "A Prelude to Hitler's Greater Germany." *American Historical Review* 59: 1 (October 1953): 43–65.

Koppes, Clayton R. "The Good Neighbor Policy and the Nationalization of Mexican Oil: A Reinterpretation." *Journal of American History* 69: 1 (1982): 62–81.

Krammer, Arnold. "Fueling the Third Reich." *Technology and Culture* 19: 3 (1978): 394–422.

Kuisel, Richard F. *Ernest Mercier: French Technocrat*. Berkeley: University of California Press, 1967.

Laqueur, Walter, ed. *Fascism: A Reader's Guide*. Middlesex: Penguin, 1976.

Larson, Henrietta M., Evelyn H. Knowlton, and Charles S. Popple. *New Horizons 1927–1950*. Vol. 3 of *History of Standard Oil Company (New Jersey)*. New York: Harper & Row, 1971.

Laux, James M. "Trucks in the West during the First World War." *Journal of Transport History* 6 (1985): 64–70.

Leach, Barry A. *German Strategy against Russia, 1939–1941*. Oxford: Oxford University Press, 1973.

Leatherdale, Clive. *Britain and Saudi Arabia, 1925–1939: The Imperial Oasis*. Frank Cass: Abingdon, 1983.

Leitz, Christian. "Arms as Levers: 'Matériel' and Raw Materials in Germany's Trade with Romania in the 1930s." *International History Review* 19: 2 (1997): 312–32.

Lemon, Edward J. "Dunsterforce or Dunsterfarce? Re-Evaluating the British Mission to Baku, 1918." *First World War Studies* 6: 2 (2015): 133–49.

Lesch, John, ed. *The German Chemical Industry in the Twentieth Century*. Dordrecht: Kluwer Academic Publishers, 2000.

Lesser, Ian. *Resources and Strategy*. Basingstoke: Macmillan, 1989.

Lesser, Ian. *Oil, the Persian Gulf, and Grand Strategy*. Santa Monica: Rand Corporation, 1991.

Levine, Alan J. "Was World War II a Near-Run Thing?" *Journal of Strategic Studies* 8:1 (1985): 38–63.

Levy, Walter J. *Oil Strategy and Politics, 1941–1981*. Boulder, Westview Press, 1982.

Liddell Hart, B.H., ed. *The Rommel Papers*. New York: Harcourt Bruce, 1953.

Liddell Hart, B.H. *The Other Side of the Hill: Germany's Generals, Their Rise and Fall, with Their Own Account of Military Events, 1939–1945*. London: Cassell, 1973.

Lifset, Robert, ed. *American Energy Policy in the 1970s*. Norman: University of Oklahoma Press, 2014.

Lind, Jennifer and Daryl G. Press. "Markets for Mercantilism? How China Secures Its Energy Supplies." *International Security* 42: 4 (2018): 170–204.

LoCicero, Michael, Ross Mahoney, and Stuart Mitchell, ed. *A Military Transformed? Adaptation and Innovation in the British Military, 1792–1945*. Helion: Solihull, 2014.

Ludendorff, Erich. *Meine Kriegserinnerungen, 1914–1918*. Berlin: Ernst Siegfried Mittler und Sohn, 1919.

Ludendorff, Erich. *Urkunden der Obersten Heeresleitung über ihre Tätigkeit, 1916/18*. Berlin: Mittler & Sohn, 1921.

McBeth, B.S. *Juan Vincente Gómez and the Oil Companies in Venezuela, 1908–1935.* Cambridge: Cambridge University Press, 1983.

McBeth, B.S. *British Oil Policy, 1919–1939.* London: Frank Cass, 1985.

McBeth, B.S. "Venezuela's Nascent Oil Industry and the U.S. Tariff on Crude Oil Imports, 1927–1935." *Journal of Iberian and Latin American History* 27: 3 (2009): 427–62.

McDonald, Andrew. "The Geddes Committee and the Formulation of Public Expenditure Policy, 1921–1922." *Historical Journal* 32: 3 (1989): 643–74.

MacDonald, C.A. "Economic Appeasement and the German 'Moderates', 1937–1939: An Introductory Essay." *Past & Present* 56 (1972): 105–35.

McMeekin, Sean. *The Ottoman Endgame: War, Revolution, and the Making of the Modern Middle East, 1908–1923.* New York: Penguin, 2015.

Mackinder, Halford. "The Geographical Pivot of History." *The Geographical Journal* 23: 4 (1904): 421–37.

Maiolo, Joseph. "Did the Royal Navy Decline between the Two World Wars?" *RUSI Journal* 159: 4 (2014): 18–24.

Manegold, Karl-Heinz, ed. *Wissenschaft, Wirtschaft und Technik: Studien zur Geschichte.* München: Bruckmann, 1969.

Mar Rubio, María del. "The Role of Mexico in the First World War Oil Shortage: 1918–1922, An International Perspective." *Journal of Iberian and Latin American Economic History* 24: 1 (2006): 69–95.

Marder, Arthur. *From the Dreadnought to Scapa Flow: The Royal Navy in the Fisher Era.* London: Oxford University Press, 1961–70, 1978.

Marder, Arthur. "The Royal Navy and the Ethiopian Crisis of 1935–36." *Journal of American History* 75: 5 (1970): 1327–56.

Martel, Gordon. "The Meaning of Power: Rethinking the Decline and Fall of Great Britain." *International History Review* 13: 4 (1991): 662–94.

Martel, Gordon, ed. *The Origins of the Second World War Reconsidered.* London: Routledge, 1999.

Maurer, John H. "Fuel and the Battle Fleet: Coal, Oil, and American Naval Strategy, 1898–1925." *Naval War College Review* 34: 6 (1981): 60–77.

Maurer, Noel. "The Empire Struck Back: Sanctions and Compensation in the Mexican Oil Expropriation of 1938." *Journal of Economic History* 71: 3 (2011): 590–615.

Mazower, Mark. *Hitler's Europe: How the Nazis Ruled Europe.* New York: Penguin, 2008.

Meier-Dörnberg, Wilhelm. *Die Mineralöversorgung der Kriegsmarine 1935 bis 1945.* Freiburg: Rombach, 1973.

Meierding, Emily. "Dismantling the Oil Wars Myth." *Security Studies* 25: 2 (2016): 258–88.

Mejcher, Helmut. "Oil and British Policy towards Mesopotamia, 1914–1918." *Middle Eastern Studies* 8: 3 (1972): 377–91.

Mejcher, Helmut. *Imperial Quest for Oil: Iraq, 1910–1928.* London: Ithaca Press, 1976.

Mejcher, Helmut. *Die Politik und das Öl im Nahen Osten.* Stuttgart: Klett-Cotta, 1980, 1990.

Mejcher, Helmut. "North Africa in the Strategy and Politics of the Axis Powers, 1936–1943." *Les Cahiers de Tunisie* 117–118: 3–4 (1981): 629–48.

Melby, Eric. *Oil and the International System: The Case of France, 1918–1969.* New York: Arno Press, 1981.

Melosi, Martin V. *Coping with Abundance: Energy and Environment in Industrial America.* Philadelphia: Temple University Press, 1985.

Meyer, Lorenzo. *Mexico and the United States in the Oil Controversy, 1917–1942.* Austin: University of Texas Press, 1977.

Michaelis, Meir. "World Power Status or World Dominion? A Survey of the Literature on Hitler's 'Plan of World Dominion' (1937–1940)." *Historical Journal* 15: 2 (1972): 331–60.
Michalka, Wolfgang, ed. *Nationalsozialistische Aussenpolitik*. Darmstadt: Wissenschaftliche Buchgesellschaft, 1978.
Mierzejewski, Alfred C. *The Collapse of the German War Economy, 1944–1945*. Chapel Hill: University of North Carolina Press, 1988.
Millet, Alan and Williamson Murray, ed. *Military Effectiveness*. Boston: Allen & Unwin, 1988.
Millet, Alan and Williamson Murray, ed. *Calculations: Net Assessment and the Coming of World War II*. New York: Free Press, 1992.
Mitchell, Timothy. "Carbon Democracy." *Economy and Society* 38: 3 (2009): 399–432.
Mitchell, Timothy. *Carbon Democracy: Political Power in the Age of Oil*. New York: Verso, 2011.
Mohr, Anton. *The Oil War*. New York: Harcourt, Brace and Co., 1926.
Monroe, Elizabeth. *Britain's Moment in the Middle East, 1914–1971*. Baltimore: Johns Hopkins University Press, 1981.
Moran, Theodore H. "Managing an Oligopoly of Would-Be Sovereigns: The Dynamics of Joint Control and Self-Control in the International Oil Industry Past, Present, and Future." *International Organization* 41: 4 (1987): 575–607.
More, Charles. *Black Gold: Britain and Oil in the Twentieth Century*. London: Continuum, 2009.
Murray, Williamson. "The Change in the European Balance of Power, 1938–1939." Ph.D. Dissertation, Yale University, 1975.
Murray, Williamson. "Munich, 1939: The Military Confrontation." *Journal of Strategic Studies* 2: 3 (1979): 282–302.
Murray, Williamson. "The Role of Italy in British Strategy, 1938–1939." *RUSI Journal* 124: 3 (1979): 43–9.
Murray, Williamson. *The Change in the European Balance of Power, 1938–1939: The Path to Ruin*. Princeton: Princeton University Press, 1984.
Murray, Williamson. "Appeasement and Intelligence." *Intelligence and National Security* 2: 4 (1987): 47–66.
Murray, Williamson. *German Military Effectiveness*. Baltimore: Nautical & Aviation Pub. Co. of America, 1992.
Murray, Williamson. *The Luftwaffe, 1933–1945: Strategy for Defeat*. Washington: Brassey's, 1996.
Murray, Williamson. *War, Strategy, and Military Effectiveness*. Cambridge: Cambridge University Press, 2011.
Murray, Williamson, MacGregor Knox, and Alvin Berstein, ed. *The Making of Strategy: Rulers, States, and War*. Cambridge: Cambridge University Press, 1994.
Neal, Larry. "The Economics and Finance of Bilateral Clearing Agreements: Germany, 1934-1938." *Economic History Review* 32: 3 (1979): 391–404.
Neumann, Franz. *Behemoth: The Structure and Practice of National Socialism, 1933–1944*. New York: Octagon Books, 1963.
Nitzan, Jonathan and Shimson Bichler. *Global Political Economy of Israel*. Pluto: London, 2002.
Nowell, Gregory. *Mercantile States and the World Oil Cartel, 1900–1939*. Ithaca: Cornell University Press, 1994.
O'Brien, Patrick. "The Costs and Benefits of British Imperialism, 1846–1914." *Past & Present* 120 (1988): 163–200.
O'Brien, Patrick. "Imperialism and the Rise and Decline of the British Economy, 1688–1989." *New Left Review* 1: 238 (1999): 48–80.

Omissi, David E. *Air Power and Colonial Control: The Royal Air Force, 1919–1939*. Manchester: Manchester University Press, 1990.

Osterhammel, Jürgen. *The Transformation of the World: A Global History of the Nineteenth Century*. Princeton: Princeton University Press, 2014.

Overy, Richard J. "Transportation and Rearmament in the Third Reich." *Historical Journal* 16: 2 (1973): 389–409.

Overy, Richard J. *War and Economy in the Third Reich*. Oxford: Oxford University Press, 1994.

Overy, Richard J. *Why the Allies Won*. New York: W.W. Norton, 1995.

Overy, Richard J. "The Four-Year Plan." *European Yearbook of Business History* (2000): 87–106.

Painter, David S. *Oil and the American Century: The Political Economy of U.S. Foreign Oil Policy, 1941–1954*. Baltimore: Johns Hopkins University Press, 1986.

Painter, David S. "Oil and the American Century." *Journal of American History* 99: 1 (2012): 24–39.

Painter, David S. "Oil and Geopolitics: The Oil Crises of the 1970s and the Cold War." *Historical Social Research* 39: 4 (2014): 186–208.

Paret, Peter, ed. *The Makers of Modern Strategy: From Machiavelli to the Nuclear Age*. Princeton: Princeton University Press, 1986.

Parker, R.A.C. "Great Britain, France and the Ethiopian Crisis, 1935–1936." *English Historical Review* 89: 174 (1974): 293–332.

Parker, R.A.C. "Economics, Rearmament and Foreign Policy: The United Kingdom before 1939—A Preliminary Study." *Journal of Contemporary History* 10: 4 (October 1975): 637–47.

Parker, R.A.C. "The Pound Sterling, the American Treasury and British Preparation for War, 1938–1939." *English Historical Review* 98: 387 (April 1983): 261–79.

Payson O'Brien, Phillips. *How the War Was Won: Air–Sea Power and Allied Victory in World War II*. Cambridge: Cambridge University Press, 2015.

Pearton, Maurice. *Oil and the Romanian State*. Oxford: Clarendon Press, 1971.

Peden, G.C. "The Burden of Imperial Defense and the Continental Commitment Reconsidered." *Historical Journal* 27: 2 (1984): 405–23.

Peden, G.C. "A Matter of Timing: The Economic Background to British Foreign Policy, 1937–1939." *History* 69: 225 (1984): 15–28.

Penrose, Edith and E.F. Penrose. *Iraq: International Relations and National Development*. London: Ernest Benn, 1978.

Perkins, John. "Coins for Conflict: Nickel and the Axis, 1933–1945." *Historian* 55 (1992): 85–100.

Perović, Jeronim, ed. *Cold War Energy*. Cham: Springer International Publishing, 2017.

Petzina, Dieter. *Autarkiepolitik im Dritten Reich: Der nationalsozialistische Vierjahresplan*. Stuttgart: Deutsche Verlagsanstalt, 1968.

Petzina, Dieter. "IG-Farben und nationalsozialistische Autarkiepolitik." *Tradition* 13 (October 1968): 250–4.

Philip, George. *Oil and Politics in Latin America*. Cambridge: Cambridge University Press, 1976.

Plumpe, Gottfried. *Die I.G. Farbenindustrie AG: Wirtschaft, Technik und Politik*. Berlin: Duncker & Humblot, 1990.

Podobnik, Bruce. *Global Energy Shifts: Fostering Sustainability in a Turbulent Age*. Philadelphia: Temple University Press, 2006.

Popple, Charles Sterling. *Standard Oil Company (New Jersey) in World War II*. New York: Standard Oil Co. (New Jersey), 1952.

Price, Christopher. *Britain, America and Rearmament in the 1930s: The Cost of Failure.* Basingstoke: Palgrave, 2001.

Ra'anan, Uri, Robert L. Pfaltzgraff, Jr., and Geoffrey Kemp, ed. *Projection of Power: Perspectives, Perceptions, and Problems.* Hamden: Archon Books, 1982.

Rabe, Stephen G. *The Road to OPEC: United States Relations with Venezuela, 1919–1976.* Austin: University of Texas Press, 1982.

Reynolds, David. *From Munich to Pearl Harbor: Roosevelt's America and the Origins of the Second World War.* Chicago: Ivan R. Dee, 2001.

Rich, Norman. *Hitler's War Aims.* New York: W.W. Norton & Company, 1973, 1974.

Richardson, Charles O. "French Plans for Allied Attacks on the Caucasus Oil Fields, January–April, 1940." *French Historical Studies* 8: 1 (1973): 130–56.

Ristuccia, Cristiano Andrea. "The 1935 Sanctions against Italy: Would Coal and Oil Have Made a Difference?" *European Review of Economic History* 4 (2000): 85–110.

Rothwell, V.H. "Mesopotamia in British War Aims." *Historical Journal* 13: 2 (1970): 273–94.

Rubin, Michael A. "Stumbling through the 'Open Door': The U.S. in Persia and the Standard-Sinclair Oil Dispute, 1920–1925." *Iranian Studies* 28: 3/4 (1995): 203–29.

Rush, Robert S. and William W. Epley, ed. *Multinational Operations, Alliances, and International Cooperation Past and Future.* Washington, DC: Center for Military History, 2006.

Salewski, Michael. *Die Deutsche Seekriegsleitung, 1935–1945.* Frankfurt am Main: Bernard & Graefe, 1970–1975.

Sampson, Anthony. *The Seven Sisters: The Great Oil Companies and the World They Shaped.* New York: Viking Press, 1975.

Santiago, Myrna I. *The Ecology of Oil: Environment, Labor, and the Mexican Revolution, 1900–1938.* Cambridge: Cambridge University Press, 2006.

Sassi, Mohamed. "The Emergence of the French Oil Industry between the Two Wars." *Business and Economic History On-line* 1 (2003). Accessed August 15, 2018. https://thebhc.org/sites/default/files/Sassi.pdf.

Sassi, Mohamed. "Evolution of the Structure of the French Oil Economy between the Wars: Toward a French Holding Company." *Business and Economic History On-line* 2 (2004). Accessed August 15, 2018. https://www.thebhc.org/sites/default/files/Sassi_1.pdf.

Scazzieri, Luigi, "Britain, France, and Mesopotamian Oil." *Diplomacy & Statecraft* 26: 1 (2015): 25–45.

Scherner, Jonas. "Bericht zur deutschen Wirtschaftslage 1943/44: Eine Bilanz des Reichsministeriums für Rüstung und Kriegsproduktion über die Entwicklung der deutschen Kriegswirtschaft bis Sommer 1944." *Vierteljahrshefte für Zeitgeschichte* 55: 3 (2007): 499–546.

Schustereit, Hartmut. "Die Mineralöllieferungen der Sowjetunion an das Deutsche Reich 1940/41." *Vierteljahrschrift für Sozial- und Wirtschaftsgeschichte* 67: 3 (1980): 334–53.

Schwarte, M., ed. *Der große Krieg, 1914–1918.* Leipzig: Johann Ambrosius Barth, 1921.

Schwendemann, Heinrich. "German-Soviet Economic Relations at the Time of the Hitler-Stalin Pact, 1939–1941." *Cahiers du Monde russe* 36: 1/2 (1995): 161–78.

Seidenzahl, Fritz. *100 Jahre Deutsche Bank, 1870–1970.* Frankfurt: Deutsche Bank, 1970.

Shulman, Peter A. "'Science Can Never Demobilize': The United States Navy and Petroleum Geology, 1898–1924." *History and Technology* 19: 4 (2003): 365–85.

Shwadran, Benjamin. *The Middle East, Oil, and the Great Powers.* New York: Council for Middle Eastern Affairs Press, 1959.

Silverfarb, Daniel. "Britain, the United States, and the Security of the Saudi Arabian Oilfields in 1942." *Historical Journal* 26: 3 (1983): 719–26.

Silverfarb, Daniel. *Britain's Informal Empire in the Middle East: A Case Study of Iraq, 1929–1941*. New York: Oxford University Press, 1986.

Simpson, Christopher, ed. *War Crimes of the Deutsche Bank and the Dresdner Bank: Office of Military Government (U.S.) Reports*. New York: Holmes & Meier, 2002.

Sluglett, Peter. *Britain in Iraq: Contriving King and Country*, London: I.B. Tauris, 2007.

Snyder, David Allan. "Petroleum and Power: Naval Fuel Technology and the Anglo-American Struggle for Core Hegemony, 1889–1922." Ph.D. Dissertation, Texas A&M University, 2001.

Snyder, Timothy. *Bloodlands: Europe between Hitler and Stalin*. New York: Basic Books, 2010.

Solberg, Carl. *Oil Power*. New York: Mason/Charter, 1976.

Steinberg, Jonathan. "The Third Reich Reflected: German Civil Administration in the Occupied Soviet Union, 1941–4." *English Historical Review* 110: 437 (1995): 620–51.

Steiner, Zara. *The Lights that Failed: European International History*. Oxford: Oxford University Press, 2005.

Steiner, Zara. *The Triumph of the Dark: European International History, 1933–1939*. Oxford: Oxford University Press, 2011.

Stern, Roger. "Oil Scarcity Ideology in US Foreign Policy, 1909–1997." *Security Studies* 25: 2 (2016): 214–57.

Sternberg, Fritz. *From Nazi Sources: Why Hitler Can't Win*. New York: Stratford Press, 1939.

Stivers, William. "International Politics and Iraqi Oil, 1918–1928: A Study in Anglo-American Diplomacy." *Business History Review* 55: 4 (1981): 517–40.

Stivers, William. *Supremacy and Oil: Iraq, Turkey, and the Anglo-American World Order, 1918–1930*. Ithaca: Cornell University Press, 1982.

Stivers, William. "A Note on the Red Line Agreement." *Diplomatic History* 7: 1 (1983): 23–34.

Stokes, Raymond G. "The Oil Industry in Nazi Germany, 1936–1945." *Business History Review* 59: 2 (1985): 254–77.

Strang, G. Bruce. "'The Worst of All Worlds': Oil Sanctions and Italy's Invasion of Abyssinia, 1935–1936." *Diplomacy and Statecraft* 19 (2008): 210–35.

Stranges, Anthony N. "Friedrich Bergius and the Rise of the German Synthetic Fuel Industry." *Isis* 75: 4 (1984): 643–67.

Stranges, Anthony N. "From Birmingham to Billingham: Synthetic Fuels in Great Britain, 1910–1945." *Technology and Culture* 26 (1985): 726–57.

Stranges, Anthony N. "Synthetic Fuel Production in Prewar and World War II Japan: A Case Study in Technological Failure." *Annals of Science* 50: 3 (1993): 229–65.

Supple, Barry. "Fear of Failing: Economic History and the Decline of Britain." *Economic History Review* 47: 3 (1994): 441–58.

Thies, Jochen. "Hitler's European Building Programme." *Journal of Contemporary History* 13: 3 (1978): 413–31.

Thomas, Georg. *Geschichte der deutschen Wehr- und Rüstungswirtschaft, 1918–1943/1945*. Boppard am Rhein: Boldt, 1966.

Tomlinson, B.R. *Political Economy of the Raj, 1914–1947*. London: Macmillan, 1979.

Tooze, J. Adam. "No Room for Miracles: German Industrial Output in World War II Reassessed." *Geschichte und Gesellschaft* 31: 3 (2005): 439–64.

Tooze, J. Adam. *The Wages of Destruction: The Making and Breaking of the Nazi Economy*. New York: Viking, 2007.

Topnik, Steven C. and Allen Wells, ed. *The Second Conquest of Latin America: Coffee, Henequen, and Oil during the Export Boom, 1850–1930*. Austin: University of Texas Press, 1998.

Toprani, Anand. "The French Connection: A New Perspective on the End of the Red Line Agreement, 1945–1948." *Diplomatic History* 36: 2 (2012): 261–99.

Toprani, Anand. "Germany's Answer to Standard Oil: The Continental Oil Company and Nazi Grand Strategy, 1940–1942." *Journal of Strategic Studies* 37: 6–7 (December 2014): 949–73.

Toprani, Anand. " 'The navy's success speaks for itself'? The German Navy's Independent Energy Security Strategy, 1932–1940." *Naval War College Review* 68: 3 (Summer 2015): 91–118.

Toprani, Anand. "The First War for Oil: The Caucasus, German Strategy, and the Turning Point of the War on the Eastern Front, 1942." *Journal of Military History* 80: 3 (2016): 815–54.

Toprani, Anand. "An Anglo-American 'Petroleum Entente'? The First Attempt to Reach an Anglo-American Oil Agreement, 1921." *The Historian* 79: 1 (2017): 56–79.

Treue, Wilhelm. "Hitlers Denkschrift zum Vierjahresplan, 1936." *Vierteljahrschefte für Zeitgeschichte* 3: 2 (1955): 184–210.

Trevor-Roper, Hugh Redwald. "Hitlers Kriegsziele." *Vierteljahrschefte für Zeitsgeschichte* 8: 2 (1960): 121–33.

Turner, Henry Ashby. "Hitlers Einstellung zu Wirtschaft und Gesellschaft vor 1933." *Geschichte und Gesellschaft* 2: 1 (1976): 89–117.

Turner, Henry Ashby. *German Big Business and the Rise of Hitler*. New York: Oxford University Press, 1985.

Ulrichsen, Kristian Coates. "The British Occupation of Mesopotamia, 1914–1922." *Journal of Strategic Studies* 30: 2 (2007): 349–77.

Van Creveld, Martin L. *Supplying War: Logistics from Wallenstein to Patton*. New York: Cambridge University Press, 2004.

Venn, Fiona. " 'A futile paper chase': Anglo-American Relations and Middle East Oil, 1918–1934." *Diplomacy & Statecraft* 1 (1990): 165–84.

Venn, Fiona. "Oleaginous Diplomacy: Oil, Anglo-American Relations and the Lausanne Conference, 1922–23." *Diplomacy and Statecraft* 20 (2009): 414–33.

Volland, Klaus. *Das Dritte Reich und Mexiko: Studien zur Entwicklung des deutsch-mexikanischen Verhältnisses, 1933–1942 unter besonderer Berücksichtigung der Ölpolitik*. Frankfurt am Main: Peter Lang, 1976.

Waters, Christopher. *Australia and Appeasement: Imperial Foreign Policy and the Origins of Appeasement*. New York: Palgrave, 2012.

Wegner, Bernd. "The Road to Defeat: The German Campaigns in Russia, 1941–43." *Journal of Strategic Studies* 13: 1 (1990): 105–27.

Wegner, Bernd. "Hitlers Besuch in Finnland: Das geheime Tonprotokoll seiner Unterredung mit Mannerheim am 4. June 1942." *Vierteljahrshefte für Zeitgeschichte* 41: 1 (1993): 117–37.

Wegner, Bernd, ed. *From Peace to War: Germany, Soviet Russia and the World, 1939–1941*. Providence: Berghahn Books, 1997.

Weinberg, Gerhard L. "22 June 1941: The German View." *War in History* 3: 2 (1996): 225–33.

Weinberg, Gerhard L. *Hitler's Second Book: The Unpublished Sequel to Mein Kampf by Adolf Hitler*. New York: Enigma, 2003.

Weinberg, Gerhard L. "Some Myths of World War II." *Journal of Military History* 75: 3 (2011): 701–18.

Wilkins, Mira. *The Maturing of Multinational Enterprise: American Business Abroad from 1914–1970*. Cambridge: Harvard University Press, 1974.

Wilkins, Mira. *The History of Foreign Investment in the United States, 1914–1945*. Cambridge: Harvard University Press, 2004.

Wylie, J.C. *Military Strategy: A General Theory of Power Control.* New Brunswick: Rutgers University Press, 1967.
Yergin, Daniel. *The Prize: The Epic Quest for Oil, Money, and Power.* New York: Simon & Schuster, 1991.
Zetzsche, Hans-Jürgen. *Logistik und Operationen: Die Mineralölversorgung der Kriegsmarine im Zweiten Weltkrieg.* Ph.D. Dissertation, Christian-Albrechts Universität zu Kiel, 1986.
Zürrer, Werner. *Kaukasien, 1918–1921: Der Kampf der Großmächte um die Landbrücke zwischen Schwarzem und Kaspischem Meer.* Düsseldorf: Droste, 1978.

Index

Note: Tables are indicated by an italic *t* following the page number. Endnotes are indicated by a note suffix (n.) following the page number.

9 780198 834601

Made in the USA
Las Vegas, NV
19 September 2026

2b0481c0-cd85-45ba-8394-4f19b693d3d1R01